Sky Watcher's Handbook

Sky Watcher's Handbook

The Expert Reference Source for the Amateur Astronomer

•

Edited by
JAMES MUIRDEN

OXFORD · NEW YORK · HEIDELBERG

W.H. Freeman and Company Limited
20 Beaumont Street, Oxford, OX1 2NQ
41 Madison Avenue, New York, NY 10010

Library of Congress Cataloging-in-Publication Data

Sky Watcher's Handbook : the expert reference source for the amateur astronomer / edited by James Muirden.
p. cm.
ISBN 0-7167-4502-X
1. Astronomy—Amateurs' manuals. 2. Astronomy—Observers' manuals. I. Muirden, James.
QB63.P7 1993 *92-32996*
CIP

Set by Keyword Publishing Services
Printed by The Bath Press Ltd

CONTENTS

Preface

AMATEUR ASTRONOMY, WHICH SEEMED TO BE PASSING THROUGH A decline in the fifties and sixties, has undergone a transformation in the seventies and eighties. 'Backyard' observers are even detecting new minor planets, while the discovery of comets, novae, supernovae and new variable stars continues apace.

Paradoxically, the technological advances in telescope guidance systems and image recording techniques, which looked set to undermine amateur work, have helped: observers with modest telescopes are now measuring cometary positions with 'professional' accuracy, and the adaptation of video methods to planetary photography is allowing detail at the very threshold of visual detection to be recorded.

Electronic communication has also strengthened the amateur's hand. A never-sleeping network of observers now spans the globe, ready to check on possible discoveries at a moment's notice and to feed the information into the professional astronomers' information service.

The highly specialized nature of the 'sharp end' of modern amateur astronomy makes it essential for the specialists themselves to contribute to a book on practical amateur astronomy. Apart from my obvious indebtedness to the distinguished amateur astronomers from many different countries who have written for this book, I should like also to thank the following individuals for their help and support: Alan M. MacRobert and his colleagues at *Sky & Telescope*, for help in contacting overseas contributors; Robin Scagell, for advising me in general terms about the content; my son Daniel for translating Chapter 14 (High-resolution lunar and planetary photography) from the original French; Michael Rodgers, who originally suggested that I should edit the book; and my wife Helen for all her help and advice.

JAMES MUIRDEN
June, 1992

CHAPTER ONE · *Observing the Sun*

PETER O. TAYLOR
AAVSO SOLAR DIVISION

IN 1801, SIR WILLIAM HERSCHEL REFERRED TO THE SUN WHEN HE wrote: 'The influence of this eminent body on the globe we inhabit is so great and so widely diffused that it becomes almost a duty to study the operations which are carried on upon the solar surface'. A lofty thought, and one to which we thoroughly subscribe, but just *what* shall we observe, and *how* can we safely study it? Perhaps we should begin with the second of these questions.

DEFENDING THE EYE

Since the first telescopic observations of the Sun were made by Galileo and Fabricius in the early seventeenth century, safety has been a prime concern. Before the harmful properties of the Sun's invisible radiation were known, the Sun was frequently viewed through blackened or coloured glass 'filters' mounted at the eyepiece. While these devices served to dim the Sun's brightness, they did not block the dangerous ultraviolet and near-infrared radiation which penetrates the Earth's atmosphere: an absolute necessity for observing the Sun without damaging the eye.

Today, amateur solar observers generally favour two methods which do meet these requirements, offering safety, convenience and low cost: the techniques of direct viewing through full-aperture filters, and of image projection. Before delving into these methods, however, let us discuss a few devices that we believe should be avoided entirely or used conditionally.

First, the *sun-cap* (eyepiece) filters that were often supplied with small telescopes are very dangerous and should NEVER be used. Ultraviolet and infrared radiation can penetrate these devices, or they may crack without warning, sending a damaging flash of sunlight into an unwary observer's eye.

In our opinion, *photographic films* are not suitable for viewing the Sun either. The once popular belief that fully exposed and developed black and white film provides a safe solar filter should be treated with caution. The metallic silver in the developed film serves as the filtering medium, and all modern colour emulsions (and some black and white films) have the silver removed during development. Thus, while these materials may look dark, they are not safe solar filters.

Welder's glass of shade number 14 is satisfactory for non-instrumental viewing, but may break suddenly if used at the eyepiece without additional filtering. It is not suitable for full-aperture use because its surface is not optically uniform and the material from which it is made is not of 'instrument quality'.

The viewing methods which have been successfully employed in the past (such as those which use light-deflecting prisms or unsilvered mirror systems) tend to be expensive to construct or require additional filtering at the eyepiece. While the best of these systems have performed well in the past, today's full-aperture filters have surpassed them in safety and quality.

DIRECT VIEWING WITH A FULL-APERTURE FILTER

Modern full-aperture solar filters are manufactured of either glass or an optical-quality plastic film, such as mylar. These materials are given special metallic coatings that block the harmful radiation from the Sun and reduce its brightness to a fraction of its original amount, affording a view that is both comfortable and safe.

Full-aperture filters also offer a unique advantage in that they block the Sun's heat *before* it enters the telescope. Heat affects optical components through unequal expansion, and degrades the resolution of the image by producing thermal currents within the tube.

Experienced solar observers generally favour direct-viewing methods over projection for seeing detail, and we agree that this is the most satisfactory way to observe sunspots and many other active solar phenomena. The filters are not expensive and are available in sizes to fit virtually any size of instrument.

Occasionally a filter's coating will become abraded, and so it should be routinely inspected. Hold the filter up to a strong light and examine it for thin areas and pinholes in its coating. A few tiny holes can safely be repaired with small dabs of black paint, but filters with more serious defects should be rejected.

Since filters made of glass generally produce a yellow-orange solar image, and the eye is more sensitive to detail when viewing in this portion of the spectrum, it is probably superior to mylar for viewing sunspots. Mylar filters are the least expensive type of solar filter, though, and offer a safety factor equivalent to glass. They generally produce a bluish image because the material blocks slightly more infrared radiation than a glass filter does. This apparent drawback can be an advantage in one respect, since a few bright phenomena such as solar white-light flares and faculae are easier to detect when viewed towards the blue region of the spectrum.

Direct viewing does not, however, lend itself easily to the process of determining the heliographic (solar) positions of sunspots or other features. The most accurate of these techniques usually require that the Sun's image be projected or photographed.

PROJECTING THE SOLAR IMAGE

Projection is certainly the safest way to observe the Sun, and it is the method of choice for those who wish to determine the locations of sunspots or other solar phenomena. An electrically-driven equatorial mounting is not required for routine observations, but it is a necessity if accurate drawings of the Sun's features or determination of their positions is to be attempted.

On the other hand, projection is not satisfactory when used with refracting telescopes of less than 60 mm aperture. In addition, owners of catadioptric telescopes should consult the manufacturer of their instrument before attempting this method, since some of the materials that are used in their construction can be damaged by concentrated heat. The secondary mirror is particularly vulnerable if it is attached with cement rather than by mechanical means.

The eyepiece that is chosen to project the image should be selected with care. Again, those with cemented components or antireflection coatings should be used with caution because heat can damage these materials. The older Ramsden or Huygenian types should have their lens-retaining rings loosened slightly to allow for expansion.

The image should be projected on to a rigid surface that is mounted perpendicular to the emergent beam. The surface should be larger than the projected image, so that the complete image of the Sun can be displayed at one time. It should be made of a smooth, flat material and painted flat white. A projection surface mounted separately from the telescope is preferred; several observers have had good results by using old music stands or similar heavy supports for this purpose. However, if the image is to be projected on a screen marked with reference lines for deriving positions, the screen must be attached to the telescope so that it shares the equatorial motion. The simple projection apparatus that is often included with small telescopes is not suitable for serious observations because its surface is generally too small, and it is subject to considerable vibration, so a home-constructed device may be necessary.

Since the Sun's angular diameter varies between 32′ 35.2″ arc and 31′ 30.8″ arc as the Sun-to-Earth distance changes throughout the year, the distance between eyepiece and surface should have some means of adjustment in order to maintain a constant image diameter. Image size can be controlled in two ways: eyepiece magnification, and distance between projection surface and eyepiece. The relationship between these aspects is defined by:

$$\text{Distance between surface and eyepiece} = \frac{107D}{M\text{-}1}$$

In the relation, M represents the magnification of the eyepiece, and D the required image diameter.

Magnification for a given eyepiece varies according to the telescope's focal length. It is obtained by dividing the focal length of the instrument by that of the eyepiece. For example, a telescope with a focal length of 900 mm requires an eyepiece with a focal length of 15 mm to produce a magnification of $\times 60$. Using such a magnification, a screen positioned about 270 mm behind the eyepiece will receive a solar image 150 mm across.

It is important to shield the projection surface from direct sunlight, so a baffle should be constructed of heavy card stock or similar material. Although its location will vary according to instrument type, the shield should be mounted so that it will completely shade the projection surface. However, such a shield only prevents direct sunlight from reaching the projection surface, and when observing out of doors additional shade may be needed to prevent light from the sky and elsewhere from reducing the contrast in the Sun's image to the point where faint features on the disk may be missed. Simple methods of increasing the contrast include holding an umbrella over the projection surface or enclosing the projection surface and the shield in a black cloth bag into which the observer's head may be inserted.

COUNTING SUNSPOTS

One of the most attractive features of observing the Sun is the variety it offers. The Sun can simply be viewed for the knowledge and enjoyment that relaxed observing in the daytime provides, an extremely rewarding pursuit in itself. Alternatively, the observer may contribute to a structured effort such as the AAVSO's Solar Division which regularly supplies data collected by its collaborators to professional astronomers. At present, there are two areas where the latter observer can contribute valuable information: the indirect detection of solar flares, and sunspot counting. The observation of flares will be treated in a later section.

Sunspot counting is truly the province of the amateur astronomer. The mid-nineteenth century discovery of the regular rise and fall of the sunspot number, now known as the *sunspot cycle* (Fig. 1.1), is generally credited to the amateur Heinrich Schwabe, and amateurs have traditionally monitored each cycle's evolution.

The *sunspot number* is not a simple count of the spots seen on the visible hemisphere, however. Sunspots occur in groups which range in size from one to well over 100 separate spots. Each day, the skilled observer first correctly divides the spots into these groups, and then counts the spots within each. The development

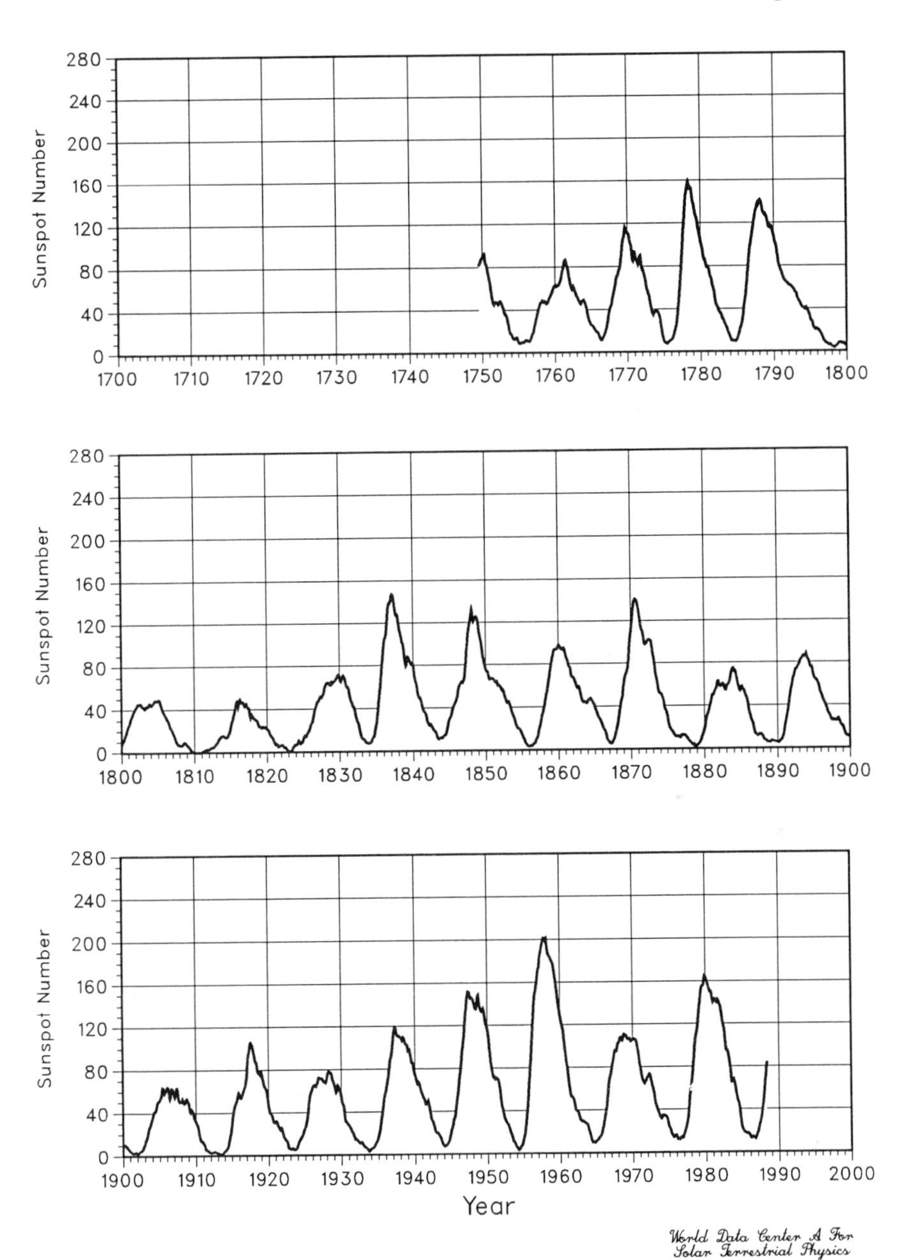

FIGURE 1.1
The sunspot cycle, 1749–1988. The smoothed monthly relative sunspot number regular rises and falls according to an average period near 11.1 years. The figure was provided by John McKinnon of the World Data Center for Solar-Terrestrial Physics, Boulder, Colorado, USA.

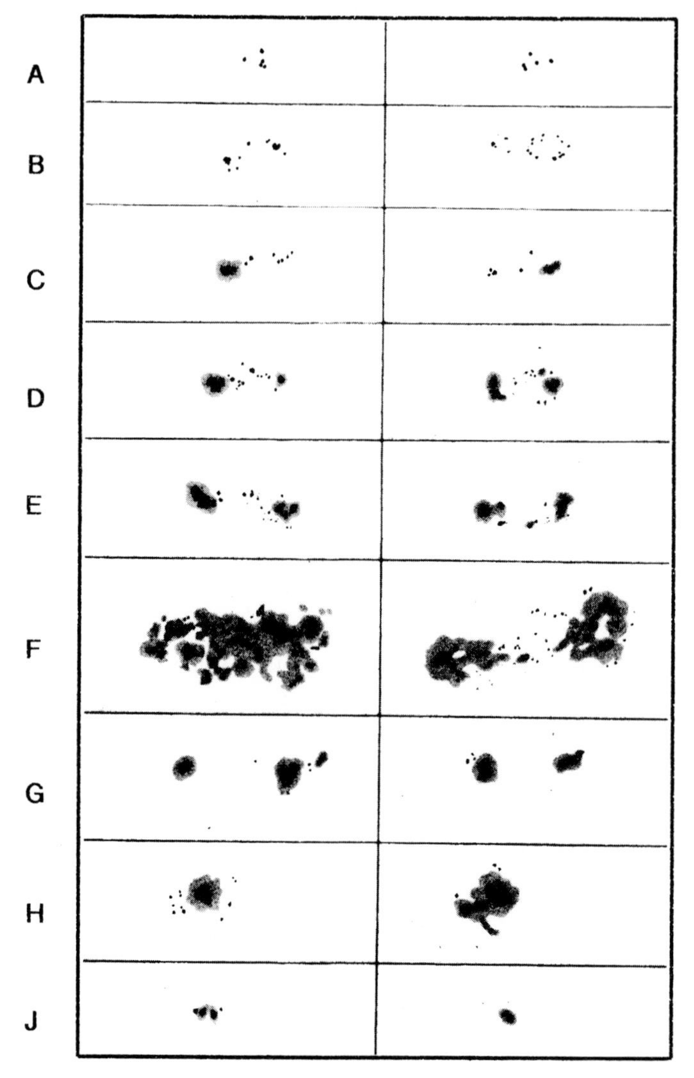

FIGURE 1.2
The classification of sunspot groups. Figure prepared from drawings contained in Waldmeier, M., *Ergebnisse und Probleme der Sonnenforschung*, 2nd edn. Geest u. Portig, Leipzig, 1955.

of sunspot groups is evolutionary in nature, and may proceed through part or all of the classes which are described below and depicted in Fig. 1.2.

A An isolated spot or small group of spots without penumbra.

B A bipolar group without penumbra in which the principal spots are separated by at least 3° of longitude.

C A bipolar group with one principal spot surrounded by penumbra.

D Both major spots have penumbra and the group is less than 10° in length.
E Similar to the D group, but with a length from 10° to 15°.
F Similar to D and E groups, but with a length which is 15° or more.
G A large bipolar group with penumbrae, but with no intermediate spots.
H A unipolar group with penumbra, with a diameter which is at least 2.5°.
J A unipolar group with penumbra, with a diameter which is less than 2.5°.

After the clusters have been determined, the number of groups (g) is then multiplied by 10 and combined with the total number of individual spots (s) according to the empirical relation:

$$\text{Relative sunspot number } (R) = 10g + s$$

Thus, a cluster such as that shown in Fig. 1.3 would receive an R value of 52 (the group value of 10 plus 42 spots).

Dividing the spots into groups can be a difficult procedure, especially when spot-complexes develop rapidly or suddenly erupt near existing groups. Since the equipment to separate them according to their magnetic qualities is not available to amateur astronomers, the observer must rely upon their experience and knowledge alone.

When an observer's estimates are analysed, each R value is reduced according to a previously determined observatory constant, or 'k-factor', before averaging with the results of others. Such factors are intended to compensate for differences in judgement, typical local seeing conditions and instrumentation, among observers.

Sunspots do not emerge at random locations. In fact, studies have shown that it is ten times more likely that a spot will appear within an active area than at

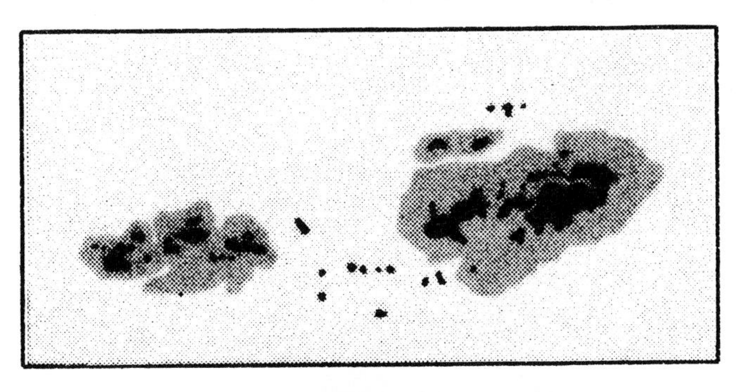

FIGURE 1.3
A sample sunspot group with an *R* value of 52 (see text).

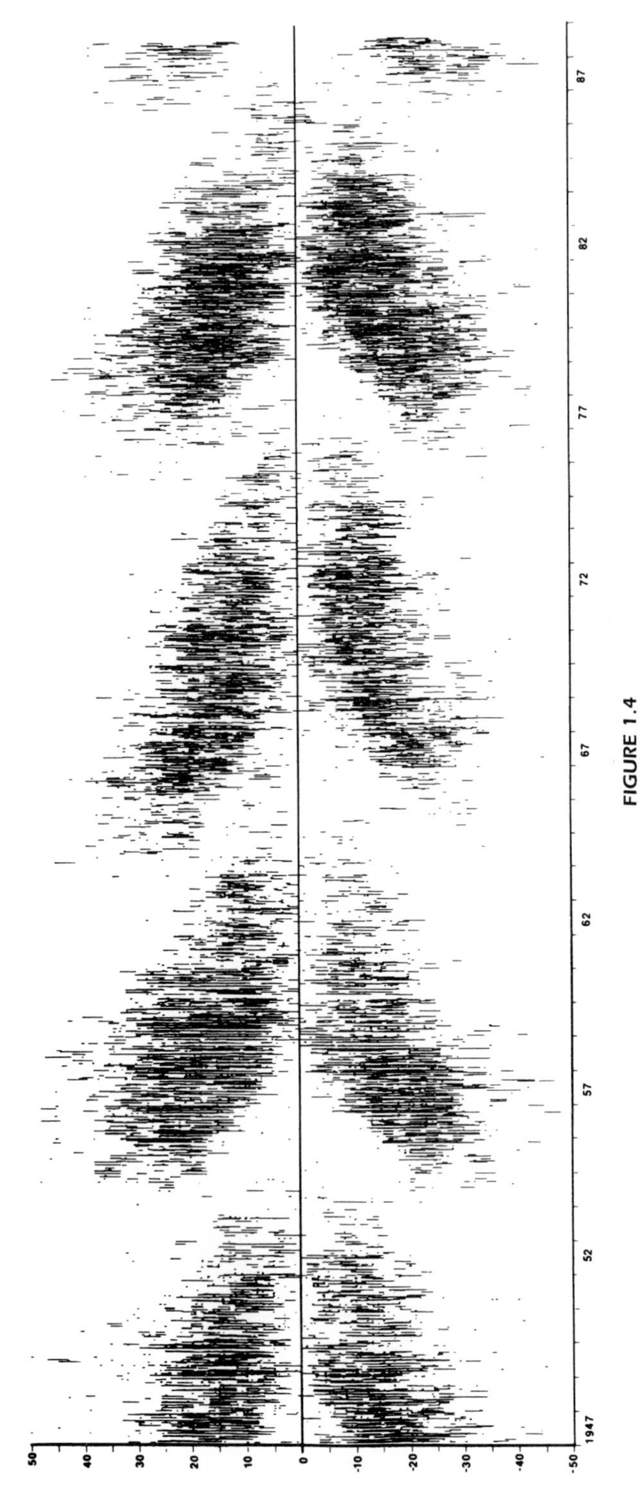

FIGURE 1.4

The 'butterfly diagram'. When the heliographic latitudes of sunspot groups are plotted against time, the characteristic shape of the outstretched wings of a butterfly appears for each sunspot cycle. The graph represents the observations of a single expert observer, Miss Hisako Koyama of Tokyo, Japan, who also supplied the diagram.

another part of the Sun. Of course, since the time of Galileo it has been recognized that spots almost always appear within 'zones' of latitude that extend 35° to the north and south of the solar equator. However, spots have been sighted in latitudes as great as 70°. The most recent examples of extreme high-latitude sunspots are two small groups which appeared in the northern and southern hemispheres during mid-1987, near latitude 60°.

In addition, when a cycle begins, its spots erupt at much higher average latitudes than do those near the end of the cycle. Typically, the first spots of a new cycle appear about 18 months *before* the minimum of the existing cycle, at an average latitude near 30°. By the end of the cycle, however, the latitude of emergence for new spots has decreased to about 7°. The graph that results when the latitudes of emerging spots are plotted against time is known as a 'butterfly diagram' (Fig. 1.4). The process, which may result from the twisting of magnetic fields deep within the Sun, is generally referred to as *Spörer's law*.

One clue to the emergence of a new spot group is the presence of the seemingly bright structures known as *photospheric faculae*, which often precede spot formation. All sunspots have associated faculae (the reverse is not true) and the two phenomena are closely related magnetically. It has been suggested that faculae appear to be brighter than the photosphere because they are several hundred kilometres higher, and are less attenuated by the Sun's atmosphere. They are often seen as whitish patches near the limb where their contrast is high, although they can be viewed all across the disk at short wavelengths.

Another common bright-appearing feature is the *light-bridge*. These elongated strips of photosphere frequently extend into the dark (umbral) areas of sunspots, and generally signal the group's impending demise. They too appear to be brighter than the surrounding photosphere, but this is an illusion produced by contrast between the thin strip of solar surface and dark umbra.

SUNSPOT SIZES

The areas of sunspot groups are normally measured in millionths of the solar hemisphere. For an image 150 mm in diameter, one square millimetre near disk centre represents an area of around 28.3 millionths. To compute a group's area, multiply the number of millimetre units within the group's border by 28.3, and, to allow for foreshortening, multiply this result by secant ρ (see below under 'Mathematical determination of positions'). Since 1000 millionths is equivalent to 3040 million square kilometres, the largest groups have areas as great as 15 thousand million square kilometres!

ORIENTING THE IMAGE

It is a simple matter to find the approximate directions on the Sun's image. Regardless of the observing method, east and west are determined by allowing the Sun to drift through the field of view with the telescope drive off. The direction of travel will be from east towards west. North and south are found by moving the front of the telescope slightly towards the north; the southern limb will disappear at the edge of the field first.

For many of those who view the Sun for enjoyment only, this simple procedure will usually suffice. However, other observers may wish to be more precise, or to measure the positions of particular features. To simplify this process, the observer who uses the direct-viewing method should obtain a low-power eyepiece (which allows the entire solar disk to be viewed at one time) equipped with a simple cross-line reticle. Then it will be necessary to make a drawing of a circle (perhaps 60 mm in diameter) which is bisected by two perpendicular lines. The observer who uses projection should prepare a similar drawing, but one which has the same diameter as the projected image, usually around 150 mm.

Direct image – When initially determining the east and west directions, rotate the eyepiece/reticle so that a sunspot tracks parallel to one of the lines as the Sun drifts through the field. Label east and west points on the appropriate bisector, and north and south on the other line. Centre the solar image in the field of view and start the drive. Any spot-groups are then indicated on the sketch by small circles or ellipses. They are located on the drawing by carefully estimating their positions relative to the crossed lines of the reticle and the Sun's limb.

Projected image – An identical process is followed when using projection. In this case, however, rotate the drawing about the optical axis until a spot tracks parallel to one of the bisecting lines. Then centre the drawing and secure it to the projection surface while maintaining its alignment. If a sheet of tracing paper is used to sketch the projected image, it can overlay a grid which is used to aid the location process. Such a device should be divided into 6 mm to 12 mm squares and oriented so that (with the drive off) a spot moves parallel to one of the horizontal lines.

HELIOGRAPHIC COORDINATES

The changing Earth–Sun geometry, due mainly to the inclinations of the axes of the Sun and Earth with respect to the plane of the Earth's orbit, causes changes in the appearance of the solar disk as we view it. Because of this, two factors are required for precise orientation: P, the position angle of the Sun's actual axis of rotation, and B_0, which can be defined either as the tilt of the Sun's north pole towards (+) or away from (−) the Earth, or as the heliographic latitude of the centre of the Sun's disk. Daily values for these quantities can be obtained from a source such as the *Astronomical Almanac* or in many astronomical handbooks, such as the *BAA Handbook*, which describe observing targets for amateur astronomers.

After observations have been concluded, measure and mark angle P on the drawing, and extend a line from it through the centre of the drawing to the opposite

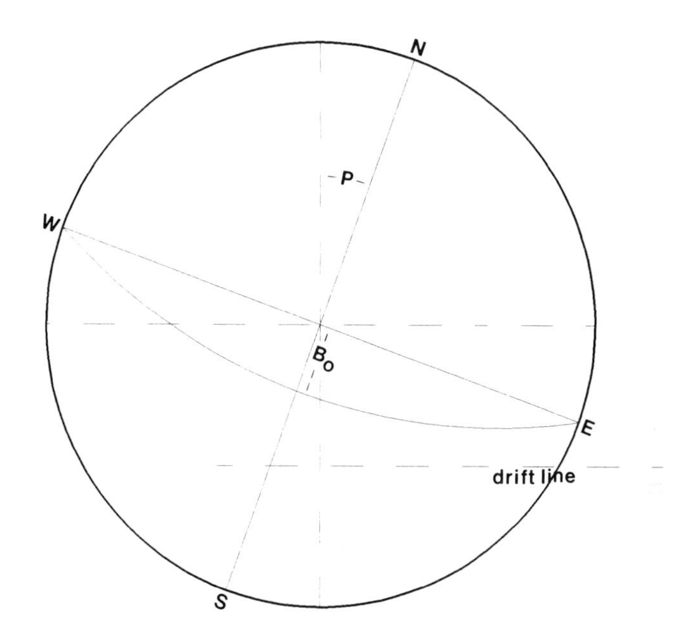

FIGURE 1.5
Orienting the Sun's image. As the Earth revolves around the Sun each year, the changing geometry of their orbital relationship affects the appearance of the solar disk as we view it. The N–S line represents the Sun's actual axis of rotation displaced from north in the sky by the amount of angle P. The equator appears to pass south of the disk centre because the Sun's north pole is tilted towards the Earth during half of each year.

edge of the circle. Positive values of P are measured east of the north point, and negative values towards the west. This line represents the Sun's rotational axis. Now draw a similar line at right angles to the first, through the centre of the circle. Refer to Fig. 1.5, and note that W and E represent true west and east on the Sun.

Select the value of B_0 for the date of observation from the table in the *Astronomical Almanac*. Since positive values show that the Sun's north pole is tilted towards the Earth by the amount of B_0, the solar equator must pass across the N–S axis south of the disk centre at distance d. On the drawing, this distance is given by

$$d = R \sin B_0$$

with R being the radius of the circle. Plot this point on the N–S line. The solar equator is represented by an ellipse passing through points W, B_0 and E.

MATHEMATICAL DETERMINATION OF POSITIONS

First proceed as outlined above to the result shown in Fig. 1.5. Then measure the distance from the centre of the drawing to a spot and call the result r. Sunspots generally occur in pairs (they are said to be 'optically bipolar') that are oriented in an E–W direction. When measuring the position of a bipolar group, use the largest spot (often the preceding, or most westerly one) as a reference point: it will often outlive the following spot. Then calculate according to

$$\sin \rho = \frac{r}{R}$$

with R again equal to the circle's radius.

Now compute B, the heliographic latitude of the spot:

$$\sin b = \cos \rho \sin B_0 + \sin \rho \cos B_0 \cos \phi$$

where ϕ is the measured angle between the north point on the solar axis and the line drawn from the drawing's origin through the centre of the spot. This procedure is shown in Fig. 1.6.

Next, determine l, the distance of the spot from the solar central meridian:

$$\sin l = \frac{\sin \phi \sin \rho}{\cos b}$$

Refer to the table in the *Astronomical Almanac* and find L_0 (the value for the central meridian at 0 h UT) for the date of the observation. A simple linear interpolation procedure can be used to refine the value of L_0 for the time of

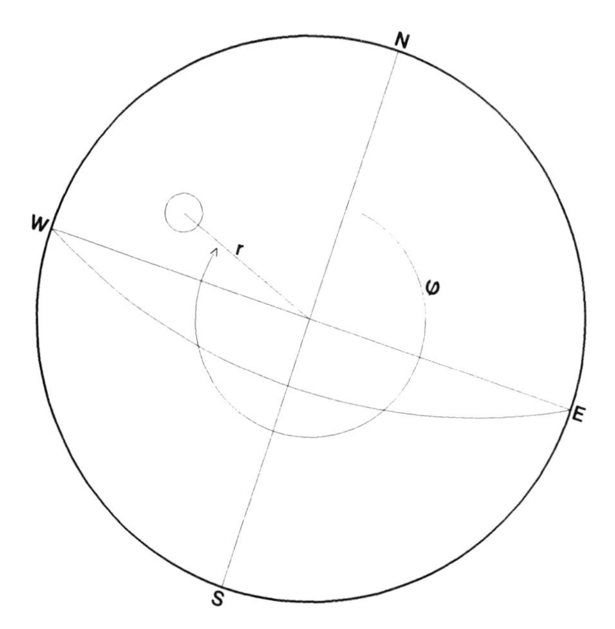

FIGURE 1.6
Measuring the position of a sunspot. One way to determine the position of a sunspot group or other feature on the Sun's disk requires that both angle ϕ and distance r be measured from a carefully prepared sketch.

observation, remembering that it *decreases* by approximately 13.2° per day. The longitude of the spot is given by

$$\text{Heliographic longitude } = L_0 - 1$$

SOLAR FLARES

Flares occur when magnetic fields (usually above sunspots) suddenly collapse and recombine into simpler structures with enormous output of energy. They are among the most energetic events in nature, with total energy emissions which can equal some 2000 million hydrogen bombs.

Detecting flares indirectly – Part of the energetic radiation that results from this process causes variations in the Earth's atmosphere that observers can monitor with simple low-frequency radio receivers. The events which result from these phenomena occur at the same time as a flare's optical appearance, and are called *sudden ionospheric disturbances* (SIDs). When the pulses which are generated by distant lightning strikes are monitored, the effect is known as a *sudden enhancement of atmospherics* (SEA). However, today more observers tune their receivers to very low-frequency radio stations since interference is minimized and their signal is more stable than that from natural sources. The latter effects are termed *sudden enhancement of signal* (SES).

The detection of solar flares by recording of low-frequency radio signals is sometimes more sensitive than that by satellite detectors. The data that are gathered in this way have long been necessary for radio propagation and flare research, and are increasingly important as we send astronauts farther into space. An adequate description of this method would consume more space than we are allotted, but the author will happily supply details of the programme to those who are interested in this type of solar observation.

White-light flares – Inexperienced observers often mistake a light-bridge for a white-light flare, but it is very easy to tell the difference: a flare changes its appearance drastically in just a few minutes, while changes in the light-bridge occur over a period of hours or days.

They are called 'white-light' flares because they extend from the Sun's atmosphere into the photospheric region and can be seen at visual wavelengths. Even though they are very rare (fewer than 90 verified sightings have occurred since the first was independently viewed by the English amateur astronomers Carrington and Hodgson in 1859), the observer should be alert for their appearance during regular observations.

If by chance the observer is fortunate enough to sight a white-light flare, the details of the observation should be carefully recorded and sent to the author or the appropriate national organization. Include its position within the group, an estimate of its brightness compared with the surrounding photosphere, and the

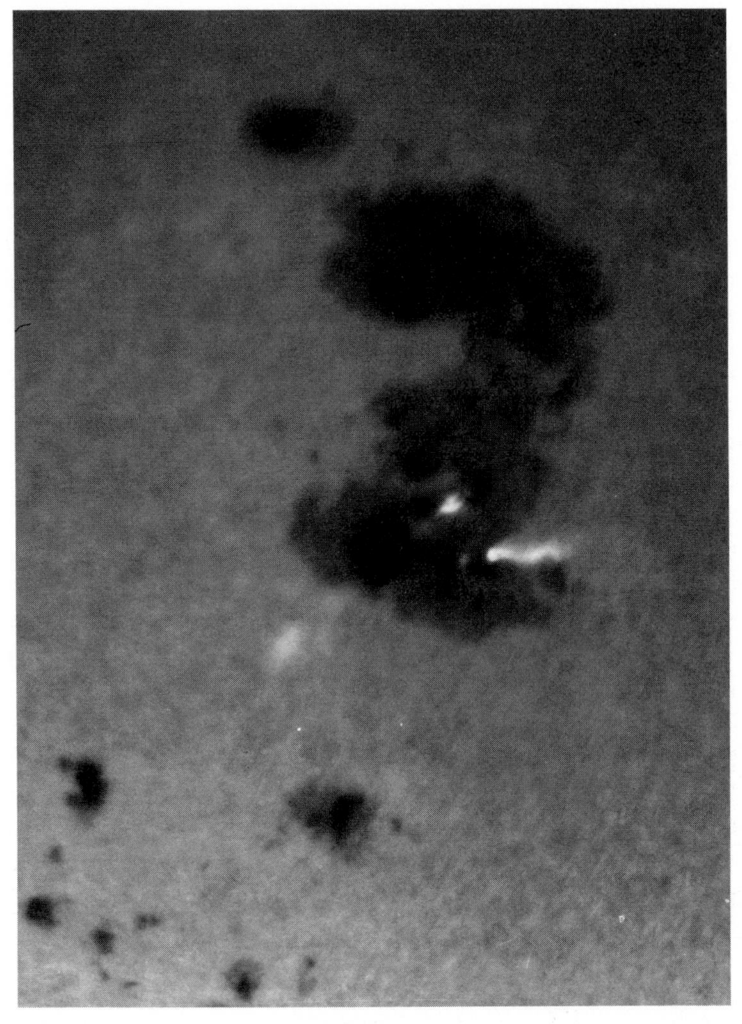

FIGURE 1.7
One of the brightest 'white-light' flares ever recorded is shown in this photograph taken on 24 April 1984. Dr Donald F. Neidig of the National Solar Observatory, Sacramento Peak, USA, provided this photograph, which shows the flare as it appeared in the normal visual spectrum (475.7 nm).

exact time and duration of the event. Accuracy and detail are very important because a verified sighting may well be included in the catalogues that list these activities. One of the most powerful of these events is shown in Fig. 1.7.

SOLAR PROMINENCES

These are gigantic eruptions of comparatively cool, dense hydrogen that extend many thousands of kilometres into space. *Hydrogen-alpha* (Hα) filters allow prominences and other spectacular solar phenomena to be seen by isolating the light of hydrogen emission in the red portion of the visible spectrum. When viewed through an Hα filter, flares appear as very bright patches, while active sunspots show a wealth of fine structure not seen by the conventional observer (Fig. 1.8). A filter's cost goes up as the width of the bandpass decreases, but narrow-band filters (about 0.6 Å or 0.06 nm) show the greatest detail. Wide-bandpass filters (4 Å

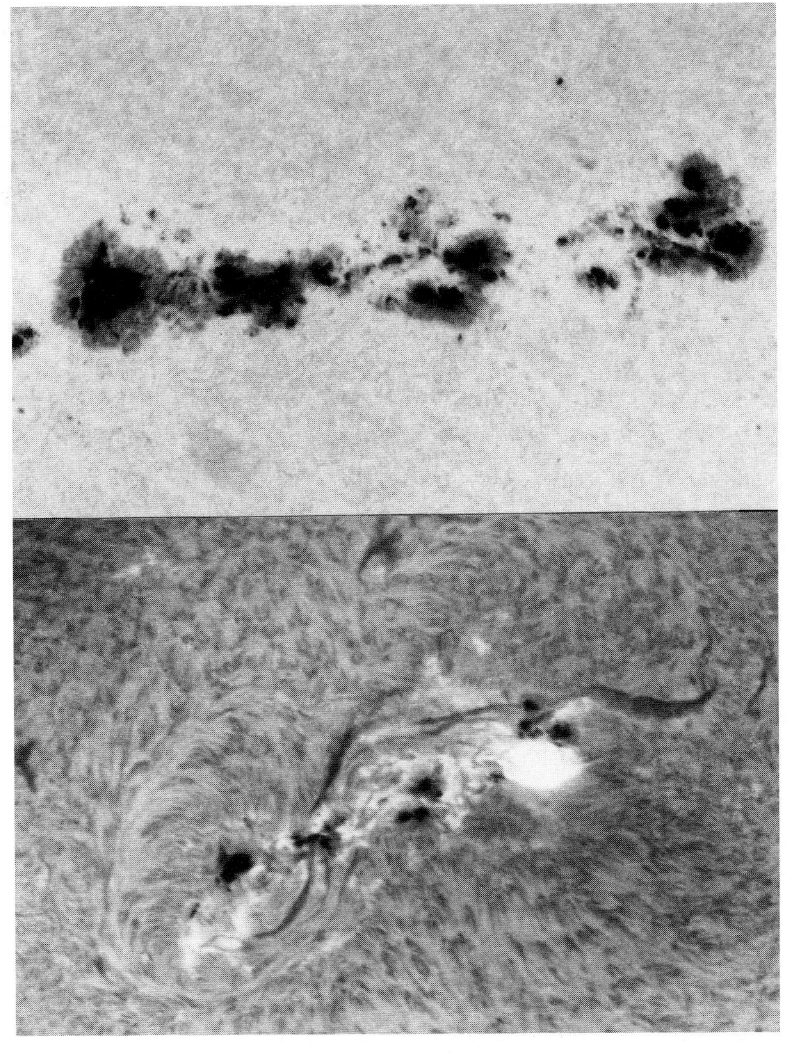

FIGURE 1.8
Differing views of the Sun. The appearance of the Sun varies considerably when it is viewed at different wavelengths. These photographs show a large sunspot group during August 1989 as seen in the normal visual spectrum (top) and when viewed through a narrow-band hydrogen-filter (bottom). Note the bright solar flare in the lower photograph. These superb photographs were taken by Professor Jean Dragesco from southern France, using a 360-mm catadioptric telescope.

or 0.4 nm) offer good views of the prominences extending from the limb, but they are not capable of showing details on the disk.

A narrow-band filter can be incorporated into an orthodox telescope fairly easily, but for viewing limb prominences with a wide-band filter it is essential to block out the solar disk, which otherwise will be too bright. An article describing a 'promscope' by the English observer H. E. Dall, who pioneered the method, will be found in reference [1].

Occasionally an observer will detect colour around a sunspot, usually as a rosy or violet tinge within the penumbra. Many of the famous solar observers of the past, Herschel, Schwabe, Schmidt and Lockyer among them, noted the effect during the course of their regular observations. Although its cause may be physiological or instrumental, other research suggests that colour effects arise from prominence activity in the sunspot area.

PROPER MOTIONS

Sunspot groups do not move around very much on the Sun; most of their apparent motion is caused by solar rotation. However, studies have shown that a group's main spots move apart by a few degrees of longitude as that group develops, then cease spreading when maximum growth is attained. Although this is an intriguing aspect, it is a difficult one for the amateur to separate from rotational movement because various latitudes of a body of gas such as the Sun rotate at different rates, a process known as *differential rotation*. This effect is illustrated for the sunspot zones by the measurements listed in Table 1.1 [2].

TERRESTRIAL EFFECTS

One of the reasons that the study of sunspots and flares continues to be so important is that the index which results from such research is highly correlated with other solar activities. Possibly the most important of these concerns the number

TABLE 1.1
THE EFFECTS OF DIFFERENTIAL ROTATION ON THE SUNSPOT ZONES

Latitude (deg)	Sidereal rotation rate (days)
0	25.1
15	25.5
30	26.5

and intensity of solar flares, which frequently contribute to the strong fluctuations in the Earth's magnetic field called *geomagnetic storms.*

The continuously expanding atmosphere of the Sun, known as the *solar wind*, rushes by the geomagnetic field, generating strong electrical currents, and flares and mass ejections intensify this effect. The best-known of the resulting phenomena are middle- and lower-latitude *aurorae*, which usually occur between 80 and 160 km above the surface of the Earth.

When a powerful flare erupts on the Sun, energetic particles stream outwards and align with the lines of force of the geomagnetic field. When they reach the Earth a day or two after the flare, the particles collide with atoms and molecules in the upper atmosphere, causing them to glow. Since the magnetic field is directed towards the Earth's magnetic poles, aurorae are much more common in high latitudes. The annual number of aurorae, which follows the yearly sunspot number but peaks slightly later, is important in that it has been used to extend the determined times of sunspot cycle maxima back to at least AD1500.

Another well-known consequence of this activity is a disruption of communications on Earth. The sudden release of energy from flares sends X-rays and other particles into space at speeds near that of light. Many of them are trapped or absorbed within the geomagnetic field and atmosphere, resulting in a change in the latter's ability to transmit radio signals. Occasionally, high-frequency radio communication becomes impossible, especially near the Earth's poles where the disruption can last for days.

SOLAR ECLIPSES

No other event in nature rivals the spectacular beauty and wonder of a total solar eclipse. The warm glow that precedes totality, the shadow of the Moon rushing towards you at thousands of kilometres per hour, the hushed expectation ... and then the awesome sight of the black disk of the Moon silhouetted against the stark white atmosphere of the Sun, the solar corona, appears. No wonder that the sight struck fear into the hearts of an unsuspecting ancient populace!

Those whose choose to view an eclipse must often travel long distances to exotic locations. Of course, for many of those who do, travel and the opportunity to talk with other enthusiasts are additional reasons for going! Details of forthcoming total solar eclipses, from reference [3], are presented in Table 1.2.

Unfortunately, there is not a lot the amateur astronomer can contribute in the way of scientific data about an eclipse. Those who have the necessary equipment and expertise may wish to make precise timings of the major stages of an eclipse from specific locations within the path of totality, and they are advised to contact the US Naval Observatory (Washington, DC 20392) for information in this regard. Otherwise, shadow bands (ripples of light and shade caused by the hair-thin solar crescent shining through unsteady air) may be seen, the flash spectrum (the sudden appearance of bright spectral lines within the chromosphere) can be photographed,

TABLE 1.2

FORTHCOMING SOLAR ECLIPSES, 1991–2010

Date	Maximum duration (min s)	Maximum width (km)	Area of visibility
1994 Nov 3	4 24	190	South America, S Atlantic Ocean
1995 Oct 24	2 10	78	Asia, Borneo, Pacific Ocean
1997 Mar 9	2 50	371	Siberia
1998 Feb 26	4 08	152	Pacific Ocean, N of South America, Atlantic Ocean
1999 Aug 11	2 23	112	Atlantic Ocean, Europe, SE and S Asia
2001 Jun 21	4 56	201	Atlantic Ocean, S Africa, Madagascar
2002 Dec 4	2 04	87	S Africa, Indian Ocean, Australia
2003 Nov 23	1 57	545	Antarctica
2005 Apr 8[a]	0 42	27	Pacific Ocean, Central America, N of South America
2006 Mar 29	4 07	189	Atlantic Ocean, Africa, Turkey, Russia
2008 Aug 1	2 28	251	N Canada, Arctic Ocean, Siberia, China
2009 Jul 22	6 40	258	Asia, Pacific Ocean
2010 Jul 11	5 20	262	Pacific Ocean, extreme S of South America

[a] Total for part of the eclipse path only.

Source: Meeus, J., Grosjean, C.C. and W. Van Der Leen 1966, *Canon of Solar Eclipses*, Pergamon Press Limited, Oxford, England.

or perhaps the onrushing shadow of the Moon can be timed; but total eclipses are best enjoyed aesthetically. Observing the partial phases of the eclipse, of course, requires an adequate solar filter, and shade 14 welder's glass is a good choice for non-instrumental viewing. No filter is employed during totality, but the observer *must not* continue to view the Sun without protection after the first bright points of sunlight – Baily's beads – begin to reappear along the trailing lunar limb.

IN CLOSING

There is an intangible but undeniable feeling that is shared by many of those who study and observe the Sun. It is beautifully stated by the renowned solar physicist Harold Zirin in his book *Astrophysics of the Sun*:

> Here before our eyes is a real star, whose remarkable phenomena we can sample in every way if we will only take the trouble. The signal-to-noise ratio is high; we can explore the world of magnetic phenomena and convection in plasmas. We can watch particles being accelerated in flares and measure the output at Earth. We can probe the interior of this star through the measurement of its oscillations, which can be carried out with exquisite precision. And with even a little telescope we can see the marvels of the sunspot cycle. I invite the reader to join the fun.

We join with Dr Zirin in his invitation ... welcome aboard!

About the Author

Peter O. Taylor has been chairman and chief analyst for the Solar Division of the AAVSO since 1980. An international network composed of well over one hundred collaborators contributes data to the sunspot and solar flare monitoring sections each month. These data are then analysed and forwarded to interested members of the professional scientific community. This effort is funded by an annual grant to the AAVSO from the National Oceanic and Atmospheric Administration. The programme also compiles information concerning aurorae and the rare solar 'white-light' flare events. Serious observers are invited to contact the programme if they would care to participate in this fascinating area of astronomy.

Mr Taylor and his wife, Pamela, reside in Georgia. His other interests include hiking, attempting to play chess, and the history of solar research.

Contact address – PO Box 5685, Athens, Ga 30604-5685, USA.

Notes and Comments

Measuring heliographic coordinates – Wilfrid Heyes, solar coordinator of *The Astronomer* magazine, writes:

> Many observers may find the use of solar graticules, such as Porter's Solar Disk and the Stonyhurst Disks, more convenient than the full mathematical determination. These graticules do not seem to be well known abroad. An excellent account of Porter's Solar Disk and its use is found in [4], and there is also an article on making Stonyhurst Disks in [5]. The advantage of a set of Stonyhurst Disks (which I use) is that they enable sunspot positions to be read off directly in latitude and longitude east or west of the Sun's meridian either from drawings or by projecting the Sun's image on to a correctly-orientated Stonyhurst Disk having the appropriate value of B_0.

> Mention may also be made of using photographs (instead of drawings) for an accurate determination of sunspot positions by measuring the position angle of the spot from the north point of the Sun's axis of rotation and the distance of the spot from the disc centre. The formulae on pages 11–13 may be applied to obtain the latitude and longitude of the spot.

On seeing this comment, the author pointed out that the AAVSO Solar Division sells (at cost plus postage) sets of 18-cm Stonyhurst Disks, sets of small (73-mm diameter) Stonyhurst Disks with instructions which explain their use with sketches made using direct viewing, and a 15-cm-diameter Porter Disk with instructions.

On the same subject, it may be worth drawing readers' attention to a very old and elegant method invented by William Carrington, a famous solar observer of the middle nineteenth century. Its main drawback – the laborious hand-calculation involved – has evaporated now that computers are here to do the work. As far as I know, this method is not described in any work currently in print. It is used with a stationary telescope, and therefore an altazimuth is as suitable as an equatorial.

Two lines AA and BB, exactly at right angles, are bisected by a third fainter line CC (Fig. 1.9). These may be drawn on a projection screen or incorporated in an eyepiece graticule. It is important that the centre of the crossed lines is on the optical axis of the telescope, and that the screen (if the projection method is used) is at right angles to the axis. The orientation is adjusted until a sunspot trails accurately along the line CC, and the telescope is arranged so that the centre of the solar image passes a little way above or below the centre of the crossed lines.

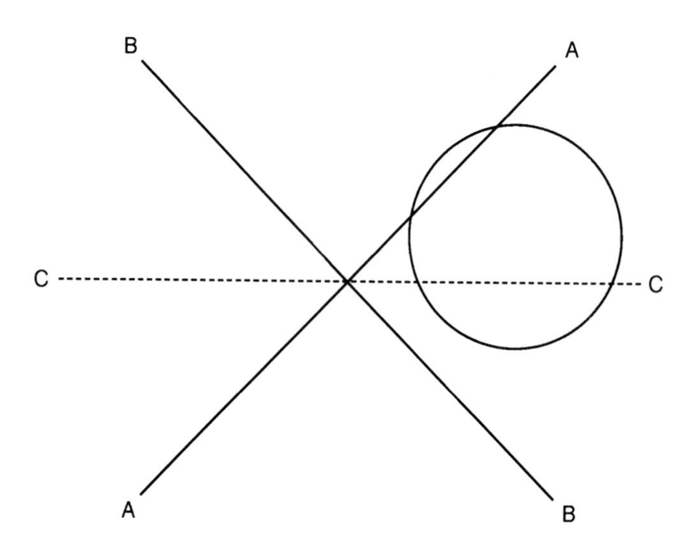

FIGURE 1.9
Carrington's method of deriving sunspot positions.

The image is then allowed to trail over the lines, and the times are recorded when the following features cross lines AA and BB:

Preceding limb (A_p and B_p)
Following limb (A_f and B_f)
The spot itself (a and b)

Six timings are therefore secured.

The values for the centre of the disk (A_c and B_c) are first derived from

$$A_c = \frac{A_p + A_f}{2} \qquad B_c = \frac{B_p + B_f}{2}$$

The calculations involve using the derived values of the spot's position angle ϕ and heliocentric angular distance from the disk centre ρ to determine its heliographic latitude B and longitude L.

Given that P = position angle of the N end of the solar axis, and i = inclination of the Sun's path to the parallel of declination passing through the disk centre (Table 1.3), then

$$\theta = \alpha + (\varepsilon \pm i - \rho)$$

where

$$\tan \varepsilon = \frac{A_f - A_p}{B_f - B_p}$$

TABLE 1.3

THE INCLINATION *i* OF THE SUN'S PATH TO THE PARALLEL OF DECLINATION PASSING THROUGH THE DISK CENTRE

		i (deg)			i (deg)
Jan	6	+0.02	Jul	9	−0.02
	14	0.03		19	0.03
	24	0.04		30	0.04
Feb	5	0.05	Aug	13	0.05
	25	0.06	Sep	8	0.06
Apr	10	0.06	Oct	13	0.06
May	2	0.05	Nov	7	0.05
	16	0.04		19	0.04
	27	0.03		29	0.03
Jun	5	0.02	Dec	7	0.02
	14	+0.01		15	−0.01
	22	0.00		21	0.00
	30	−0.01		30	+0.01

and

$$\tan\alpha = \frac{a - A_c}{b - B_c}\,\frac{1}{\tan\varepsilon}$$

while

$$\frac{r}{R} = 2\sec\alpha\,\frac{b - B_c}{B_f - B_p}$$

Then

$$\sin B = \cos\theta\cos B_0\sin\rho + \sin B_0\cos\rho$$

where B_0 = heliographic latitude of the centre of the disk; and $\sin l = \sin\theta\sin\rho\sec B$ with $L = L_0 - l$, where L_0 = heliographic longitude of the centre of the disk, and l = longitude of the spot measured from the central meridian.

With direct keying-in of the contact times, values of B and L could be obtained immediately, with further runs permitting a check on accuracy.

Magnetometer observations – 'Disturbances in the geomagnetic field are closely tied to the condition of the solar wind, to the interplanetary magnetic field, and ultimately to solar activity', wrote Ron Livesey in *Sky & Telescope*, October 1989. In this article he described a simple swinging magnet carrying a small mirror to deflect a light beam along a scale. This 'jam-jar magnetometer', illustrated in Fig. 1.10, has served as an early-warning detector for auroral displays, as well as permitting regular monitoring of solar activity.

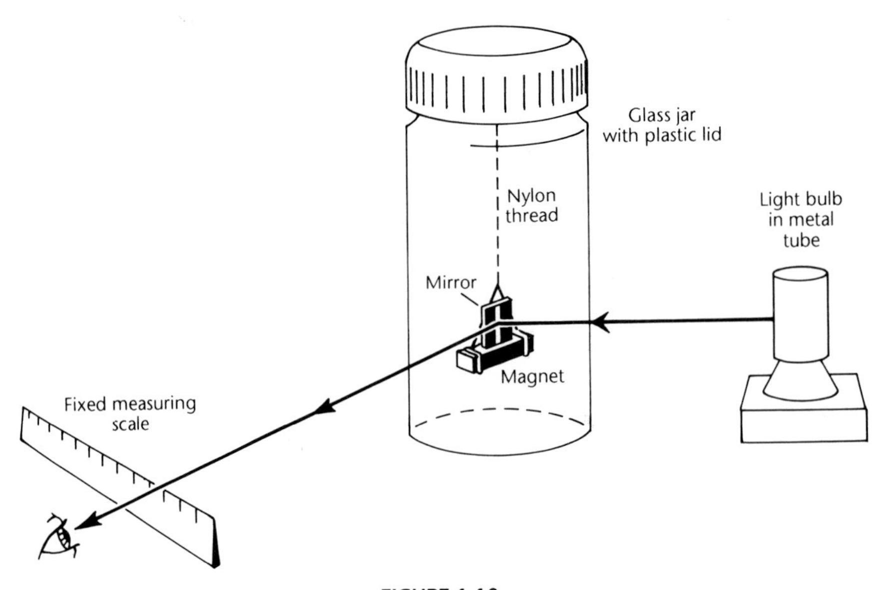

FIGURE 1.10
The 'jam-jar' magnetometer.

The illustration is almost self-explanatory, but two points should be noted. The mirror is blackened except for a vertical central strip about 1 mm wide. This produces a well-defined, narrow image at the end of the measuring scale, whose position can be read off easily. The magnet needs to be damped to prevent excessive swinging. Pieces of cardboard glued to the back and sides of the magnet achieve this by increasing the air drag.

Regular hourly readings are enough to monitor normal activity, but during solar storms variations may be observed almost continuously. For further information, consult the article [6] – Ed.

Bibliography

Astronomical Almanac, US Government Printing Office, Washington, DC, and HMSO, London.

Noyes, R.W., *The Sun, Our Star*. Harvard University Press, Cambridge, Mass., 1982.

Zirin, H., *Astrophysics of the Sun*. Cambridge University Press, Cambridge, 1988.

References

1. Dall, H.É., *B.A.A. Journal*, 77, 2 (1967).
2. Menzel, D.H., *Our Sun*. Country Life Press, Garden City, NJ, 1949.
3. Meeus, J. *et al.*, *Canon of Solar Eclipses*. Pergamon Press, Oxford, 1966.
4. Muirden, J., *The Amateur Astronomer's Handbook*, 3rd edn, p. 126. Harper & Row, London, 1987.
5. Mayne, A., *B.A.A. Journal*, 78(5), 356 (1968).
6. Livesey, R.J., *Sky & Telescope*, 78(4), 426 (October 1989)

CHAPTER TWO · *Observing the Modern Moon*

JOHN E. WESTFALL
EXECUTIVE DIRECTOR, ASSOCIATION OF LUNAR AND PLANETARY OBSERVERS

OUR MOON IS AN APPROXIMATELY SPHERICAL BALL OF ROCK 3476 KM in diameter, with a negligible atmosphere. It is rotationally locked in an orbit about 384 000 km from the Earth. These bald facts have several implications. The Moon is one of the two largest-appearing objects in the sky, over half a degree across. It is the only body which remains about the same angular size as it goes through a complete sequence of phases from New to Full (the *lunation*). The same lunar hemisphere always faces the Earth. It is about a hundred times nearer than the nearest distance any major planet can approach us, and is thus the only celestial body on which we can readily see relief features from the Earth. Finally, it is the only extraterrestrial body on which men have walked: the extensive results of its study from space and on its surface both restrict and benefit the amateur astronomer who wishes to contribute to lunar science. For the more casual observer, our nearest neighbour provides a constantly changing panorama of features to view and record.

LIGHTING AND VIEWING CONDITIONS

The fact that the same hemisphere of the Moon is always turned towards us, combined with its changing phases, means that its appearance is determined by the position of its *terminator* (the sunrise/sunset line) as it moves across the disk. Before Full phase we see the sunrise terminator; after Full, the sunset.

The *age* of the Moon, measured in days, is a fairly crude measure of lunar lighting. *Solar colongitude* (C) measures the lunar longitude of the sunrise terminator, measured to the lunar west continuously from 0° through 360°. To about ±7°

accuracy, the colongitude is 270° at New Moon (0 days), after which it increases by about 12.2° per day, reaching 360° (or 0°) at First Quarter (7 days), 90° at Full Moon (14 days), and 180° at Last Quarter (21 days). The position of the terminator determines which lunar areas will be in night and invisible, which will have a low sun angle to bring out their relief features, and which will have a high Sun to accentuate their tonal detail.

The *selenographic solar latitude* (b.) has a much smaller effect on the visibility of features: the Moon has almost no seasons because its equatorial plane is tilted only 1.5° to the ecliptic. For all but the polar regions shadows fall about east–west and relief features oriented in that direction are less clearly illuminated than those aligned more north–south.

Colongitude and solar latitude accurately define the solar lighting at any moment for the whole Moon, and may be used to compute the Sun's altitude for any point on its surface (see the section on slope and elevation measurement). The Sun is not all that shines on the Moon, however, and near New Moon one often sees the night portion faintly illuminated by second-hand sunlight reflected from the Earth, a phenomenon known as *Earthshine*.

The Moon's periods of rotation and revolution are the same – 27.3 days – which is why we see only one side. On the other hand, the Moon's elliptical orbit means that its angular orbital velocity varies, causing a ±7° east–west wobble called the *libration in longitude* (l'). Similarly, the tilt of the Moon's axis to its orbit creates a ±6° north–south wobble, the *libration in latitude* (b'). The effect of these librations is to cause the areas near the limb to be alternatively hidden and exposed. Thus, there is 41% of the Moon always turned toward the Earth, 41% always invisible, and 18% occasionally visible.

PERSPECTIVE AND POSITIONS

The Moon appears disk-like to us, and we tend to forget that we are actually looking at a globe. Near the apparent centre of the disk we are looking down on the Moon's features and consequently see them in true perspective. The nearer we approach the limb, the more foreshortened the features appear, making their interpretation more difficult and their appearance more affected by libration. Besides this, near the limb an object's elevation affects its apparent position, high features being displaced away from the disk centre and low features being displaced towards it.

When observing or simply describing the Moon's features we need to use a standard direction and coordinate system. North and south are simple: we use the Moon's poles. East and west are more uncertain. When we look at the Moon projected on to the sky, with north up, celestial east is to the left. However, were we standing on the Moon facing north, lunar east would be to our right, just as on the Earth. Until 1961, lunar observers used the celestial direction system. In that year, the IAU (International Astronomical Union) recommended that the lunar system (north up, east to the right) be used, and the 'IAU system' is now the dominant one. The IAU also recommended that maps, drawings, and photographs be oriented with north up. However, the normal inverted telescopic view in our Northern Hemisphere is south up, and that is the convention used in the illustrations in this chapter (Fig. 2.1).

Positions on the Moon are referred to by either of two interchangeable means. The first, called *direction cosines*, superimposes a square Cartesian grid upon the apparent disk, centred on the disk centre at mean libration. One axis (η) is aligned to lunar north; the other, called ξ, is positive to the lunar east. This two-dimensional system is convenient for mapping the appearance of the Moon at mean libration, and can be made more useful by adding a third dimension (ζ) positive towards the Earth. When using a 'true' map projection, predicting solar lighting, or making slope or altitude measurements, then latitude (ϕ) and longitude (λ) are the more convenient position system, and this is the one used in this chapter (in Fig. 2.1, for example). These angular coordinates are used just as they are on the Earth and the other bodies of our solar system.

LUNAR AMATEUR ASTRONOMY

Some familiarity with the facts and conventions described above is necessary for all but the most casual lunar observing. For more serious lunar work, there are two paths that the amateur can take. The first we might call 'Moon appreciation'. The Moon's proximity and lack of atmosphere allow us to see more detail than with any other celestial object, with features somewhat analogous to the terrestrial ones with which we are familiar. Add to this the fact that the appearance of all these features is always different owing to changing lighting and libration, and it is clear that simply viewing the Moon in an appreciative and intelligent manner can be an inexhaustible source of fascination.

There is, though, another direction that the serious amateur can take: contributing to lunar science by increasing our knowledge about the Moon. Here, the role of the Earth-based observer is more restricted than before the Space Age. For example, there is little need for the telescopic charting of the Moon's features except in some special cases. Many of the lunar 'mysteries' were cleared up by the Lunar Orbiter, the Apollo missions, and other space missions. On the other hand, the Moon was visited for only a brief period and only a small portion of it was studied in any detail. The serious lunar observer must be familiar with the results of the lunar missions, if only to determine what gaps in our knowledge remain to be filled. Several such gaps are described later in the sections on observing methods and special projects.

THE MOON'S SURFACE FEATURES

Since the Moon has no significant atmosphere, and because we are currently in no position to explore its underlying geology directly, its surface features alone are available for study. These exist in greater variety than many suppose: there are plenty of craters, but there is much else as well. For example, the Moon has two 'personalities': when the Sun is low for a lunar region, *relief features* are accentuated; however, when the Sun is high it is the *albedo features* that are clearest. When particular lunar features are cited in this chapter, they (or their location) can be found in Fig. 2.1.

Albedo Features

As with most solid bodies of the solar system, the albedo (tonal) features are all that we see from a distance: a mosaic of regions of different reflectivity. To the naked eye, these create the pattern that some call 'The Man in the Moon'. These relatively dark patches, called *maria* ('seas': *mare* in the singular form) occupy

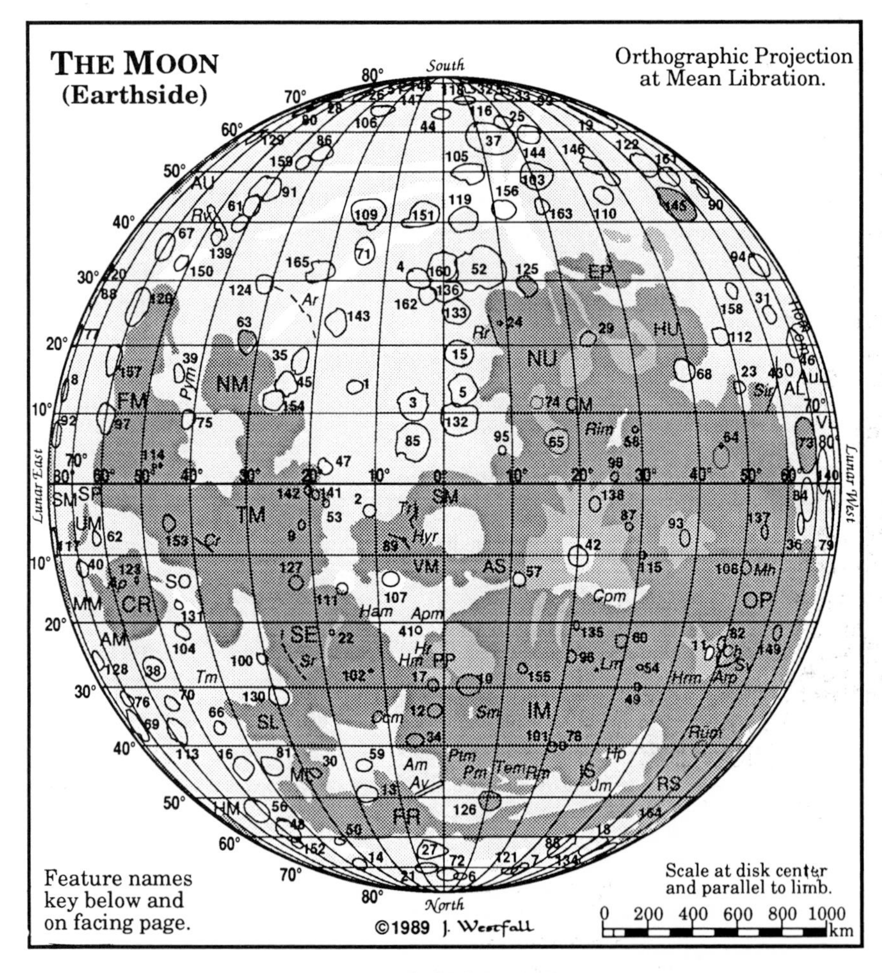

FIGURE 2.1
Map of the Moon. Features are shown by numbers for craters and abbreviations for other categories (see accompanying key).

Craters

1. Abulfeda, 14S14E
2. Agrippa, 04N10E
3. Albategnius, 11S04E
4. Aliacensis, 31S05E
5. Alphonsus, 14S03W
6. Anaxagoras, 74N10W
7. Anaximenes, 72N44W
8. Ansgarius, 13S79E
9. Arago, O6N21E
10. Archimedes, 30N04W
11. Aristarchus, 24N48W
12. Aristillus, 34N01E
13. Aristoteles, 50N17E
14. Arnold, 67N36E
15. Arzachel, 18S02W
16. Atlas, 47N44E
17. Autolycus, 31N01E
18. Babbage, 60N57W
19. Bailly, 67S60W
20. Barnard, 29S86E
21. Barrow, 71N08E
22. Bessel, 22N18E
23. Billy, 14S50W
24. Birt, 22S08W
25. Blancanus, 64S22W
26. Boguslawsky, 73S43E
27. W. Bond, 65N04E
28. Boussingault, 70S55E
29. Bullialdus, 21S22W
30. Bürg, 45N28E
31. Byrgius, 25S65W
32. Cabeus, 85S36W
33. Casatus, 73S30W
34. Cassini, 40N04E
35. Catharina, 18S24E
36. Cavalerius, 05N67W
37. Clavius, 58S14W
38. Cleomedes, 28N56E
39. Colombo, 15S46E
40. Condorcet, 12N70E
41. Conon, 22N02E
42. Copernicus, 10N20W
43. Crüger, 17S67W
44. Curtius, 67S04E
45. Cyrillus, 13S24E
46. Darwin, 20S69W
47. Delambre, 02S18E
48. de la Rue, 59N53E
49. Delisle, 30N35W
50. Democritus, 62N35E
51. Demonax, 78S59E
52. Deslandres, 32S05W
53. Dionysius, 03N17E
54. Diophantus, 28N34W
55 .Drygalski, 80S80W
56. Endymion, 54N56E
57. Eratosthenes, 14N11W
58. Euclides, 07S30W
59. Eudoxus, 44N16E
60. Euler, 23N29W
61. Fabricius, 43S42E
62. Firmicus, 07N63E
63. Fracastorius, 21S33E
64. Flamsteed, 04S44W
65. Fra Mauro, 06S17W
66. Franklin, 39N48E
67. Furnerius, 36S60E
68. Gassendi, 18S40W
69. Gauss, 36N79E
70. Geminus, 34N57E
71. Gemma Frisius, 34S13E
72. Goldschmidt, 73N03E
73. Grimaldi, 05S68W
74. Guericke, 12S14W
75. Gutenberg, 09S41E
76. Hahn, 31N74E
77. Hecataeus, 22S79E
78. Helicon, 40N23W
79. Hedin, 03N76W
80. Helmholtz, 68S64E
81. Hercules, 47N39E
82. Herodotus, 23N50W
83. J. Herschel, 62N41W
84. Hevelius, 02N68W
85. Hipparchus, 06S05E
86. Hommel, 55S33E
87. Hortensius, 06N28W
88. Humboldt, 27S81E
89. Hyginus, 08N06E
90. Inghirami, 48S69W
91. Janssen, 45S42E
92. Kästner, 07S79E
93. Kepler, 08N38W
94. Lagrange, 33S72W
95. Lalande, 04S09W
96. Lambert, 26N21W
97. Langrenus, 09S61E
98. Lansberg, 03S27W
99. le Gentil, 74S76W
100. le Monnier, 26N30E
101. LeVerrier, 40N21W
102. Linné, 28N12E
103. Longomontanus 50S22W
104. Macrobius, 21N46E
105. Maginus, 50S06W
106. Manzinus, 68S27E
107. Manilius, 14N09E
108. Marius, 12N51W
109. Maurolycus, 42S14E
110. Mee, 44S35W
111. Menelaus, 16N16E
112. Mersenius, 22S49W
113. Messala, 39N60E
114. Messier, 02S48E
115. Milichius, 10N30W
116. Moretus, 71S06W
117. Neper, 09N84E
118. Newton, 77S17W
119. Orontius, 40S04W
120. Petavius, 25S60E
121. Philolaus, 72N32W
122. Phocylides, 53S57W
123. Picard, 14N54E
124. Piccolomini, 30S32E
125. Pitatus, 30S14W
126. Plato, 51N09W
127. Plinius, 15N24E
128. Plutarch, 24N79E
129. Pontécoulant, 59S66E
130. Posidonius, 32N30E
131. Proclus, 16N47E
132. Ptolemaeus, 09S02W
133. Purbach, 26S02W
134. Pythagoras, 63N62W
135. Pytheas, 20N21W
136. Regiomontanus, 28S01W
137. Reiner, 07N55W
138. Reinhold, 03N23W
139. Rheita, 37S47E
140. Riccioli, 03S73W
141. Ritter, 02N19E
142. Sabine, 01N20E
143. Sacrobosco, 24S17E
144. Scheiner, 60S28W
145. Schickard, 44S55W
146. Schiller, 52S40W
147. Schomberger, 77S25E
148. Scott, 82S45E
149. Seleucus, 21N66W
150. Stevinus, 32S54E
151. Stöfler, 41S06E
152. Strabo, 62N54E
153. Taruntius, 06N46E
154. Theophlius, 11S26E
155. Timocharis, 27N13W
156. Tycho, 43S11W
157. Vendelinus, 16S62E
158. Vieta, 29S56W
159. Vlacq, 53S39E
160. Walter, 33S01E
161. Wargentin, 50S60W
162. Werner, 28S03E
163. Wilhelm, 43S21W
164. Xenophanes, 57N80W
165. Zagut, 32S22E

Mare Units

(With approximate center latitudes and longitudes in degrees.)

AL. Lacus Aestatis, 15S69W
AS. Sinus Aestuum, 11N07W
AM. Mare Anguis, 23N68E
AU. Mare Australe, 50S90E
AuL. Lacus Autumnae, 15S80W
CM. Mare Cognitum, 10S20W
CR. Mare Crisium, 15N60E
EP. Palus Epidemiarum, 32S28W
FM. Mare Fecunditatis, 10S50E
FR. Mare Frigoris, 60N00E
HM. Mare Humboldtianum, 55N80E
HU. Mare Humorum, 25S40W
IM. Mare Imbrium, 40N20W
IS. Sinus Iridum, 45N32W
MM. Mare Marginis, 15N85E
MS. Sinus Medii, 00N00E
ML. Lacus Mortis, 45N27E
NM. Mare Nectaris, 15S35E
NU. Mare Nubium, 20S15W
OP. Oceanus Procellarum, 20N50W
PP. Palus Putredinis, 27N00E
RS. Sinus Roris, 50N60W
SE. Mare Serenitatis, 25N15E
SM. Mare Smythii, 00N85E
SO. Palus Somni, 15N44E
SL. Lacus Somniorum, 39N30E
SP. Mare Spumans, 00N65E
TM. Mare Tranquillitatis, 05N30E
UM. Mare Undarum, 07N70E
VM. Mare Vaporum, 14N04E
VL. Lacus Veris, 10S85W

Non-Crater Relief Features

Ap. Promitorium Agarum, 14N65E
Am. Montes Alpes, 47N00E
Av. Vallis Alpes, 49N03E
Ar. Rupes Altai, 22S23E
Apm. Montes Apenninus, 20N00E
Arp. "Aristarchus Plateau," 28N50W
Cpm. Montes Carpatus, 15N25W
Ccm. Montes Caucasus, 35N09E
Cr. Rupes Cauchy, 09N37E
Ch. "Cobra Head," 25N49W
Com. Montes Cordillera, 20S80W
Hm. Mons Hadley, 27N04E
Hr. Rima Hadley, 25N03E
Ham. Montes Haemus, 19N11E
Hrm. Montes Harbinger, 27N41W
Hp. Promitorium Heraclides, 41N34W
Hyr. Rima Hyginus, 08N06E
Jm. Montes Jura, 48N35W
Lm. Mons La Hire, 28N26W
Mh. "Marius Hills," 12N55W
Pm. Mons Pico, 46N09W
Ptm. Mons Piton, 41N01W
Pym. Montes Pyrenaeus, 15S41E
Rr. Rupes Recta, 22S08W
Rm. Montes Recti, 48N20W
Rv. Vallis Rheita, 40S48E
Rim. Montes Riphaeus, 08S28W
Rom. Montes Rook, 22S85W
Rüm. Mons Rümker, 41N58W
Sv. Vallis Schröteri, 25N51W
Sr. "Serpentine Ridge," 25N25E
Sir. Rima Sirsalis, 15S61W
Sm. Montes Spitzbergensis, 35N05W
Tm. Montes Taurus, 28N42E
Tem. Montes Teneriffe, 48N13W
Tr. Rimae Triesnecker, 05N05E

about 29.6% of the lunar near side, but only 1.3% of the far side. The largest is an *oceanus* (Oceanus Procellarum), and it should be noted that, like most of the categories of lunar formation, these features have a Latin class name. The major maria are mapped in Fig. 2.1, and represent relatively smooth, low-lying, young basaltic lava plains.

The remainder of the Moon's surface consists of the *highlands* or *terrae* (singular, *terra*). Lighter in tone than the maria, the highlands are also rougher and more weathered. For the most part they live up to their name, being higher than the maria. The maria–highlands distinction is clear at all phases, but the maria are seen in their fullest extent near full phase. Smaller dark features that are maria in every respect but their size often have the names *lacus* (lake), *palus* (marsh) or *sinus* (bay).

The keen eye may also discern light patches or streaks on the parts of the Moon that have a high sun. With even a pair of binoculars, these *ray systems* become obvious. They tend to form bright patches (*nimbuses*) centred on certain craters, sometimes accompanied by radial streaks that extend large distances from the central crater. Some major ray systems, such as those centred on the craters Tycho, Copernicus, Kepler, Aristarchus and Proclus are indicated in Fig. 2.1. With a more powerful instrument we see that even tiny craters often have light nimbuses around them. More rarely, a crater has a dark nimbus: that immediately surrounding Tycho is the largest-scale example.

Maria, ray systems and nimbuses form only the more obvious tonal features. Actually, under a high Sun, at all scales one finds a continuous range of shades of grey. Crater rims tend to be bright: their floors are often dark. A few craters have dark radial bands within them, or sometimes dark spots, both varying in intensity as the Sun angle changes. As with, for example, Mars, tonal features are occasionally associated with relief features, but frequently they are not.

RELIEF FEATURES

Amongst the Moon's relief features, the largest specimens are the *basins*, or *impact basins* as most lunar scientists feel that they are caused by the impacts of asteroids, comets, or both. They are roughly circular, often with two or more concentric rings (*multi-ring basins*), but their rims are often weathered and incomplete. The interior is depressed in relation to the exterior. The smaller specimens are about 300 km across; the diameter of the largest one is about 2000 km. Because the crust is thinner on the side of the Moon facing the Earth, several of the earthside basins have been flooded with lava and form the circular maria: Maria Crisium, Nectaris, Serenitatis, Imbrium and Humorum are clear examples. Figure 2.2 shows a stylized profile of a basin, along with several other recognized forms of lunar relief.

In their general morphology, the smaller basins merge with the larger *craters*. The famous craters of the Moon are circular, elliptical or polygonal depressions, with their interiors lower than the surrounding terrain and their rims elevated slightly above the outside surface and higher above the interior. The modern convention is to use the term *crater* for all such features, regardless of their size or detailed morphology. An older usage is to call the larger craters *walled plains*. This reflects the fact that they have flat interior *floors* with a ring of mountains circling them. Examples of such features are Endymion, Plato, Clavius, Schickard and Bailly. With the walled plains, the interior floor may have craters, hills and other features, but no clear-cut *central mountain*. As one proceeds down the size scale, dropping below about 100 km in diameter, the floor becomes smaller in relation to the rim diameter and central mountains or central peaks become more common. These are sometimes called *ringed plains*, and include Petavius, Theophilus, Eratosthenes, Copernicus and many others. When the crater rim diameter drops below about 30 km, central peaks disappear. An old term for this smallest group of craters is *craterlets*. When a craterlet's rim is flush with the outer surface, the term *craterpit* applies.

The descriptions above fit well-defined craters. However, like all forms of lunar relief, craters exhibit various degrees of weathering. The chief agency of lunar weathering is the continual influx of meteorites; but temperature changes, solar radiation and the solar wind also play a role. Thus an older crater may be fragmentary, with a broken rim like Pitatus or Fracastorius, or have very subdued relief. Craters can also be inundated with mare lava, forming *rings* such as the

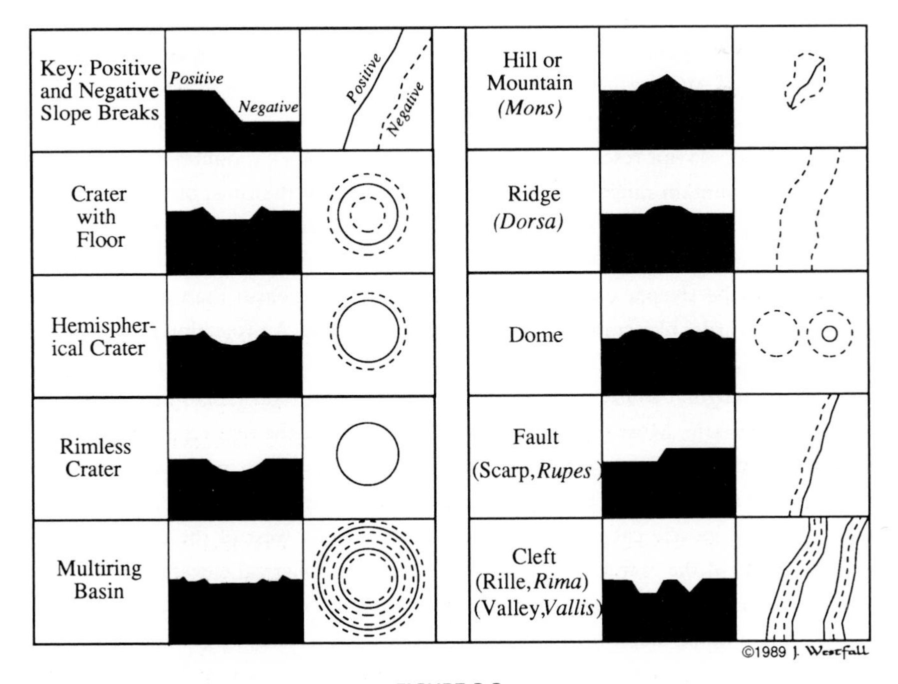

FIGURE 2.2
Types of lunar feature. Profile and plan views are shown.

Flamsteed Ring. In the extreme case, one has a barely-discernible outline called a *ghost crater*. Other crater forms include the *summit craters* on some domes and peaks, and merged rows of craters called *cataenae* or *crater chains*.

Most lunar scientists hold that the great majority of craters are *exogenous*: they were formed by the impact of meteorites, asteroids and even comets. Craters so formed are called *primary craters*. In addition, the explosive formation of craters threw out clumps of material (*ejecta*) that often, when landing, created *secondary craters*. For example, the rows of craterpits east of Copernicus are secondary craters caused by the formation of that crater. A significant minority of craters have locations or form patterns that are unlikely to be due to random impacts. Examples of these are the summit craters and crater chains, which are suspected of being caused by 'volcanism' or, better, *endogenous processes*.

All lunar depressions, including craters, are examples of *negative relief features*. This category also includes elongated depressions, usually called *rimae* (singular, *rima*) such as the aforementioned crater chains represented by the Hyginus Rille. Winding elongated depressions tend to be called *vallis*, a loose category that includes the *graben* (depression between two parallel faults), of which the Vallis Alpes (Alpine Valley) is a clear example, and large crater chains like the Vallis Rheita. Quite often such valleys radiate from the major basins, indicating that they are formed from large secondary craters.

Elevations constitute the Moon's *positive relief features*. However, there is one form, the *rupes*, which is neither positive nor negative, being a fault with the terrain on one side lower than on the other. The two best-known examples are the Rupes Recta (Straight Wall) and the Rupes Cauchy.

The largest positive relief features on the Moon are its mountain ranges, the *montes*, which do not resemble the familiar folded or block mountains of the Earth. The lunar mountain ranges are the rims, or portions of the rims, of the basins. For example, the Mare Imbrium Basin is rimmed by the Montes Harbinger, Jura, Alpes, Caucasus, Apenninus and Carpatus. The terrain of such ranges is often confused, but tends to be steeper on the portion facing into the basin than on the portion sloping outwards: like crater walls but on a larger scale. A projection of the montes into the maria receives the fitting name of *promitorium*. A large number of individual mountains (singular *mons*) are found, both in the mountain ranges and as isolated peaks in the maria. Most such peaks are not named, but the first type is represented by Mons Hadley in the Apennines, and the second by Mons Pico in the Mare Imbrium.

Smaller relief features are rarely named. Small elevations, particularly if conical in profile, are loosely called *hills*, like the Marius Hills west of the crater Marius. In some parts of the maria are found low, often hemispherical elevations: the *lunar domes*. A well-known *dome field* lies near the craters Hortensius and Milichius.

The remaining major form of positive relief is the *wrinkle ridges* which meander in parts of some of the maria. The prominent example in eastern Mare Serenitatis has the nickname 'Serpentine Ridge'. These often parallel the rims of the mare basins.

Thus we have listed over two dozen types of features on the Moon's surface, each of which can take many forms. Further investigation will show a number of minor forms of features, some hybrids and some that defy classification.

EQUIPMENT AND RECORDS

The Moon being the most accessible astronomical body, a variety of instruments and methods of observing it can be used. Instrumental requirements for making useful drawings are a small telescope at least 100 mm aperture, with good optics and preferably an equatorial mounting. The more 'technological' forms of observations – photography, video making, CCD imaging, photometry and measuring of elevations – work best with a somewhat larger instrument on a clock-driven equatorial mounting: fine slow-motion controls in both right ascension and declination are then very helpful. For drawing, the only accessories one needs are a small set of high-quality eyepieces (Plössls or orthoscopics work well), supplemented by an achromatic Barlow if the telescope has a short focal ratio. When atmospheric conditions permit it, magnifications in the range of 1–2 per millimetre aperture should be available. In addition to or in place of eyepieces, the other modes of observing the Moon require their own specialized equipment, described below.

Whatever form one's lunar observing takes, observations lose most of their value unless accompanied by adequate records. At a minimum, record everything pertinent that could not be computed later. This would at the least include the date and UT to one minute; the telescope aperture and type; the magnification (if a visual observation); the atmospheric seeing, specifying whether the ALPO 1–10 or the Antoniadi I–V scale is used); and the atmospheric transparency (for lunar work, expressed best in the ALPO scale ranging from 0 for worst to 5 for perfect). Optional but useful are the colongitude at the time of observation; and for observations of the polar regions, or whenever the lighting is critical, also give the solar latitude. If the observations are submitted to someone else, include your name and full address. The more specialized forms of observing, as will be seen, require some specific data.

DRAWING THE MOON

This, the oldest form of lunar observation, is still used because with a given instrument and atmospheric conditions the eye can detect finer detail than photographic film or the video camera are able to do. Drawings of the Moon need not be attractive, but they should be accurate. The most common working medium is soft pencil and white paper, held on a clipboard and illuminated at the telescope by a low-wattage red light. More talented souls sometimes use charcoal, chalk, or crayon to shade in their drawings. The person who draws the Moon need not be an artist, but should be able to portray each feature seen in correct-scale size and shape, and in the correct position relative to other features. Refinements such as inking shadows in block or showing shading gradations are not always needed, and in any case can often be done after the observing session.

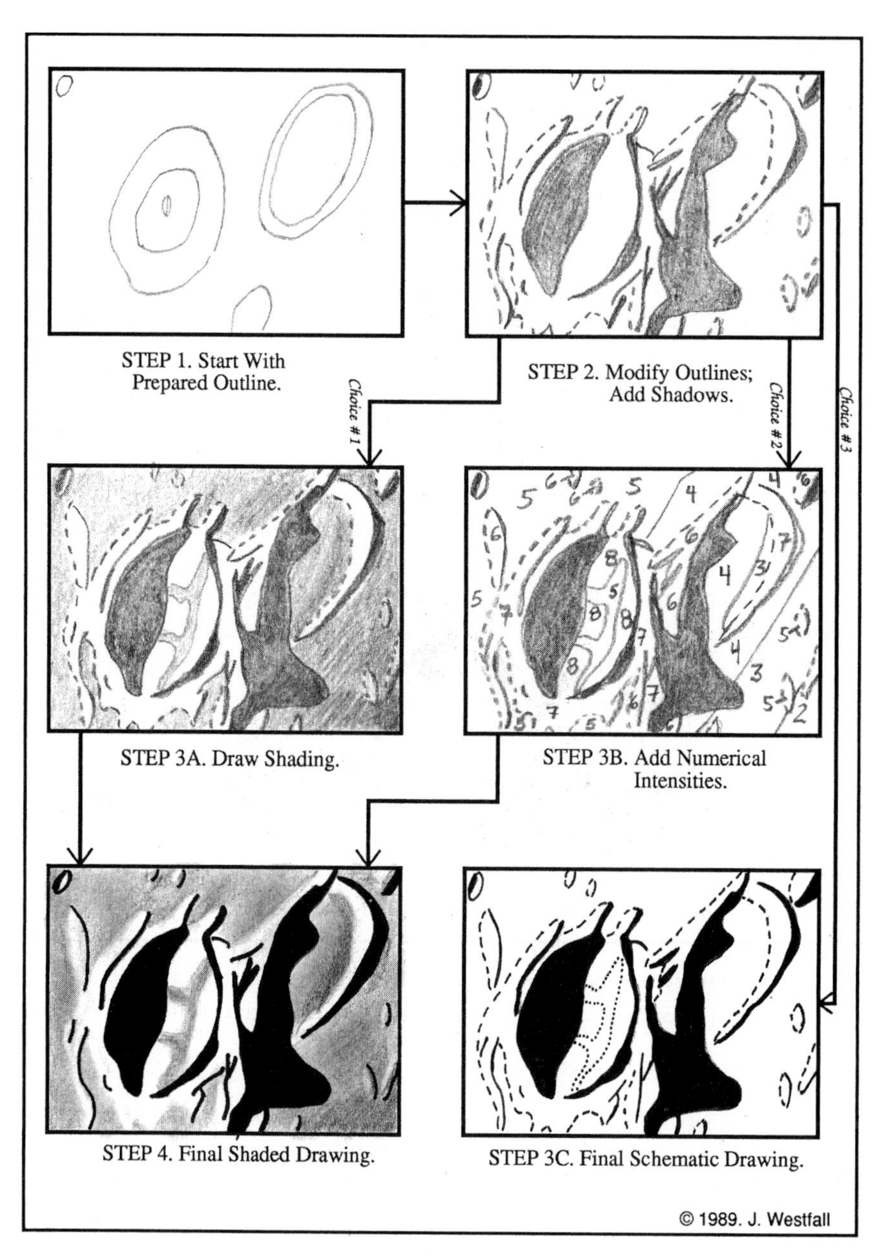

FIGURE 2.3
Preparing a lunar drawing. The different stages are shown. The crater Aristarchus is on the left, with Herodotus on the right, as drawn by the writer on 12 December 1959 (02.35–03.25 UT), with a 508-mm refractor at ×350. Colongitude 53.5°. Seeing fair, transparency good. South at top. See also text.

Figure 2.3 shows the four steps of making a lunar drawing, not counting the initial selection of the area to be drawn. The first step is to use an Earth-based photograph, or an accurate orthographic-projection (Earth-perspective) map to trace with thin lines the outlines of major features on to the blank paper. The Moon holds considerable detail, so make your drawing at a large scale of at least 0.5 mm and preferably 1 mm per kilometre (in other words, 1 : 2 000 000 or 1 : 1 000 000). If working elsewhere than near the centre of the disk, it is best to trace outlines from a photograph taken under librations similar to those at the time of observation.

Step 2 involves drawing in heavier lines the actual feature outlines as seen at the telescope, correcting their positions for libration if necessary. Follow this by drawing the outlines of black shadows and the position of the terminator if it falls within the drawing area.

The third step can take several forms, depending on one's talents and on what one wishes to portray. The simplest method (Step 3C) is to avoid all shading, but to indicate the bases of elevations with dashed lines, adding solid lines for those crests not casting black shadow. Another way to avoid representing shading at the telescope is to outline areas of different tone and to indicate their grey levels on a 0 (shadow) to 10 (white) scale. This is Step 3B. The final approach is actually to execute the shading at the telescope (Step 3A). Each of these three approaches is shown in a different portion of Fig. 2.3.

Step 4 follows the telescope session, but as soon as possible after it. Here, shadows are inked in black and smudges are erased. If the telescopic drawing simply shows line work, these may now be inked. If tonal zones are shown by numbers, they can be converted to actual shading. If one has done shading at the telescope, the shading strokes can be smoothed, either with an artist's stub, a tissue or cloth or even a finger. Necessary data should then be entered on the same sheet of paper as the drawing. However the drawing is finished, it should finally be covered with plastic, cellophane or spray fixative.

LUNAR PHOTOGRAPHY

Chapter 14 of this book addresses the problems of lunar and planetary photography, but the following notes may also be found useful.

Photographs of the Moon do not show as fine detail as does the eye under the same conditions, nor do they have the tonal range of the eye. The advantage of a photograph is that it shows a large area simultaneously, is objective, and has accurate relative positions. The basic equipment most lunar photographers use is a telescope with an accurate motor drive, preferably adjustable to the true 'lunar rate' at the time of observation; a means to enlarge the Moon's image, such as an eyepiece-projection system or a Barlow lens, and a 35-mm single-lens reflex camera. The last should have a removable lens, a cable release, a 'bulb' or B shutter setting, a reflex mirror that locks up to avoid mirror slam during the exposure, and a clear viewing screen with a magnifier for precise focusing.

Several factors influence the choice of film type. The Moon, frankly, is not very colourful, so black-and-white film is usually more than adequate. On the other hand, colour film is convenient if slides are required. Areas of the Moon under high sunlight show weak contrast, implying that a film, or processing, giving high contrast should be used. However, low-sun areas have very high contrast, requiring low-contrast film or developing in order to show them properly. One result of these conditions is that, except very near full phase, one cannot properly show both high-sun and low-sun areas on the same frame. One Kodak product recommended by most lunar photographers is Technical Pan 2415 film, which is suitable for all black-and-white work because of its extremely high resolution and its adaptability to speed and contrast modification by proper development. Kodak colour slide films that are often used are Kodachrome 64 or Ektachrome 100, both good for tonal colour rendition, or Kodachrome 200 for low-light situations. A special-purpose colour slide film, Kodak Photomicrography 2483, has extremely high resolution, contrast and colour separation: it is capable of routinely recording the subtle colours of the Moon, but must be exposed precisely and is very slow (ISO 16).

Assuming that the seeing is good, the focus is sharp, and there is no vibration during the exposure, the most common difficulty in lunar photography is proper exposure. Any table or formula is only a first start, and there is no substitute for experimentation, but *as a start* the writer uses this rule-of-thumb formula:

$$\text{Exposure (seconds)} = K \times \frac{(\text{focal ratio})^2}{\text{film ISO speed}}$$

The factor K depends on the Moon's phase and the sun angle on the area. It varies with phase as follows:

Full Moon	0.005
Gibbous (10 or 18 days)	0.01
Quarter (7 or 21 days)	0.02
Thick crescent (5 or 23 days)	0.04
Thin crescent (3 or 25 days)	0.1

The Full Moon value applies to the entire disk; for the other phases, the value is for medium-sun areas (solar altitude 10–20°). For terminator areas K should be at least doubled, while for high-sun areas it may be halved.

There are three basic optical systems which are used for lunar photography. The simplest and quickest (for example, to be used when waiting in line at a telescope) is the *afocal* method. Here, the camera lens is retained and focused at infinity. The telescope eyepiece is also retained, and should also be focused at infinity, either by viewing with one's eye relaxed (retaining glasses if worn) or, better, by focusing on the camera's focusing screen. The camera lens is simply held to the eyepiece and the exposure made. The effective focal ratio is that of the telescope multiplied by the factor (camera focal length/eyepiece focal length). One problem with this method is that the film is looking through *three* optical systems, which does not promote the sharpest image. Another difficulty is in positioning the camera at the eyepiece, although a tripod or special couplings can be used.

In terms of resolution on the film, the sharpest lunar photographs are taken at the telescope's prime focus, with the camera lens removed. A simple T-adaptor couples the camera to the eyepiece tube. This also makes for a relatively bright lunar image. The drawback is that the image scale is small and thus the film resolution is not equal to the telescope's potential resolution.

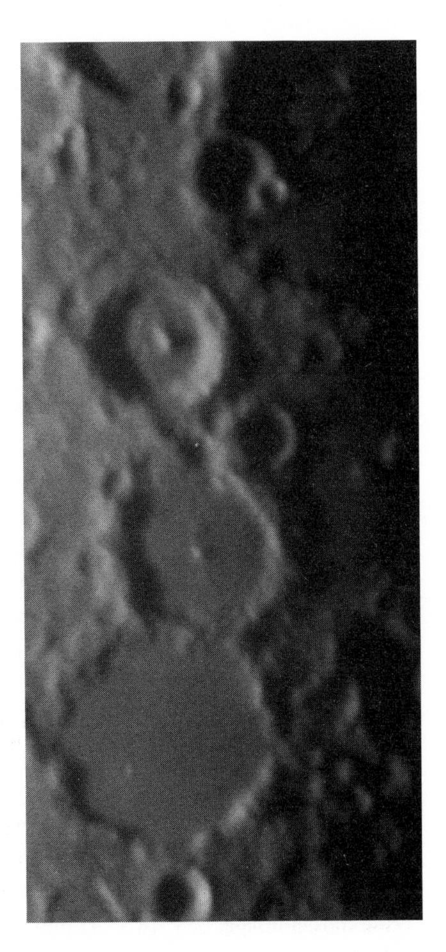

FIGURE 2.4
Lunar photography. Two photographs by the writer of the First Quarter Moon, taken within 4 minutes of each other and showing the effect of image enlargement. Both were obtained on 24 January 1972 (02.09–02.13 UT) with a 254-mm Cassegrain reflector on Plus-X film, and enlarged the same amount from the original negative. The left-hand view was at the Cassegrain focus at f/16.5, exposure 1/15 s, showing the area from Ptolemy at the bottom to the south limb at the top. The right-hand photograph was enlarged with a Barlow lens to f/55.6, exposure 1.5 s, showing the Ptolemy–Arzachel area. Colongitude 8.0°.

Thus we usually wish to enlarge the Moon's image on the film. This entails placing another optical component between the telescope objective or mirror and the film. The most popular component is a high-quality eyepiece, and the method is called *eyepiece projection*. Several firms offer couplers that link the telescope, the eyepiece, and the camera (with lens removed). By changing the eyepiece and the eyepiece–film spacing a considerable range of magnification is available. The chief difficulty is that eyepieces are not designed for projection, and may create a curved field of sharpest focus. Barlow (negative) lenses, which extend the focal length, avoid this problem, but have more limited magnification ranges and often suffer from internal reflections.

Figure 2.4 shows two lunar photographs, taken through the same telescope and enlarged the same amount compared with the original negatives. The left frame was taken at prime focus and looks the sharper of the two. The right frame, using Barlow amplification, does not look as sharp as the other but actually has a higher resolution of lunar detail because of its larger scale.

VIDEO IMAGING

A small but increasing number of amateur astronomers are finding the Moon a suitable subject for video imaging. Compared with film photography, videography has the advantage of inexpensively creating 30 frames per second. This makes it possible to search through many frames to find the one with the best seeing. Also, the image can easily be recorded on tape and replayed or copied. It is also easy to enter such data into a video digitizer for computer processing. Another advantage is that focusing, framing and exposure are monitored in 'real time'.

The optical systems used for videography are analogous to those used for conventional photography. If the video camera lens can be removed, one has the most versatile system, allowing the prime-focus, eyepiece-projection or Barlow-amplification methods to be used. If the lens cannot be removed, only the afocal method is possible, but the zoom lenses on most video cameras allow considerable flexibility here. The writer's video equipment is perhaps typical: a black-and-white CCD camera with removable lens. The CCD element is quite small, about 6 mm × 8 mm, so that the entire Moon cannot be fitted on to the frame with focal lengths longer than about 600 mm. The unit is quite sensitive, with Earthshine and the umbra of a lunar eclipse being captured with an f/3.64, 500-mm focal length system. Except for thin crescents and the nearly-full phases, the system gives well-exposed images at f/16.5. At gibbous and full phases, images may be obtained with Barlow amplification at f/50. If one requires well-exposed images at other phases, it is likely that a specialized system that allows longer integration times, and which might have to be refrigerated, would have to be used. Initial experiments with CCD digital lunar imaging have been very promising. This requires that a computer be used in conjunction with the camera.

It is always interesting to view the videotape on a monitor, but one often wishes to have 'hard' copy. If the image has been read into computer memory, a laser printer can provide a paper copy, but with some loss of detail and contrast. Ironically, the best inexpensive method of obtaining a high-quality paper copy appears to be by photographing the monitor screen. The exposure time and aperture setting for this are best found by an exposure meter, but is well within most film-camera capabilities. Exposure times of at least 1/4 second should be used to prevent inadvertently capturing the scan pattern of the monitor. For further information the reader is referred to Chapter 17.

PHOTOMETRY AND COLORIMETRY

These techniques measure the intensity and colour of the light of the Moon or of some selected portion of it. Except during lunar eclipses (see below) there is little scientific purpose in measuring the light of the entire Moon. Instead, the photometric study of a specific lunar site can tell us about the albedo and surface structure of that point. Unlike the case with lunar relief, the study of tonal variations and colours of the Moon is still extremely incomplete.

Lunar photometry can be done with a conventional photoelectric photometer. The writer has used his for a number of lunar projects without any damage to the instrument. However, if the photometer is used at the prime focus, the Moon is often too bright for the unit's range. In this case, use a Barlow lens to increase the effective focal ratio, which also has the advantage of reducing the area of lunar surface sampled by the photometer. For example, the writer uses a Barlow lens when doing lunar photometry, giving an f/80 system with an angular sample or 'footprint' of about 6″ arc (about 12 km on the Moon). One needs to compare one's brightness measurements with a comparison standard that is also an extended source. Fortunately, the phase-dependent brightness functions of several hundred lunar sites have been measured, so these can be used conveniently for comparison. A series of brightness measures for the same site for at least half a dozen well-spaced lighting conditions allows us to determine the normal (vertical Sun) albedo as well as two other photometric parameters – site roughness and soil compaction.

Figure 2.5 shows an example of the results of a lunar photometry project, comparing the relative brightness of two sites on the floor of the crater Plato in the B (blue) band during the course of a lunation.

The catalogue of comparison sites lists brightness in the B band. One can also make simultaneous readings in other colours: for example V (visual), R (red), or even U and I (ultraviolet and infrared, respectively). These can then be reduced to colour indices, such as B–V. This is probably the simplest way to do quantitative lunar colorimetry. Some observers can detect the Moon's faint hues visually, but their subjective descriptions often disagree. It is more reliable to look for colours by viewing alternately through different coloured filters: the Wratten tricolor series

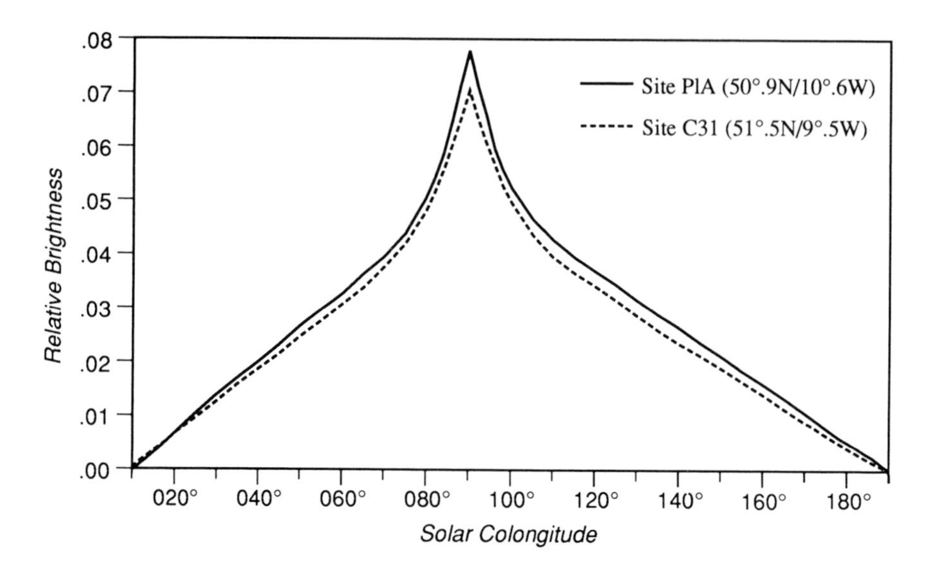

FIGURE 2.5
Lunar photometry. Graph of apparent brightness of two sites on the floor of the crater Plato from local sunrise (near colongitude 10°) to sunset (near colongitude 90°). 'P1 A' is the light patch on Plato's south-west floor, while 'C31' is the area of the central craterlet. Units are in terms of a completely-reflecting surface under vertical solar illumination. Note the brightness surge at Full (colongitude 90°), known as the *opposition effect*. Differences between the two curves are due to differences in albedo and surface compaction.

of No. 25 (red), No. 58 (green) and No. 47 (blue) is a good combination. The colour differences between different regions on the Moon can also be recorded on colour film, particularly Kodak Photomicrography 2483, mentioned above.

When studying lunar colours, a reflecting telescope is best, because of its lack of chromatic aberration. It should be used at a magnification of no more than about ×0.6 per millimetre of aperture. The Moon should be high in the sky and the atmospheric transparency quite good. Most coloured areas are best seen near Full phase, but sometimes the terminator appears as red, brown or green. Several areas change their hues over the course of a lunation. There are a number of lunar areas that consistently show distinctive colours, although their exact hues are debatable. A few of these areas are: the Aristarchus Plateau (north-west of Aristarchus: yellow, brown or red); Mare Crisium (green, yellow); Mare Humorum (green, yellow or red); Sinus Roris (yellow or red); Mare Serenitatis (yellow or red); and Mare Tranquillitatis (green, yellow or red).

SLOPES AND ELEVATIONS

The third dimension of the Moon's surface can be measured by several means. The simplest approach is to estimate the slopes of features by noting the time when their shadow either ceases to be pure black and turns dark grey (in the lunar morning) or changes from dark grey to black (lunar afternoon). This timing can be converted to the amount of slope in the direction the Sun's rays were falling. To do this, calculate the solar colongitude (C) and latitude (b.) for the time of observation, and then measure on a map or photograph the lunar latitude (ϕ) and longitude (λ). The altitude of the Sun, A, which is equal to the slope, is given by the formula

$$\sin A = \sin \mathrm{b.} \sin \phi + \cos \mathrm{b.} \cos \phi \cos(\lambda + C) \quad (1)$$

If the point's latitude and longitude are not known, they may be found from the (ξ, η) coordinates (see above) as follows:

$$\sin \phi = \eta \quad (2)$$

$$\sin \lambda = \xi \sec \phi \quad (3)$$

To determine an actual altitude, not only must the shadow be black, but its length must be measured. There are several ways of doing this. The simplest is visually to estimate the shadow's length (L) as a proportion (P) of the *unforeshortened* diameter of a crater of known diameter (D). Then:

$$L = P \times D \quad (4)$$

The shadow length may be measured more accurately if one has a stopwatch. If the telescope drive is left on at the *sidereal* rate, the Moon will drift eastwards across the field at, very roughly, 1″ arc for every 2 seconds of time. Using an eyepiece with a cross-wire, time how many seconds (T) it takes for the shadow to drift across the wire, and how long (T') it takes for a crater of known diameter. Make sure that the drive has no periodic errors and that the comparison crater's unforeshortened axis is parallel to the direction of drift. If so, the factor P in equation (4) is equal to T/T'.

Naturally, many shadows may be measured easily from photographs of known scale. This may be subject to systematic error, however, depending on the exposure and resolution of the photograph. The most accurate method of shadow measurement is to use a filar eyepiece micrometer. If the screw constant of the device has been found for the telescope being used, one may easily find the shadow's apparent length in seconds of arc. If not, the factor P is found by measuring the comparison crater as well. If one has found the shadow's length in seconds of arc, this needs to be converted to units of the lunar radius (L') by dividing L by S', the apparent lunar *semi-diameter* (radius) at the place of observation. The simplest way of finding S' is to use an ephemeris computer program. If the shadow's length, L, has been found in kilometres, it needs to be converted to units of the Moon's radius, L', by dividing by the lunar radius of 1738 km.

The final angle we need to calculate is the phase angle (g) – the angle between the Sun and the Earth as seen from the Moon:

$$\cos g = \sin b' \sin \mathrm{b.} + \cos b' \cos \mathrm{b.} \sin(l' + C) \quad (5)$$

The result from equation (5) may be in error by up to 1° if the Moon is near the horizon. To decrease the maximum error to about 0.25°, b' and l' should be the *topocentric librations* for the observer, rather than the *geocentric librations*.

From the quantities already found, the height H of the feature is found from:

$$H = L' \sin A \operatorname{cosec} g - 0.5(L' \cos A \operatorname{cosec} g) \tag{6}$$

Equation (6) gives the height in units of the lunar radius; to convert to metres, multiply by 1 738 000. This will be a *relative* height; the difference in absolute altitude between the two ends of the shadow. Thus it is important to record the coordinates of both shadow ends.

OBSERVING PROJECTS

Armed with one or more of the above techniques, there are several forms of lunar study to which the observer may contribute. These days, though, the value of the contribution depends on the observer's knowledge, including that of the results of the space missions, as well as his or her dedication and ingenuity. Studies need not be confined to personal telescopic observations: there are observatory photographs, and most of the many lunar images from orbiting vehicles still await study and measurement.

Topographic Studies

One result of the space missions is that there is no scope for *general* amateur mapping of the Moon. There certainly is room for several special projects, though. There are several areas where the space photography is of low quality: the most extensive of these is the southern and south-western limb marginal zone. In addition, only a few lunar areas were photographed well from space at low sun angles, so there remains the need for study of *low-relief features*, such as domes and wrinkle ridges. Making drawings and long-exposure photographs, and measuring elevations, are good techniques for investigating the low-sun areas near the terminator. One little-investigated type of feature in mare areas is large-scale *terminator deformations*, where extensive raising or lowering of the Moon's surface has caused the terminator to deviate from its normal elliptical shape. Such features are best shown by photographing the terminator with unusually long exposures.

Selected Areas

This project concentrates observing resources on one or a few areas. One important aspect of such studies is to determine, from lunar sunrise to sunset, the usual pattern of tonal changes with sun angle that every lunar area experiences in its own way. Drawings and photographs can be used here, but the results are most reliable if photoelectric photometry can be done as well. With the regular tone–sun angle relationship established, we have a firm base for detecting unusual tonal changes. Naturally, the slope angle and direction of slope of the terrain affect the brightness of an area, so a programme of slope and elevation measures is an important aspect of a selected area study. As an example, the Selected Areas Program of the Lunar Section of the ALPO is studying the craters Alphonsus, Aristarchus–Herodotus, Atlas, Copernicus, Plato, Theophilus and Tycho. Given enough 'observer power', this list could and should be extended.

Lunar Transient Phenomena (LTPs or TLPs) – This refers to short-term localized lunar changes, rather than the permanent topographic changes reported in the past but now largely discounted by the evidence of space photography. Transient events, though, continue to be reported and assume several forms:

1. Temporary colours, often red but sometimes blue or other hues.
2. Bright flashes or streaks, particularly in shadows or on the night hemisphere.
3. Extended foggy or cloudy patches.
4. Areas of apparent obscuration, perhaps implied by the absence of normally-visible features.
5. Deviation of a shadow from pure black.
6. Changes in the brightness, size or position of tonal features that do not fit the regular lunation pattern.

Unfortunately, most LTP reports are serendipitous, rather than the result of patrols. Thus most events are neither confirmed nor objectively recorded. Simultaneous patrols by independent observers are needed. For visual work, a thorough knowledge of the area being monitored is essential. Colour events are most reliably detected by colour filters, particularly the rapid red–blue alternation technique known as 'moon blink'. Photography helps give objective status to a report, and videotaping is a very useful way of monitoring an area for an extended period. A very promising approach is photoelectric photometry, particularly if done in different colour bands.

Where to look for LTPs is a matter of taste. Statistically, one should probably pick randomly-selected locations, or at least avoid the places where everyone else is looking (except of course in an organized simultaneous-observation programme). On the other hand, one may want to improve one's changes by picking areas of previous LTP reports. Such areas include Alphonsus, Aristarchus–Herodotus (the 'Cobra Head'), Atlas, Conon, Copernicus, Mare Crisium, Eratosthenes, Eudoxus, Gassendi, Grimaldi, Hercules, Kepler, La Hire, Linne, Menelaus, Messier-Messier A, Petavius, Philolaus, Pico, Piton, Plato, Posidonius, Proclus, Riccioli, Schickard, Palus Somnii, Theophilus, Tycho, Werner and many others.

Lunar Eclipses

An eclipse of the Moon occurs when it passes through the Earth's shadow. In a *penumbral* eclipse only the shadow's penumbra is involved – the part where the

Sun's direct light is partly but not completely blocked by the Earth. The shadow *umbra* is the central part of the shadow where no direct sunlight reaches; the light rays which illuminate the Moon here have been refracted and scattered in the Earth's atmosphere, usually resulting in a red, orange or yellow hue. In a partial eclipse, part but not all of the Moon enters the umbra. In total eclipses, the entire Moon is immersed in the umbra near mid-eclipse. Naturally, a partial eclipse will have two penumbral phases on either side of the partial phase. A total eclipse follows the sequence penumbral – partial – total – partial – penumbral. Figure 2.6 shows the Moon's path relative to the Earth's shadow during each of the three forms of eclipse.

A typical lunar eclipse lasts several hours, which is fortunate because there is plenty to do. Every eclipse is unique and somewhat unpredictable owing to changing conditions in the Earth's atmosphere. Even unillustrated and qualitative verbal descriptions of the size, darkness and colour of the different portions of the umbra

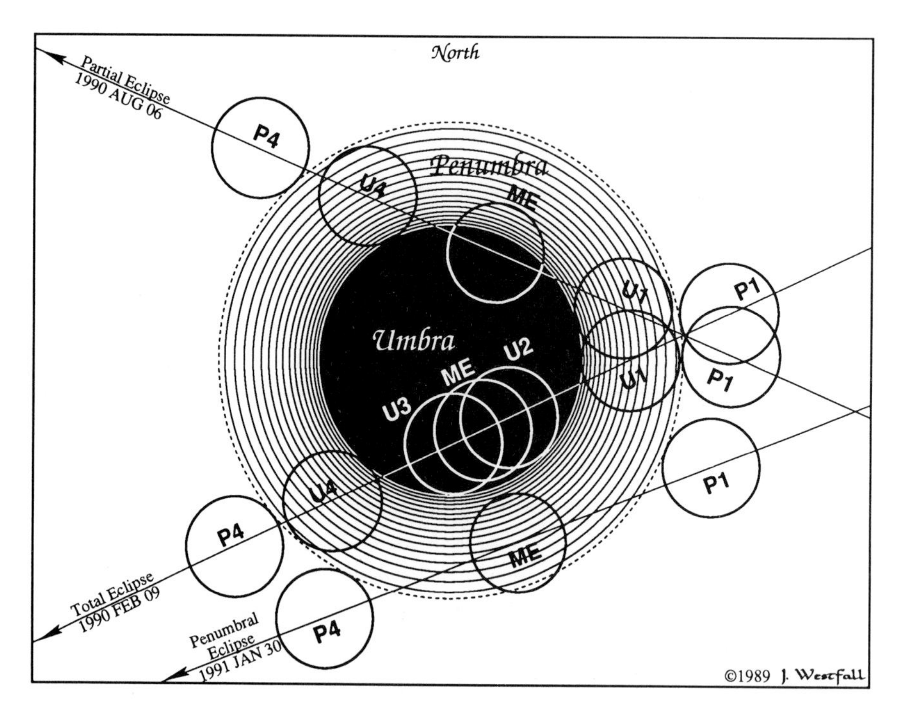

FIGURE 2.6

Lunar eclipses. Motion of the Moon through the penumbra and umbra of the Earth's shadow during the total eclipse of 9 February 1990, the partial of 6 August 1990, and the penumbral of 30 January 1991. The Moon's position is shown for the times of contact of its limb with the penumbra and umbra, abbreviated as follows: P1 = first penumbral contral (beginning of penumbral phase); U1 = first umbral contact (beginning of partial phase); U2 = second umbral contact (beginning of totality); ME = mid-eclipse; U3 = third umbral contact (end of totality); U4 = fourth umbral contact (end of partial phase); and P4 = fourth penumbral contact (end of penumbral phase). The second and third penumbral contacts have not been noted.

and penumbra are valuable. Such reports are enhanced by a series of whole-disk drawings, with colours named or actually drawn in, at selected times during the eclipse. A standard way of describing the darkness of a total eclipse is the *Danjon scale* (L), which is a numerical ranking scale intended to be applied near mid-eclipse as follows:

$L = 0$	Very dark eclipse, Moon almost invisible, especially at mid-eclipse.
$L = 1$	Dark eclipse, grey or brownish colouration; details distinguishable only with difficulty.
$L = 2$	Dark red or rust-coloured eclipse, with a very dark central umbra and the outer edge of the umbra relatively bright.
$L = 3$	Brick-red eclipse, usually with a bright or yellow rim to the umbra.
$L = 4$	Very bright copper-red or orange eclipse, with a bluish very bright shadow rim.

When necessary, intermediate ratings such as $L = 2.5$ may be used. All the forms of observations described up to this point are easy to make with the naked eye, but binoculars are also appropriate.

Eclipse Photography

Lunar eclipse photography is not hard to do; a tripod-mounted telephoto lens of 400-mm focal length or above will show the penumbra and umbra quite well. To photograph the Moon during totality will require fast film and a fairly fast lens in order to prevent the exposure being long enough to cause blurring (in other words, no more than 2 seconds with a 400 mm lens). If the camera is mounted 'piggy-back' on a motor-driven telescope, much longer exposures are possible. Prime-focus eclipse photography using a motor-driven telescope is also quite feasible, although a 'lunar rate' drive is recommended for long exposures during totality. Exposure times, especially during totality, are partly a matter of judgement, but some starting values for an f/8 optical system are:

Uneclipsed Full Moon	1/250 s with ISO 64 film
Moon deep in penumbra	1/60 s with ISO 64 film
Moon 0.1–0.5 in umbra	1/30 s with ISO 400 film
Moon 0.5–0.75 in umbra:	
Portion in penumbra	1/30 s with ISO 400 film
Portion in umbra	1 s with ISO 400 film
Moon 0.75 in umbra to totality	1 s with ISO 400 film
Beginning or end of totality	1–2 s with ISO 400 film
10 minutes after commencement or before end of totality	2–4 s with ISO 400 film
Mid-eclipse	4–8 s with ISO 400 film

It is also possible to videotape an entire lunar eclipse. The writer has used an f/3.64 optical system to record the umbral interior for an $L = 2$ eclipse, although the lens then had to be stopped down for the partial and penumbral phases.

Lunar Eclipse Photometry

This is concerned with finding the brightness of the eclipsed Moon, or of particular areas of it, during the eclipse; this information translates into the amount of dimming in particular areas of the umbra and penumbra. Estimates of the Moon's magnitude should be done throughout an eclipse. During the course of a total eclipse, the Moon typically changes in brightness by a factor of 10 000 or more. During totality, the Moon is usually dim enough to be compared directly with bright stars or planets if the observer is sufficiently near-sighted for the star or planet's blurred disk to be about the same size as the Moon. As with all such photometry, the altitude of the comparison object should be about the same as that of the Moon, otherwise a *differential extinction correction* will have to be applied.

In another method, the Moon can be viewed with reversed binoculars, to make it both fainter and more star-like. Assuming a 25% light loss in the binoculars, the light loss of the Moon in magnitudes varies according to the magnification of the binoculars as follows: ×6, 4.2; ×7, 4.5; ×8, 4.8; ×10, 5.3; ×11, 5.5; ×14, 6.0; ×20, 6.8. Another method for making the Moon appear more stellar is to view its reflection in a convex reflector. Finally, those with photoelectric photometers can make spot readings, as opposed to whole-disk photometry: a series of spot measures on a few bright craters throughout an eclipse enables one to map the light distribution throughout the umbra and penumbra.

Other Eclipse Observations

Another factor which changes from one eclipse to the next is the size of the Earth's umbra. This dimension can be measured accurately by timing the four instances when the umbra edge touches a limb of the Moon and when it crosses craters. Accuracy to 0.1 minute is needed here, and the list of recommended craters is Aristarchus, Aristoteles, Copernicus, Eudoxus, Grimaldi, Kepler, Manilius, Menelaus, Plato, Plinius, Proclus, Pytheas, Taruntius, Timocharis and Tycho.

LTPs are sometimes reported during lunar eclipses. Look for *eclipse-induced changes* in the list of LTP sites presented above. For example, light nimbuses have been reported to change their size during eclipses. With a photoelectric photometer, by making a series of readings through B, V and R filters, *lunar luminescence* may be detected, particularly during the penumbral phases.

An average of 2–3 lunar eclipses happen each year, about half being visible from a particular location. The lunar eclipses due between 1991 and 2000 are listed in Table 2.1.

TABLE 2.1
FORTHCOMING LUNAR ECLIPSES, 1991–2000

Mid-Eclipse	Eclipse		
UT Date and Time	Type[a]	Mag.[b]	Where Visible[c]
1991 Dec 21 10^h 33^m	Partial	0.088	**A:** No. American except SE; NW Asia **B:** SE No. America; Latin America **E:** Cent., S & E Asia; Australasia
1992 Jun 15 04 57	Partial	0.683	**A:** USA and Canada except W and NW; Latin America **B:** Africa; W Europe **E:** Rem. USA and Canada; New Zealand
1992 Dec 09 23 44	Total	1.271	**A:** Africa; Europe; W Asia; NE Canada; E Brazil **B:** Rem. Asia; NW Australia **E:** Rem. No. & So. America
1993 Jun 04 13 01	Total	1.561	**A:** Japan; Australasia **B:** No. and So. America except E **E:** Asia except W; E Africa
1993 Nov 29 06 25	Total	1.087	**A:** No. America; So. America except E **B:** E So. America; Africa except E; Europe; Middle East **E:** NE Asia; NE Australia; New Zealand
1994 May 25 03 31	Partial	0.242	**A:** SE & E USA; SE Canada; Latin America; W Africa **B:** Rem. Africa; Europe; Middle East **E:** Rem. No. America except Alaska

[a]Eclipse type – 'Pen.' = penumbral eclipse.
[b]Mag. – The *magnitude* of the penumbral or umbral eclipse; ranging from 0.0 for the shadow just grawing the Moon's limb, to 0.5 for the Moon being half-covered, to 1.0 for the Moon barely being entirely covered; 1.0 and above indicates that the Moon is entirely in the penumbra for a penumbral eclipse, and that the eclipse is total for an umbral eclipse.
[c]*Where visible* – **A** means that all the eclipse is visible; **B**, that the beginning is visible, but not the end; **E**, that the end is visible, but not the beginning. 'Rem.' means remainder.

TABLE 2.1 continued

FORTHCOMING LUNAR ECLIPSES, 1991–2000

Mid-Eclipse	Eclipse		
UT Date and Time	Type[a]	Mag.[b]	Where Visible[c]
1994 Nov 18 06 44	Pen.	0.881	**A:** No. America; So. America except E **B:** Europe; W Africa; E So. America **E:** NE Asia; NE Australia; New Zealand
1995 Apr 15 12 18	Partial	0.112	**A:** Pacific Basin; Australasia; NE Asia; Alaska **B:** No. America; So. America except E **E:** Rem. Asia except W
1995 Oct 08 16 04	Pen.	0.827	**A:** Asia except SW; Australia **B:** New Zealand; Alaska; W Canada; NW USA **E:** SW Asia; Africa; Europe
1996 Apr 04 0010	Total	1.379	**A:** Europe; W Middle East; Africa; W Brazil **B:** Asia except NE; W Australia **E:** Rem. So. America; No. America except Yukon, Alaska
1996 Sep 27 02 54	Total	1.240	**A:** E. No. America; So. America; NW Africa; SW Europe **B:** Rem. Africa and Europe; W and SW Asia **E:** Rem. No. America
1997 Mar 24 04 40	Partial	0.918	**A:** No. America except NW; So. America **B:** Africa; Europe; Middle East **E:** NW No. America; New Zealand

[a]Eclipse type – 'Pen.' = penumbral eclipse.
[b]Mag. – The *magnitude* of the penumbral or umbral eclipse; ranging from 0.0 for the shadow just grawing the Moon's limb, to 0.5 for the Moon being half-covered, to 1.0 for the Moon barely being entirely covered; 1.0 and above indicates that the Moon is entirely in the penumbra for a penumbral eclipse, and that the eclipse is total for an umbral eclipse.
[c]*Where visible* – **A** means that all the eclipse is visible; **B**, that the beginning is visible, but not the end; **E**, that the end is visible, but not the beginning. 'Rem.' means remainder.

TABLE 2.1

FORTHCOMING LUNAR ECLIPSES, 1991–2000

Mid-Eclipse	Eclipse		
UT Date and Time	Type[a]	Mag.[b]	Where Visible[c]
1997 Sep 16 18 46	Total	1.190	**A:** Asia except NE; Middle East; E Africa; W Australia **B:** NE Asia; Rem. Australasia **E:** Rem. Africa; Europe; E So. America
1998 Mar 13 04 21	Pen.	0.707	**A:** No. America except Alaska; So. America; SW Europe; NW Africa **B:** Rem. Africa and Europe; Middle East **E:** Alaska
1998 Aug 08 02 25	Pen.	0.122	**A:** S and E No. America; So. America; Africa; Europe except E and NE **B:** Middle East; E and NE Europe **E:** Rem. USA; Cent. Canada
1998 Sep 06 11 10	Pen.	0.813	**A:** W No. America; NE Asia; Australia except W; New Zealand **B:** Rem. No. America; So. America except E Brazil **E:** Rem. Australia; Rem. Asia except W
1999 Jan 31 16 19	Pen.	1.005	**A:** Asia except SW; Australia **B:** New Zealand; W and NW No. America **E:** SW Asia; Europe; Africa except W

[a]Eclipse type – 'Pen.' = penumbral eclipse.
[b]Mag. – The *magnitude* of the penumbral or umbral eclipse; ranging from 0.0 for the shadow just grawing the Moon's limb, to 0.5 for the Moon being half-covered, to 1.0 for the Moon barely being entirely covered; 1.0 and above indicates that the Moon is entirely in the penumbra for a penumbral eclipse, and that the eclipse is total for an umbral eclipse.
[c]*Where visible* – **A** means that all the eclipse is visible; **B**, that the beginning is visible, but not the end; **E**, that the end is visible, but not the beginning. 'Rem.' means remainder.

TABLE 2.1 continued

FORTHCOMING LUNAR ECLIPSES, 1991–2000

Mid-Eclipse	Eclipse		
UT Date and Time	Type[a]	Mag.[b]	Where Visible[c]
1999 Jul 28 11 33	Partial	0.396	**A:** Australia except NW; New Zealand **B:** No. America except NE; So. America except E Brazil **E:** Rem. Australia; Asia except W
2000 Jan 21 04 44	Total	1.325	**A:** No. and So. America; W Europe; NW Africa **B:** Rem. Africa and Europe; W Asia **E:** NE Asia
2000 Jul 16 13 56	Total	1.769	**A:** Japan; Australasia **B:** No. America except NE; W and S So. America **E:** Asia; E Africa

[a]Eclipse type – 'Pen.' = penumbral eclipse.
[b]Mag. – The *magnitude* of the penumbral or umbral eclipse; ranging from 0.0 for the shadow just grawing the Moon's limb, to 0.5 for the Moon being half-covered, to 1.0 for the Moon barely being entirely covered; 1.0 and above indicates that the Moon is entirely in the penumbra for a penumbral eclipse, and that the eclipse is total for an umbral eclipse.
[c]*Where visible* – **A** means that all the eclipse is visible; **B**, that the beginning is visible, but not the end; **E**, that the end is visible, but not the beginning. 'Rem.' means remainder.

Lunar Occultations

The Moon frequently passes in front of stars and occasionally in front of planets. The latter are interesting to watch, but stellar occultations by the Moon can give useful scientific information. For this purpose, one needs to know the latitude and longitude to 1″ arc, and the elevation to 10 metres. Visually, with a small telescope, time to 0.1 second precision the moment when a star appears or disappears at the Moon's limb. To record the times, use a tape recorder and a radio that can receive short-wave time signals.

Once the raw timing is corrected for the observer's *personal equation*, the result is applicable to finding the accurate position of the Moon and the profile of its limb. From a given site, most such events are total occultations, where there is a single disappearance and a single reappearance. If the observer is willing to travel to observe, there is occasionally the opportunity to observe a *grazing occultation.* Here, the star just skims the Moon's southern or northern limb, blinking on and off as it passes behind lunar valleys and mountains. Grazing events are best studied by observing teams, and some have employed video cameras to record the event.

Total-occultation timings provide information about the position of the Moon, stellar diameters, and the presence of unseen companion stars. Grazing occultations also help map the detailed profile of the Moon.

About the Author

John Westfall, a native of San Francisco, resides there with his wife Elizabeth, his two sons (sometimes), two dogs, six telescopes and five computers.

Involved in amateur astronomy for 45 years, he is Executive Director of the ALPO, an international group of about 700 amateur and professional observers of solar-system objects. He also edits their quarterly journal *The Strolling Astronomer.* When his duties and the San Francisco fog and earthquakes allow him to observe, his chief interests are lunar and planetary mapping, video imaging and photoelectric photometry.

Professionally, he is Professor of Geography at San Francisco State University, where he has taught since 1968. He likes to teach courses on remote sensing, surveying, statistical and computer methods, and environments of the future.

Amateur astronomy allows little time for other hobbies. He does like to travel, which includes eclipse-chasing; photography; the history of astronomy and exploration; and reading science fiction and the works of H. P. Lovecraft.

Contact address – Association of Lunar and Planetary Observers, 8930 Raven Drive, Waco, TX 76712, USA.

Bibliography

Alter, D. *Pictorial Guide to the Moon*, 3rd edn. T. Y. Crowell, New York, 1979.

The Astronomical Almanac. Annual joint publication by the UK and USA Nautical Almanac Offices, Washington and London.

BAA, *Guide to observing the Moon.* Enslow Publishers, Hillside, NJ; Aldershot, UK, 1986.

Cadogan, P. *The Moon — our Sister planet.* Cambridge University Press, Cambridge, 1981.

Cherrington, E.H. *Exploring the Moon Through Binoculars and Small Telescopes.* Dover Publications, New York, 1984.

Corliss, W. *The Moon and the Planets.* The Sourcebook Project, Glen Arm, Md, 1985.

Dobbins, T. *et al., Introduction to Observing and Photographing the Solar System.* Willmann-Bell, Richmond, Va, 1988.

French, B. M. *The Moon Book.* Penguin Books, Harmondsworth, UK, 1977.

Genet, R. (ed.), *Solar System Photometry Handbook.* Willmann-Bell, Richmond, Va, 1983.

Graham, F.G. and Westfall, J.E. *ALPO Lunar Eclipse Handbook.* ALPO Lunar Section, San Francisco, 1989.

Guest, J. and Greeley, R. *Geology on the Moon.* Wykeham Publications, London, 1977.
Heiken, G. *et al., Lunar Sourcebook.* Cambridge University Press, Cambridge, UK, 1991.
Hill, H. *A Portfolio of Lunar Drawings.* Cambridge University Press, Cambridge, UK, 1991.
Moore, P. *New Guide to the Moon.* W.W. Norton & Co., New York, 1976.
Price, F. *The Moon Observer's Handbook.* Cambridge University Press, Cambridge, UK, 1988.
Roth, G.D. (ed.), *Astronomy: a Handbook.* Sky Publishing, Cambridge, Mass, 1975.
Rükl, A. *Atlas of the Moon.* Hamlyn, 1991.
Schultz, P.H. *Moon Morphology.* University of Texas Press, Austin, 1976.
Sidgwick, J.B. *Observational Astronomy for Amateurs.* Enslow Publishers, Hillside, NJ, 1982.
Westfall, J. *Lunar Photoelectric Photometry Handbook.* ALPO Lunar Section, San Francisco, 1984.
Westfall, J. (ed.), *ALPO Solar System Ephemeris.* ALPO San Francisco, published annually.
Wilhelms, D.E. *The Geologic History of the Moon.* US Government Printing Office, Washington, 1987.
Wilkins, H.P. and Moore, P. *The Moon.* Macmillan, New York, 1955.

Notes and Comments

The value of occultation observations – I remember attending a BAA meeting as a very young member when a device called the Markowitz Moon Camera was described. We were told that this would make occultation observations redundant because of the precision with which it could photograph the Moon in a star field. Not surprisingly the feeling was 'There goes another field of useful observation.'

Therefore I was interested to read David Dunham, Director of IOTA, writing in *Sky & Telescope* for November 1988 as follows:

> Until the high accuracy of space-based astrometry is proven, we should not fall into the trap of the late 1950s, when interest in lunar occultations plummeted due to the development of the Markowitz dual-rate Moon camera. This device was predicted to determine the Moon's position better than occultation timings. It took nearly a decade to show that occultations were still superior.
>
> Occultation data are still competitive for several measurements. For example, when trying to determine the deceleration of the Moon's motion along the ecliptic due to tidal forces in the Earth–Moon system, the results are proportional to time squared. Hence, the long span of occultation data

gives them an advantage over the individually more accurate but much more recent laser-ranging data.

High-speed observations using CCDs have allowed very close double stars to be discovered as the lunar limb obscures or uncovers one component before the other. Star diameters, and even layers in stellar atmospheres, have been measured in this way!

It will be surprising if amateurs with the appropriate equipment do not find a whole new field of work here. The use of video cameras should also lead to improved accuracy in occultation timings.

Shadow measurement and eclipses – The author's further comments on these topics should be noted. Regarding the use of shadow measurement to obtain the heights of lunar features, he points out that both ground-based and space-based photographs appear to have systematic errors when used for this purpose, so the visual approach actually seems superior for all but the smaller features. Amateurs have the additional advantage that they can measure shadows under a wide range of lighting, enabling them to draw profiles, for example.

On timing of the passage of the umbra across craters during an eclipse, he points out that this traditional activity can give useful results if the craters are spread over a large area of the Moon, and if many viewers time both the entries and exits of craters from the umbra. The umbral enlargement and ellipticity do appear to vary significantly between eclipses, and we need to see if this is related to mid-eclipse magnitude, terrestrial stratospheric aerosols, and solar activity.

Lunar transient phenomena – The ALPO and the BAA both conduct regular searches, although their study went 'out of fashion' in the early 1970s. This was due partly to exaggerated reports that gave the subject a bad name, and partly to the difficulty of explaining them in a convincing way (a bad mark for scientific method!).

The ALPO search technique is carefully defined. Each observer is assigned four features where LTPs have been reported previously, as well as a control feature with no suspected events and a known epicentre for moonquakes. Each of these six locations is then assigned several reference points (such as the four cardinal points on its wall if it is a crater, as well as the central peak), which are checked against a comparison point a little north or south of the feature, usually on a plain.

The relative brightness of these points is then checked at least twice during each observing session, with at least 10 minutes elapsing between each check.

The results are then analysed to produce graphs of albedo against lunar phase throughout the lunar day, and any abnormal brightenings or darkenings will be revealed. – Ed.

CHAPTER THREE · *Observing Venus*

RICHARD McKIM
OUNDLE, NORTHAMPTONSHIRE, ENGLAND

To the memory of my late friend, J. Hedley Robinson, Director of the BAA Mercury & Venus Section from 1965 to 1979

VENUS IS THE SECOND PLANET OUT FROM THE SUN, ORBITING THE central star in 224.7 days in a near-circular orbit at a distance of 108 million km. It is observable from Earth for most of the time except when it is near conjunction; it outshines all the other stars and planets visible in the dawn or dusk sky, and it is often visible in full daylight if one has a clear sky and knows just where to look. Venus has an equatorial diameter of 12 104 km, and a mass 0.815 that of Earth's. Thus the planet's density, surface gravity and escape velocity closely resemble those of our own world. Telescopically the planet exhibits a phased disk and little else, for its surface is forever hidden beneath impenetrable clouds. Together with Mercury and Mars, Venus is a rocky world like Earth; but here the similarity ends.

In examining Venus in the light of the Space Age one is forcibly reminded that 'beauty is only skin-deep'. Venus, the Roman goddess of love, outwardly the beautiful, brilliant star of the morning or evening sky is a very different place in close-up. To dip beneath the thick clouds of this planet — the Earth's twin in size — is literally to step into Hell. The spacecraft that have visited Venus reveal it to be an exceptionally hot world, with a dense, poisonous atmosphere. Man can never visit the surface of Venus in person.

THE RUNAWAY GREENHOUSE

The Venusian atmosphere consists mostly of carbon dioxide, a molecular compound with the chemical formula CO_2 (96.5% of the lower Venusian atmosphere consists of this gas, the rest is nitrogen). Water vapour is present at a low concentration, the amount being altitude-dependent. The presence of such a large amount of CO_2 on Venus has led to a 'greenhouse effect' orders of magnitude greater than the problem which faces Man on planet Earth. In a domestic greenhouse, solar energy strikes the ground and is mostly absorbed. Infrared radiation is emitted by the ground, and the carbon dioxide in the air inside the greenhouse absorbs this infrared radiation, raising the temperature. The infrared radiation cannot pass out through the glass of the greenhouse. Molecules absorb radiation when the energy of the radiation coincides with transitions in the natural stretching or bending frequencies of the chemical bonds present. Carbon dioxide is not the only 'greenhouse gas'; although the Earth's most important atmospheric constituents — oxygen and nitrogen — do not absorb any infrared radiation of the frequencies radiated by the Earth, water vapour, methane, and many products resulting from man's lifestyle and industry are all absorbers.

Increased use of fossil fuels increases the CO_2 level in our atmosphere, which can then act as one gigantic greenhouse. Unfortunately, only photosynthesis removes CO_2 on a large scale, and the destruction of the world's rainforests in a myopic attempt by the governments of some developing nations to sustain the world's still-growing population can only spell disaster if allowed to continue. The mean abundance of carbon dioxide in dry air on Earth is presently only about 0.03% by volume. On Venus then, there is a 'runaway greenhouse'.

Much carbon dioxide is tied up on Earth in crustal carbonate rocks like limestone and dolomite. If these are heated to high enough temperatures they break down to generate yet more CO_2. Venus today is a planet with a mean surface temperature of 735 K (462°C), far hotter than would be expected merely from Venus being closer to the Sun than the Earth. This is well above the melting point of lead, or the boiling point of sulphuric acid. Finally, the Venusian atmosphere is so dense that the surface pressure is between 92 and 95 times that of Earth's.

HOW FAST DOES IT SPIN?

The visible clouds of Venus consist of an aerosol of droplets of a solution of sulphuric acid in water of about 85% strength (the particles having diameters of ca. 2 μm) and it is these clouds which give the planet such a high albedo. The clouds occur in strata some 47–65 km above the surface, and are overlain by a transparent mist of tinier sulphuric acid droplets between altitudes of 70 and 90 km. From Galileo's early telescopic observations up till the present day, no observer has ever been able to see much in the way of details in the Venusian clouds: sometimes a brighter area here, sometimes a darker patch there. Three and a half centuries of visual observation failed to establish a consensus rotation period. The most prolific observers spent whole lifetimes in studying the planet without solving this basic question.

Many observers tried to estimate the rotation period from shifts in elusive markings. Perhaps the closest was Dr W.H. Steavenson in 1924, using a 150-mm refractor, who suggested 8 days; Professor W.H. Pickering had earlier thought 68 hours to be correct, but his rotational axis was nearly coincident with the orbital plane, which was rather an odd conclusion to have drawn. But in 1927 Dr Frank Ross at Mount Wilson Observatory photographed the planet in the near-ultraviolet (UV) below 400 nm, and to his surprise found prominent bright and dark areas all over the disk. His photographs were not sufficiently systematic for him to deduce a rotational period, but it was clearly less than the duration of the planet's year. Curiously, there the problem was allowed to remain.

In 1957 C. Boyer, then an amateur astronomer working with a 260-mm reflector, took a series of almost daily UV photographs, and found that the patterns of cloud markings reappeared every 4 days. This important discovery was confirmed by work with professional instruments in the next decade, and the space probe images by Mariner 10 (1974), Pioneer Venus Orbiter (1978) and Galileo (1990) have enabled the motions of individual clouds to be studied in detail.

The UV markings of Venus have characteristic shapes, and Boyer has described these himself [1]:

> We discovered that there are two distinct kinds of dark markings and nicknamed them Y and ϕ, according to their shapes. Both lie on or near the equator, with the 'tails' of their long axes following in the rotational motion. By studying the recurrence of these features, we were able to refine the rotation period: 3.995 days (synodic).

The periods of individual features show some fluctuations, and the characteristic patterns have been known to disappear temporarily. Sulphur dioxide is a UV-absorber, and may largely account for the characteristic dark markings in UV images. Photochemical reactions convert it to sulphur trioxide in the higher atmosphere, which then reacts with water vapour to form an aerosol of sulphuric acid.

A different kind of technique was needed to determine the rotational period of the solid surface. Earth-based radar established in the 1960s that Venus turns on its axis very slowly, once in every 243 days, so that its 'day' is actually longer than its 'year'! The axial inclination is 178° with respect to the orbital plane. (Spectroscopic techniques had been applied but had yielded conflicting results for reasons which remain obscure. In the 1890s, Percival Lowell working at Flagstaff thought that the period must be very long, but later work confirmed the rapid rotation of the upper atmosphere).

THE SOLID SURFACE

Radar maps made from Earth in the 1970s (at Arecibo and Goldstone) showed what appeared to be craters and mountainous areas on Venus. The spectacularly successful Soviet Venera landers have returned images of the surface rocks. Many broken rocky plates with surprisingly sharp edges can be seen on these images; the surrounding surface of the planet is smooth but fractured. The onboard radars of the US Pioneer Venus Orbiter and Veneras 15 and 16 confirmed the presence of craters and mountains, building up a detailed map of very nearly the whole planet. In recent years the Magellan probe has produced spectacular high-resolution radar images of the surface.

The majority of the surface is quite flat, but two 'continental' areas named Aphrodite Terra and Ishtar Terra together cover some 10% of the planet and stand out above the mean surface level. Ishtar Terra is the larger area, the size of Australia, containing Maxwell Montes (i.e. the Maxwell Mountains). Maxwell Montes is the highest observed formation on the surface, rising to 11 km and apparently of volcanic origin. Other features on the planet also show strong signs of volcanic origin, while there are lava flows and sunken trenches identifiable in the images. Small domes appear elsewhere, and there are impact craters as well as the volcanic features. There is no definite evidence of present-day volcanic activity, but fluctuations in the sulphur dioxide concentration recorded between the Soviet Vegas 1 and 2 could be explained by such contemporary outbreaks.

As with the Viking Mars landers, the Soviet Venera and Vega landers determined the chemical composition of the surface, finding it variable from site to site but generally resembling terrestrial tholeitic basalts, an igneous rock associated with volcanoes.

Much more could be said about the surface of the planet, but rather more has been said about the atmosphere for that is the only part of the planet which can be studied telescopically!

THE SYNODIC CYCLE

In eight terrestrial years Venus makes 13 orbits of the Sun, very nearly, so there is an 8-year cycle during which Venus goes through a series of eastern and western elongations. Not all elongations are equally favourable; for northern hemisphere observers a favourable elongation would probably count as one where the declination at dichotomy or inferior conjunction was well north of the celestial equator. The phases of the planet can be seen with binoculars, but a telescope is needed to see more.

Some drawings of Venus by the author in 1988 are given in the illustrations, and show most of the features that an observer would normally expect to see. Figures 3.1a–f were all drawn with my 216-mm Newtonian reflector and 75-mm refractor. It is usual in planetary drawings to black-round the background, but I have not done so in these pictures, for Venus is often seen against a bright twilight sky, or indeed in full daylight with the Sun above the horizon!

The Gibbous Phase

Figure 3.1a shows the planet at a gibbous phase as drawn on 14 February 1988. Superior conjunction had been on 23 August 1987, some 6 months earlier, and I had begun to observe the planet seriously the following February. In Fig. 3.1a the

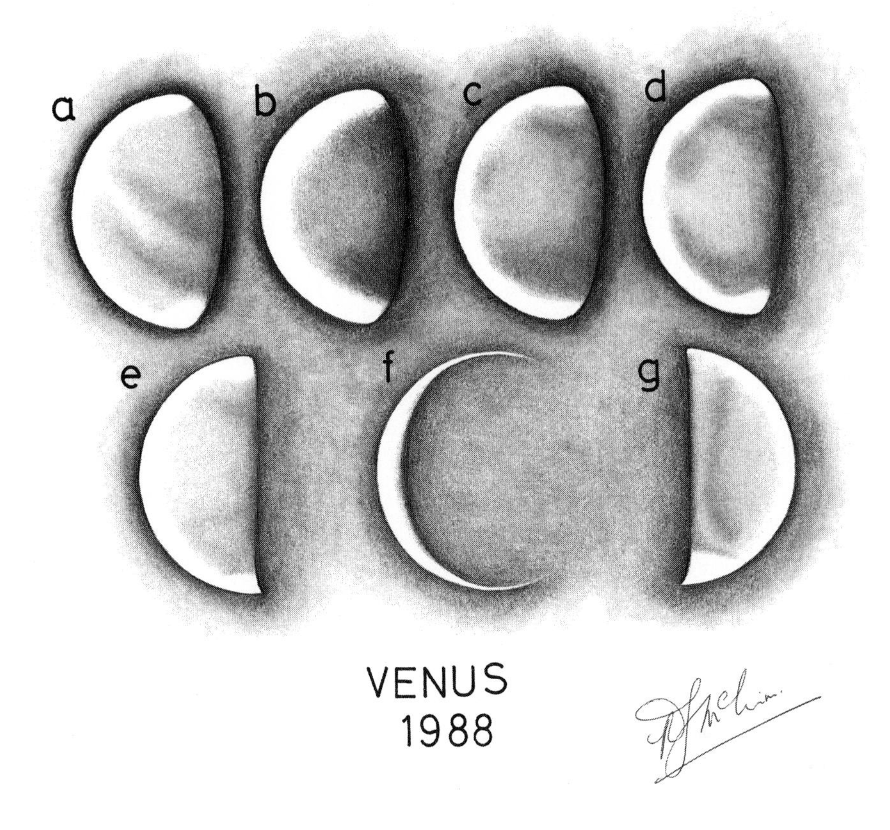

FIGURE 3.1
Drawings of Venus by the author, 1988. South is up. See text for details. (a) Feb 14 d 17 h 30 m. No filter and Wratten 25 (red) filter, 216-mm reflector, ×232. (b) Feb 14 d 17 h 40 m. As (a), but with Wratten 47 (blue-violet) filter. (c) Feb 17 d 17 h 00 m. As (a), no filter. (d) Feb 25 d 16 h 30 m. As (a), no filter and Wratten 25 (red) filter. (e) Mar 23 d 18 h 15 m. As (a), no filter, Wratten 15 (yellow), 25 (red) and 58 (green) filters. (f) May 28 d 20 h 35 m. No filter, 75-mm refractor, ×120. (g) Aug 22 d 09 h 55 m. No filter, 360-mm refractor, ×140.

limb is a little brighter than the mid-disk, and there is a falling-off in illumination at the terminator. The cusps are marked by brighter areas, seen as enlargements of the limb brightening. Such areas are often referred to as cusp caps. Also in the drawing there are two vague bands of shading running diagonally across the disk, representing clouds with lower albedo than their surroundings. The phase was observed (by measuring the sketch) to be 67% , against the ephemeris figure of 72% . The terminator is illuminated very obliquely and is rather faint, accounting for the fact that the observed phase is less than predicted. This effect was discovered by the German astronomer Schröter, some two centuries ago who found by observation that the terminator when supposed to be straight was in fact slightly crescentic. Today the phenomenon is known generally as 'Schröter's effect' and is manifest in observed dichotomy being early in an evening elongation or late at a morning elongation.

Filter Observations

It is interesting to observe Venus with narrow-passband colour filters, and the Kodak 'Wratten' series of gelatine (or glass) filters is recommended. They should be mounted in the drawtube or over the end of the eyepiece. On 14 February 1988 I first observed Venus in white light (Fig. 3.1a), and then with different filters. With a red (No. 25) filter the shadings seemed like Fig. 3.1a; Fig. 3.1b was then drawn with the blue-violet (No. 47) filter. The terminator shading is shown as being heavier and more extended in blue light, while the limb brightening is widened. There seemed to be some dusky shadings present, but they did not appear quite the same as in white or red light. Unfortunately the seeing was by then deteriorating, and observations with other filters were less useful. Such is often the case. Picking up Venus just before it becomes visible to the naked eye at sunset is a good time to start work; the image is neither too bright nor too faint, but soon it becomes dazzlingly bright and the seeing worsens with the progressively decreasing altitude. Blue-filter drawings are important because the darker shadings, so prominent in the UV photographs, are just visible in blue or violet light (though not in white light or yellow light).

Amateur astronomers first experimented seriously with colour filters in the 1940s, although it had been known for several decades that a red or yellow glass would bring out more detail on Mars, for example. In the BAA, the late J. Hedley Robinson, A.W. Heath and the late V.A. Firsoff tried out different filters from 1956 onwards [2]. Finally, it is sometimes said that a green filter (such as a Wratten 58) will highlight the bright areas such as the cusp caps, but I have personally been unable to confirm this.

The Apparent Diameter

Near superior conjunction, both the elongation from the Sun and the disk diameter are quite small. As Venus draws away from the Sun in an evening (or eastern) elongation, it becomes easier to observe. Its altitude above the horizon during twilight is improved, and so are the seeing conditions, generally speaking. At greatest elongation the planet is some 46° from the Sun and setting several hours after it. The phase gradually decreases towards inferior conjunction, while at the same time its angular diameter increases considerably. (The angular diameter is 10″ arc at superior conjunction and 64″ arc at inferior conjunction). The limb remains bright, and the terminator shading tends to become more obvious around the period of dichotomy or half-phase. Figures 3.1c and 3.1d were made 8 days apart on 17 and 25 February 1988: there is a general similarity in the positions of the darker shadings, after two rotations of the upper atmosphere. Unfortunately no observation could be made on 21 February. The cusps are quite bright in both drawings.

Dichotomy and the Cusps

Figure 3.1e shows Venus approximately at dichotomy — or half phase — on 23 March. The cusps are bright and there are polar 'collars' to each. Pioneer Venus Orbiter imagery shows these collars, and the bright areas at the cusps, to be quite real, and not due to any contrast effects. Indeed, the collars are now known to represent the extent of the polar vortices imaged by the spacecraft. The S cusp is blunted and the N somewhat extended. This difference in behaviour of the two cusps is quite typical near dichotomy. Terminator irregularities will be most noticeable, with the obliquely illuminated terminator now seen to best advantage. Spacecraft data prove these irregularities simply to be the origins of dark shadings starting out from the terminator.

Figure 3. 1f, drawn on 28 May shows the planet approaching inferior conjunction. The crescent is very narrow (it being impossible to distinguish albedo variations across the disk) and the horns are extended beyond the semidiameter. Near inferior conjunction, if the planet passes very close to the Sun it is sometimes possible to see it as a complete ring of light; unfortunately I have never quite managed this! The extension of the cusps arises from the reflection and/or refraction of the light from the Sun in the dense CO_2 atmosphere. The degree of extension seems to be merely dependent on the angular separation between Venus and the Sun.

Of course, following inferior conjunction the cycle of phases repeats itself. Figure 3.1g shows the darkest marking I remember seeing, with the 360-mm Arcetri Observatory refractor in Florence, Italy on 22 August 1988. It appeared to be a dark vertical band, parallel to the terminator and was confirmed by my Italian host, Marco Falorni. (As luck would have it, there were no filters to hand and so it could not be studied at different wavelengths!) Note also that the cusps are unequally bright and unequally rounded. Sometimes Venus appears quite blank in excellent seeing in white light. I have observed Venus since 1973 and find that the dark shadings on mid-disk are never easy to see, and even harder to draw realistically. However, the cusps are often bright, and subject to change in area and shape from day to day. Often there are dusky collars visible to each.

After August 1988, Venus was again approaching the Sun in the sky and ultimately reached superior conjunction on 4 April 1989, taking us back to our starting point. The synodic period is on average 584 days long.

AN OBSERVATIONAL PROGRAMME

Why observe Venus nowadays? Most types of traditional observation of the planet must now be regarded as largely recreational, for no scientific goals will be achieved by merely drawing the planet occasionally in white light. However, it is worthwhile to continue to monitor the planet on account of the occasional unusual features which Venus throws up. These are limb and terminator deformities, and the Ashen Light.

First of all, of course, we must find the planet. In a twilight sky it is glaringly obvious. With the Sun above the horizon, setting circles are extremely useful. Set the telescope on the Sun, and offset the instrument by a distance equal to the difference in RA and declination between the Sun and Venus. Use a pre-focused wide-field eyepiece giving a real field of $\frac{1}{2}$–1°. Daylight observation presents a problem because the telescope will be heated by the Sun, leading to tube currents and poor seeing conditions. The most successful observers have used a movable sunshade to keep direct sunlight off the telescope tube during daylight work on Venus or Mercury. Magnification depends on the circumstances. I use ×232 in twilight on my 216-mm Newtonian, though ×130 gives a brighter image and is preferred for daylight work. It is rarely possible to usefully employ a higher power with Venus unless the altitude is very high at the time of observation. Near inferior conjunction the disk will of course be relatively large.

Drawings

Disk drawings of the planet may be made with any aperture, though for serious work a 75-mm refractor or 150-mm reflector would be the minimum necessary. Drawings should be made to a diameter of 50 mm. The phase should be sketched in first, followed by any bright areas at the limb, cusps and terminator. Then any faint shadings can be added. An HB pencil is probably the most generally useful; it should be employed lightly on a pre-drawn outline, and the drawing worked up into a finished state by softening the outlines of the features with an artist's stump or a pad of cotton wool. It is conventional to black-round the background, but for observations such as that in Fig. 3.1f this is unrealistic, and a light pencil background is preferred for artistic effect.

Intensity Estimates

There is a scale of relative intensities in use for Venus, which may be useful instead of verbose descriptions of the subtle features seen. The following is the BAA scale:

Intensity	*Feature*
0	Brilliant white areas such as limb brightening and cusp caps.
1	Bright areas near the limb or at the cusps.
2	The general disk background.
3	The general terminator shading and normal dusky shadings.
4	Unusually dark shadings sometimes seen near the terminator of bordering the cusps.
5	Very dark shadings, usually seen only near the terminator; this would be an exceptional darkness for a Venusian feature.

Drawings should also be made with filters, though the filters vary in density and some will be unsuitable for small apertures. The Wratten No. 15 can be used with all apertures as a standard yellow filter. For small instruments the blue filter chosen should be the No. 44A, but for larger instruments the denser No. 47 (or 49) is preferred. A No. 25 makes a useful red filter, with the denser No. 29 preferred for large instruments. The No. 58 green can also be tried. Polaroid filters can be useful in darkening the sky background for daylight observations.

Phase Measurements

Several amateur groups monitor the so-called Schröter effect, which has been mentioned earlier. Strictly speaking, one should use only phase data acquired micrometrically at the telescope (or by using a finely ruled eyepiece reticule), or measured from photographs with the appropriate filters (see later). A compromise for the visual observer is a set of phase comparison blanks, consisting of a series of phased outlines of the planet to a diameter of about 25 mm, which may be held at arm's length for comparison with the image in the field of view. One simply writes down the phase of the one which corresponds most nearly to the telescopic view, and then draws an identical phase on one's sketch.

Many papers have appeared on this subject. The height of the clouds of Venus may vary somewhat with the solar cycle or for other long-term climatic reasons, and the shadow profile at the terminator and hence the observed phase may be affected from one elongation to the next. It has to be said, however, that the accumulation of much of this 'phase anomaly' data has thrown no light upon the structure of the Venusian atmosphere!

V.A. Bronshten, recounting amateur work in the Soviet Union over many years, has remarked [3] that the observed phase when estimated visually is open to error. Using 'model planets' suitably illuminated he observed that with a gibbous disk the phase is generally underestimated, while with a crescent the phase is often exaggerated. Figure 3.2 presents my own observations of the phase in white light for the 1980 evening elongation, compared with the ephemeris from the *BAA Handbook*. It will be seen that apart from the obvious fact that observed dichotomy was early, the observed curve (O) lay generally beneath the calculated curve (C). But at small phases the reverse was true. Figure 3.3 is my (O – C) diagram for the same elongation, which shows what we may term the 'Bronshten effect'.

Several observers in the United Kingdom repeated these experiments with model planets and obtained similar results. This suggests that visual estimation of phase may be unreliable. However, the (O – C) discrepancy at dichotomy does vary from elongation to elongation from a day or two to about two weeks, so it is hard to believe that the more general anomaly can be entirely accounted for by psychophysiological effects in the human eye and brain.

Limb and Terminator Deformations

In any event, a more fruitful line of work is to monitor the limb and terminator for any large deformities, representing high cloud seen in profile or casting a shadow.

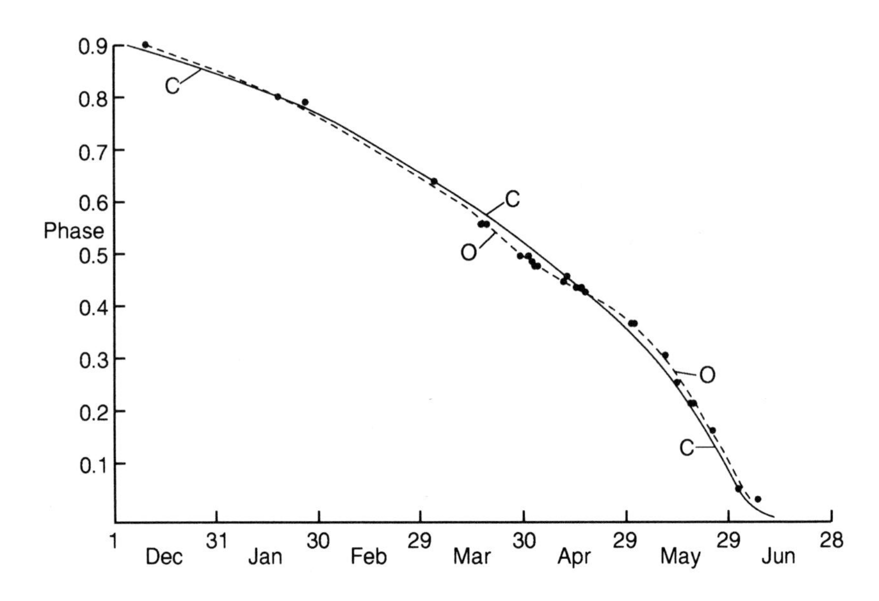

FIGURE 3.2
The observed (O, ------) and calculated (C,————) phase curves for Venus at the 1980 eastern elongation. Observations by the author from measured drawings made without a filter using 200-mm and 320-mm refractors and a 216-mm reflector.

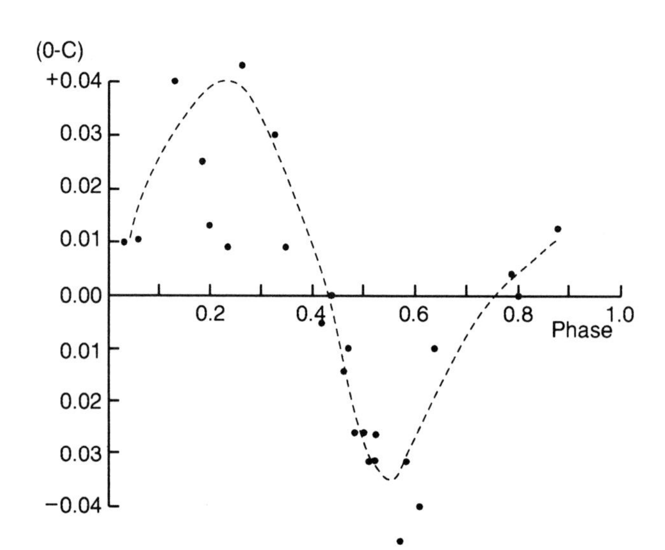

FIGURE 3.3
The variation of the observed minus calculated (O-C) phase as a function of phase for the 1980 eastern elongation of Venus.

FIGURE 3.4
A copy of a drawing made by H. McEwen, 19 February 1913, with a 125-mm refractor, ×144. Note the large 'bite' in the terminator.

Deformations of the terminator, and unequal rounding of the cusps are moderately common. The most famous incident in BAA history occurred on 15 February 1913, when there was a deep notch in the terminator, independently described by Frank Sargent at Bristol and Henry McEwen, Director of the Mercury and Venus Section at Cambuslang, Glasgow (see Fig. 3.4). McEwen saw it again on the 19th: exactly one rotation of the atmosphere later!

Limb deformities are rare, and even more rarely are they corroborated by independent evidence. The Reverend T.H. Foulkes, observing beneath the clear skies of Malta in February 1897 once saw a small 'hump' on the limb, while J.D. Greenwood found a similar feature in March 1985. These two instances are perhaps the only ones ever to have appeared in the *BAA Journal* during the century of the Association's existence.

Unusual Markings

Unusually dark shadings may be seen in white or yellow light from time to time; some were seen in 1924 and 1978, while I saw one in 1988 (see earlier). Such features are certainly worth reporting, but the sensible observer would immediately notify other observers to obtain confirmation at the same hour, just as he or she ought to do for *any* unusual planetary feature. Sometimes, in perfect seeing, the disk can appear broken up into a chequered pattern of small spots and streaks representing small-scale cloud structure. The late Robert Barker, who used a 300-mm reflector, sometimes reported this effect in the 1920s and 1930s, though his interpretation that he was seeing the ground beneath the clouds was, of course, unsound. A.J. Hollis had a similar sighting in 1988, with a 360-mm Cassegrain, near the region of the terminator.

The Ashen Light

The 'Ashen Light' has been left till last to describe, for it is undoubtedly the most controversial phenomenon Venus can display. I have personally never seen it distinctly, though I did record a suspicion of it once under poor seeing conditions. The appearance is similar to our Moon seen with the naked eye when it is a very young (or very old!) crescent. Sometimes the Earth's clouds reflect enough light back into space to feebly illuminate the dark part of the Moon. The effect with Venus is of course not due to any Venusian satellite, but might arise from some effect akin to terrestrial airglow.

When Venus is a narrowing crescent, set against a darkening sky, it is a good time to search for any 'ashy' glow of the unilluminated hemisphere. Being a rather feeble glow, it is never visible against a daylight sky, but when seen under favourable conditions at dawn or dusk it has been variously described as grey or reddish by different observers. Independent observations have been made by experienced observers, and yet the phenomenon has not been accepted by the professional community until recently. Observations of the effect with refractors must remain open to doubt, but there is no reason to doubt observations with reflectors. If an observer suspects the Ashen Light, he should check that it remains visible when the bright crescent of the planet is blocked out by a suitably shaped occulting bar in the field of view. Constructional details for an occulting bar can be found elsewhere [4]. If the Ashen Light disappears when the illuminated disk is hidden then it is illusory, perhaps due to internal reflection in the eyepiece or some other cause. Both the Wratten 58 green and No. 35 purple filters have been recommended for use in conjunction with an occulting bar when searching for the Ashen Light, but I have not found them to be of any use.

In 1988 there was a worldwide 'Ashen Light' campaign when amateurs were encouraged to watch Venus carefully through the crescent phases. Although the planet did not put on any strong display of the Ashen Light, there were enough corroborated observations in April–May to give further credence to its reality. The sightings were correlated with Pioneer Venus Orbiter data on the solar wind, but the investigators found no evidence that the Ashen Light owed its origin to a stream of particles from the Sun.

In 1983 D.A. Allen and J. Crawford made a series of infrared photographs of Venus and found cloud patterns on the unlit hemisphere; these clouds rotated in 5.4 days and are believed to be lower-lying than the UV features. They are thought to be illuminated by solar radiation scattered through the atmosphere from the day side of the planet. Could such clouds sometimes be dimly seen in white light, giving rise to the 'Ashen Light'? The Galileo probe showed that these features lay in the lower clouds at about 50 km altitude. The dark IR clouds are regions of increased sulphuric acid concentration. Infrared work is not a job for the amateur as it requires a very dry atmosphere and large aperture.

PHOTOGRAPHIC WORK

In photographing Venus it is best to insert a yellow filter, such as the Wratten No. 15, into the lightpath to sharpen up the image. With a refractor, the use of filters in photographing the planets is essential. Using a yellow filter gives the planet an appearance very close to that in white light. In an effort in the 1940s to reproduce the subtle shadings observable visually in the Venusian clouds, one French astronomer repeatedly copied his yellow-filter negative images to yield high-contrast positive prints. But most amateurs who want to carry out a research project would be better advised to photograph the planet with red and blue colour filters, or better still to experiment with the UV end of the spectrum.

Filters in Photography

It has been found, both visually and photographically, that the phase in blue light is less than the phase in red. This may be due to the greater terminator shading in blue, which must arise from the greater scattering of the higher frequency blue light (scattering by a planetary atmosphere varies with the fourth power of frequency of the light). The difference in phase ('R – B') is thought to differ from elongation to elongation. Other factors may determine the observed (or photographed) phase: inadequate exposure of the dim terminator, for instance, will lead to an underestimated phase and scientifically useless data. At least the disk of the planet will appear better defined against a blue sky background in red or yellow light. Observers should use the Wratten No. 25 or No. 29 red filters or the No. 47 or 49 blue-violet ones for this work (29 with 49 or 25 with 47). The blue-green No. 44A is too broad-band for satisfactory results. With all planetary photography one should aim to get an acceptably large image and the shortest possible exposure; further advice is given in Chapter 14.

Boyer's original UV work used a 260-mm reflector with an effective focal length (using a Barlow lens) of 10 metres, and a Wratten No. 34 filter to cut out visible light. An exposure of 1 second was suitable on Pan F film. More modern emulsions would probably accept a shorter exposure. This however, presents a problem with focusing! It will be necessary to find the UV focus by experiment. Modern UV glass filters are available, and it is desirable that they should have a peak transmittance near 370 nm; the Schott filters UG2 and UV5 may be mentioned here. Photography should preferably be done with a reflector with an aluminized mirror. Amateurs are experimenting with CCD-imaging of Venus. It is now possible to record the UV-markings with suitable filters, despite the fact that CCDs tend to be red-sensitive.

Some observers have managed to see the UV markings of Venus more clearly following cataract operations. The human crystalline eye lens absorbs UV light effectively, so its removal might be recommended to potential Venus observers! Ewen Whitaker has described his own experiences in this field [5].

TRANSITS OF VENUS

Venus normally just misses the Sun as seen from Earth at inferior conjunction, but if it is near a node at such times it will transit the Sun's disk, appearing as a black spot. It is surrounded by a brighter circle of light due to its atmosphere. The planet can also be seen as a dark spot for a short distance off the disk, projected on the inner solar corona. Such transits used to be of great importance for determining the solar parallax, but they are of less importance today. No-one now living can have observed the last transits of 1874 and 1882. The transits occur in pairs and must take place in June or December. The next pair occur on 8 June 2004 and 6 June 2012. In the United Kingdom the 2004 event will be the more favourable, and only the egress of the planet will be visible at the second transit.

MERCURY: A POSTSCRIPT

Much of what has been said about the *technique* of observing Venus applies equally to Mercury. This small, airless rocky world broadly resembles the Moon in its topography and composition. Its smaller elongation from the Sun makes it a challenge to observe. However, an excellent albedo map has been drawn by J.B. Murray and A. Dollfus from high-resolution drawings and photographs from the Pic du Midi Observatory, and it is unlikely to be improved upon except by the Hubble Space Telescope (whose use in planetary observation is unlikely to be extensive). In any case, Mariner 10 mapped 50% of the planet from close quarters in 1974–5, showing the accuracy of the existing chart. Observations of Mercury are of no scientific value, but it is satisfying to catch the planet in the twilight, and to glimpse the slightly lighter and darker areas on its surface. Its transits are more frequent than Venus, and the next two will take place on 6 November 1993 and 15 November 1999. Occasionally Venus and Mercury approach each other closely enough in the sky to be observed in the same low-power field of view. At such times their brightness and colour may be compared. On 1 June 1980 I made such a comparison: Mercury was dull grey with a yellow or pink tinge and Venus was brilliant silvery white. Mercury was then gibbous and Venus showed a fine crescent.

CONCLUSION

Venus is a frustrating world to study seriously. Amateur work is limited, but even after the various space probe missions our knowledge concerning the frequency of and mechanism for the Ashen Light is rather sketchy. Committed amateurs can still play a useful role in observing Venus if they are presented to concentrate their efforts in the directions illustrated.

ORGANIZATIONS

The two most active amateur organizations concerned with receiving Venus observations are the Mercury and Venus Section of the BAA [6] and the Venus Section of the ALPO [7]. Interested observers will find the addresses of these organizations elsewhere in this book. In Germany there is the Planetary Section of the Bund der Sternfreunde, and in Australia there is the National Association of Lunar and Planetary Observers.

Recently the Unione Astrofili Italiani has created a Venus observing group. Other organizations (such as the Société Astronomique de France and the Oriental Astronomical Association in Japan) have planetary sections which are (rightly!) primarily concerned with Mars and Jupiter. Most issue drawing blanks or printed report forms to save time at the telescope, or can supply filters or other observational aids.

About the Author

Richard McKim, BSc, PhD, C CHEM, MRCS, FRAS, lives in Oundle, Northamptonshire, England, and has been an amateur astronomer since 1972. He joined the BAA in 1973 and observes with his home-made 216-mm and 300-mm Newtonian reflectors. He is now Director of the BAA Mars Section, Assistant Director of the Jupiter Section, a committee member of the Saturn Section, and was BAA Papers Secretary during 1986–1990.

He has carried out original research into the historical observations of W.R. Dawes (the 'eagle-eyed' nineteenth-century English observer of double stars, faint satellites, and planetary markings — Ed.), seasonal variability in the atmosphere of Saturn, and the Martian dust storms. His latest project is the writing of a biography of the famous Greek planetary observer E.M. Antoniadi. He was awarded the BAA Merlin Medal in 1985.

He edited a historical Memoir telling the story of the BAA from 1940 to 1990 that was published for the Association's Centenary. He is also Vice-President of the Groupement Internationale des Observateurs des Surfaces Planétaires, and a member of the ALPO. For his planetary observations he has been allowed to work with the large professional telescopes of the Meudon and Pic du Midi observatories.

Richard McKim is a chemistry teacher at Oundle School, and an ex-assistant Housemaster. He has a doctorate in chemistry from Cambridge University.

Contact address – British Astronomical Association, Burlington House, Piccadilly, London W1V 0NL, UK.

References

1. C. Boyer, The four-day rotation of Venus' atmosphere, *BAA Journal*, **83**, 363–367 (1973).

2. The first such paper to appear was: J. Hedley Robinson, The atmosphere of Venus, 1956, Spring, *BAA Journal*, **66**, 261–264 (1956).
3. V.A. Bronshten, The Schröter effect in the USSR, *BAA Journal*, **81**, 181–185 (1971).
4. A.C. Curtis, 'Ashen Light' observations of Venus, and a special occulting bar, *BAA Journal*, **74**, 229–234 (1964).
5. E.A. Whitaker, Visual observations of Venus in the UV. *BAA Journal*, **99**, 296–297 (1989).
6. For a summary of many years of work by the BAA, see *Report on the planet Venus*, BAA Memoirs, **41**, 1974.
7. The ALPO programme is given in: J.L. Benton, *ALPO Journal*, **30**, 239–245 (1984).

Notes and comments

Venus in daylight – A study of the observations contributed to any amateur group will almost certainly reveal an overwhelming preponderance of crescent views. Yet far more of the planet's surface can be seen in the quarter of its orbit between superior conjunction and elongation, when it is gibbous. Why is this phase so neglected?

One answer is presumably because its disk appears much smaller. But even at superior conjunction the diameter is 10″ arc — a size at which keen Mars observers are certainly still active. It reaches a diameter of 15″ arc at a phase of about 75%, and this is larger than Mars at an aphelic opposition. In terms of elongation from the Sun — another factor affecting observation — there is no difference between a phase of 75% and 25%.

I suspect that in addition to the question of size there is the question of brightness (mag. −3.5 at 75%, mag. −4.3 at 25%). Observers perhaps think that the fainter object will be harder to find, and do not try. In fact, if you work it out area for area, the gibbous disk has a *greater intensity of illumination* than the crescent one. This means that although the planet appears smaller, the contrast against the sky is higher. It is certainly true that the planet will be harder to sweep up using a low-power finder that shows it only as a starlike point, but a few back-and-forth excursions with the main instrument at a power of ×40 or so should locate the small disk as easily as the large one.

Having located it, does the smaller disk size make a great difference? Again, comparing the situations of the Venus and Mars observers, it surely does not. In general, Venusian markings (or at any rate the ones that amateurs record) are broad and of low contrast. The observer is not straining after fine Martian-type detail. If the major and even minor markings on Mars can be recorded on its small disk at an unfavourable apparition, surely useful work can be done on Venus when its disk is no smaller? — Ed.

CHAPTER FOUR · *Observing Mars*

MASATSUGU MINAMI
DIRECTOR, ORIENTAL ASTRONOMICAL ASSOCIATION MARS SECTION

THE PLANET MARS WAS A MYSTERIOUS 'STAR' TO ANCIENT astronomers, since it moved very rapidly against the celestial background and glared particularly brightly, with a reddish colour, when it came to opposition every 2 years and 2 months. It was particularly bright at this time every 15 or 17 years. The name 'Mars' is a Roman translation of the Greek Αρης, which was the name of a warlike god, and the symbol ♂ for Mars is composed of a spear and a shield. When it shines most brilliantly in the sky, at its closest oppositions, it is in the neighbourhood of α Scorpii; the fact that both objects have a similar ruddy hue must have led to the name Antares (anti-Ares) being given to the star. In China, Antares was called 大火 (great fire), and Mars itself is called 火星 (fire star), pronounced *Huoxing* in Chinese and *Kasei* in Japanese.

The planet was also puzzling to observers in Old China because of its mysterious behaviour and sinister colour, and it was called 熒惑, the Chinese characters expressing the dazzling and puzzling characteristics of the planet Mars.

THE MOTION OF MARS

Mars completes an apparition every 780 days, but the reason it appears larger and brighter every 15 or 17 years, when it comes to opposition in Scorpius or Sagittarius, is the considerable eccentricity of its orbit. The apparent diameter of the planet at a perihelic opposition is 1.8 times larger than at an aphelic opposition: the maximal diameter ranges from 25.1″ arc down to 13.8″ arc at these extreme oppositions (see Fig. 4.1).

It is worth remembering that the unusually high ellipticity of the orbit of Mars led Johannes Kepler to discover his three laws of planetary motion at the beginning of the seventeenth century, and these in turn inspired the discovery by Isaac Newton of the classical equation of motion which governs universal matter as well as the celestial bodies. These discoveries might have been considerably delayed if Kepler had not been able to use Tycho Brahe's observations of celestial motions of Mars, and if Mars itself had moved in a less elliptical orbit.

The last approximately perihelic oppositions occurred on 10 July 1986 and on 28 September 1988, and the apparent diameters (hereafter denoted by δ) on those days were 23.3″ and 23.8″ arc, respectively. These diameters were the largest since 1971, and Mars showed good and detailed views — three drawings obtained in 1986 and 1988 are shown in Figs 4.2, 4.3 and 4.4. Note that the latitudes of the disk centres (hereafter denoted by ϕ) differ in 1986 and 1988. The south polar caps are also in different season and show different sizes.

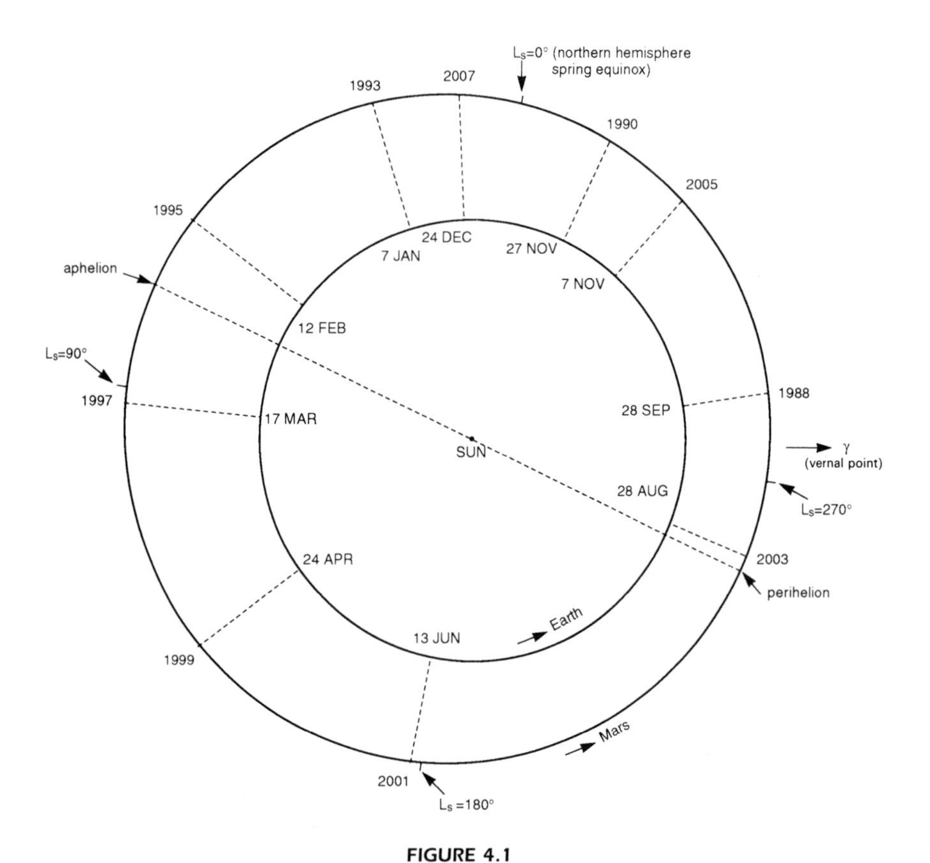

FIGURE 4.1

Oppositions of Mars, 1988–2003. Oppositions of Mars are shown in this heliocentric chart of the orbits of Mars and the Earth. The dotted line through the Sun is the line of apsides for Mars. Perihelion occurs at about $L_s \times 250°$, while aphelion is at about $L_s \times 70°$.

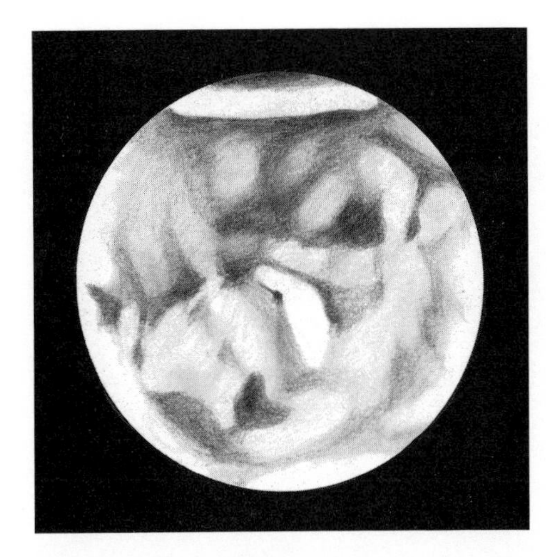

FIGURE 4.2

A drawing of Mars, 1986: 6 July, 17.15 GMT; $\omega \times 61°$, $\phi \times 7°S$, $L_s \times 201°$, $\delta \times 22.7''$, $< 5°$. 250mm refractor, ×420. This view was obtained at the perihelic apparition of 1986. The south polar cap is still large, but slightly shadowy at the southern limb. The big eye-like marking is Solis Lacus, and the preceding dark area is composed of Mare Erythraeum, Sinus Aurorae and Margaritifer Sinus. Sinus Meridiani is partly seen. The 'canal' to the north of Solis Lacus is called Agathodaemon, the site of the huge canyon 'Valles Marineris'. Nilokeras is the claw-like dark marking. (M. Minami, Taipei.)

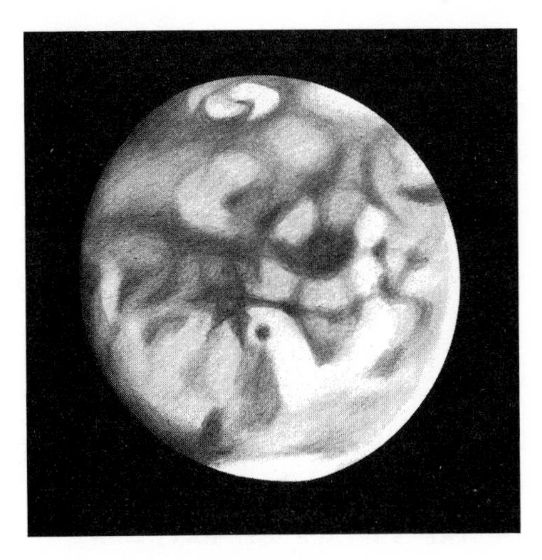

FIGURE 4.3

Drawings of Mars, 1988: 25 August, 17.20 GMT; $\omega = 65°$, $\phi = 20°S$, $L_s = 259°$, $\delta = 21.1°$, $\iota = 27°$; 250mm refractor, ×420. The south polar cap is now smaller with a rift inside it. Solis Lacus is still dark and large. Note that the value of ϕ is considerably different from the image in 1986, and so Nilokeras is now difficult to see. M. Sirenum is appearing at the western (right-hand) limb. Phasis looks slightly different compared with 1986. Note that the evening limb has a noticeable defect of illumination because the time is one month before opposition. (M. Minami, Taipei.)

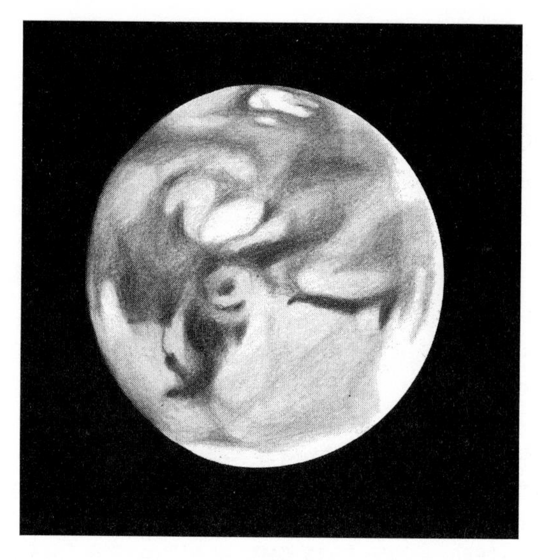

FIGURE 4.4
Drawings of Mars, 1988: 10 September, 19.10 GMT; $\omega = 308°$, $\phi = 20°S$, $L_s = 269°$, $\delta = 23.3''$, $\iota = 15°$; 250mm refractor, ×420, ×540. M. Tyrrhenum and the dark Syrtis Major appear on the afternoon side. A spot to the south of Syrtis Major is the centre of a crater called 'Huygens'. The north-western part of Hellas is dusty. The bar-like marking on the morning side is S. Sabaeus; the right nail of Fastigium Aryn is, however, covered by a morning mist. A white detachment of the south polar cap is sometimes called a 'mountain of Mitchel' (Novus Mons). The season is just at the summer solstice of the southern hemisphere. (M. Minami, Taipei.)

The opposition in 1990 occurred on 27 November 1990 with $\delta = 18.1''$ arc. The following list gives the maximum disk diameter at succeeding oppositions:

7 January 1993	14.9″ arc
12 February 1995	13.9″ arc
17 March 1997	14.2″ arc
24 April 1999	16.2″ arc
13 June 2001	20.8″ arc
28 August 2003	25.1″ arc

The opposition in 2003 will be rare in occurring so near perihelion, and the diameter will be the maximum possible.

OBSERVERS OF THE SURFACE OF MARS

A new epoch for planetary astronomy began at the beginning of the seventeenth century with the invention of simple kinds of telescope that could readily reveal the shapes and surface detail of some of the planets. Galileo observed that Mars was not always perfectly round, but sometimes showed a slight phase or defect of illumination, but the first definite record of surface markings must be attributed to

Christiaan Huygens' drawing made on 28 November 1659, on which a dark wedge-shaped feature, nowadays called Syrtis Major, is clearly depicted. The apparent diameter of Mars was then about 17″ arc, and he used an object glass with a focal length of about 6.8 metres. Huygens also found that Mars rotates in somewhat more than 24 hours, and at the perihelic apparition of 1672 he made out the south polar cap. It is sometimes said that G.D. Cassini discovered both polar caps at the aphelic apparition of 1666, but since it is impossible to see both the south and north caps at the same time Cassini must have seen only the north cap, believing that bright Hellas, or something like the polar hood, was the southern cap.

During the eighteenth and nineteenth centuries there were many excellent European observers of Mars, such as Herschel, Schröter, Dawes, Beer, Mädler, Secchi, Lockyer, Proctor, Phillips, and others. Special mention must be made of Giovanni V. Schiaparelli, observing from Milan, who drew several maps of Mars based on observations starting in 1877. He is famous for detecting several 'canals' using a 220-mm refractor, but more important was his establishment of a Martian longitude system, measured westwards, and his naming of the markings using Latin nomenclature based on old terrestrial geography, Greek mythology and other sources — for example, Hellas, Hesperia, Mare Sirenum, Cerberus, Deucalionis Regio, Hellespontus, Nix Olympica, and so on. These names have been retained.

The convention that the Martian prime meridian is fixed at what was then known as Fastigium Aryn dates back to the days of Beer and Mädler, but it was Schiaparelli who decided that the longitude should run from east to west. This means that the longitude of the central meridian as presented to the Earth, which is usually denoted by ω, increases as time passes, and it can be determined by consulting the *Astronomical Almanac, BAA Handbook*, or some other source, and using simple addition.

The existence of the artificial canals, described in exaggerated form by Lowell in the late nineteenth century, was based on optical illusion.

Another observer deserving separate mention is a former Director of the BAA Mars Section, E.M. Antoniadi, who made a study of previous Martian topographic observations and drew up a detailed map with more than 500 named markings. He himself began his research with the 1888 apparition, but from 1909 onwards he used the 830-mm refractor at Meudon, Paris, and published his great book *La Planète Mars* in 1930. His map should, however, be regarded as a historical index map, some of the markings on which he had failed to verify himself, so that their existence is doubtful. It was Antoniadi who suggested the name Sinus Meridiani for the area of Fastigium Aryn.

SEASONS ON MARS

The surface markings on Mars are essentially stable, but long before Schiaparelli's time it was known that seasonal and secular changes occur. Markings can change in both colour and intensity; there are sometimes morning and evening

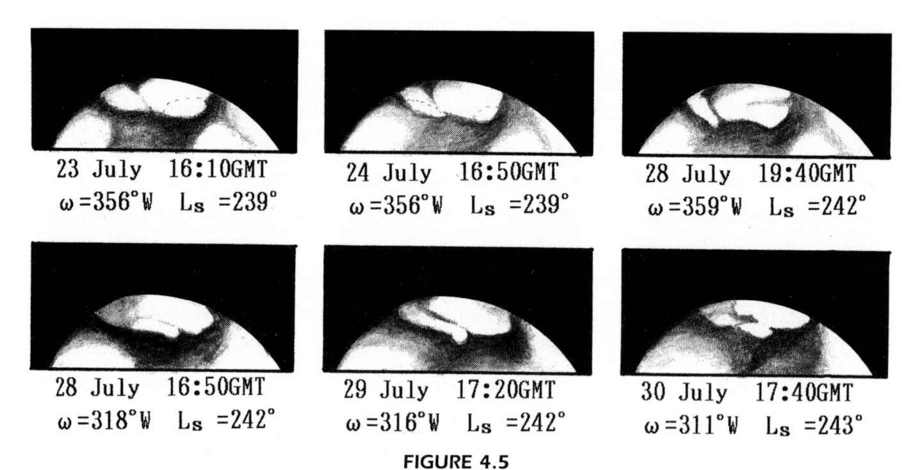

FIGURE 4.5
The south polar cap gradually thawed away, showing several rifts and shadows inside. Here the south polar region was caught from similar angles between $L_s = 239°$ and $L_s = 243°$ just when Novus Mons was detached from the south polar cap in 1988. (M. Minami, 250mm refractor, Taipei.)

hazes, and of course the polar caps show their annual regression. The existence of four seasons on our neighbour world is due to the fact that the inclination of the equator with respect to the orbit is about 25.2°. To indicate the Martian season, it is customary to use the areocentric longitude of the Sun, L_s, which, like the Earth's, is measured in the orbital plane from its ascending node with the plane of the Martian equator. Thus (for the northern hemisphere) the vernal equinox corresponds to $L_s = 0°$, the summer solstice (and therefore the winter solstice of the southern hemisphere) to $L_s = 90°$, the autumnal equinox to $L_s = 180°$, and the winter solstice to $L_s = 270°$.

The north polar cap is seen to regress from around $L_s = 0°$ to 150°, while the south cap starts shrinking at about $L_s = 150°$, reaching its minimum size at $L_s = 330°$. The south cap, when at its greatest extent, is $1\frac{1}{2}$ times larger than the north cap. This follows from the fact that northern midwinter occurs when the planet is near perihelion, while southern midwinter is at aphelion.

In Fig. 4.5, views of the south polar region observed in 1988 are shown: the peninsula called Novus Mons was detaching from the cap itself, a rare event to be seen every 15 or 17 years.

In the 1990s, in contrast, we will be able to watch the north polar cap: its formation and its recession and its disappearance. Especially in the 1992/93 apparition, we will be faced with the rare moment when the deposit polar cap makes its appearance from under a haunting polar hood.

THE MARTIAN ATMOSPHERE

It has already been mentioned that some of the surface features, although basically fixed, are subject to secular changes. For example, the famous dark marking known as Solis Lacus changes in shape and intensity from time to time, while a once dark and broad 'canal' known as Thoth–Nepenthes is nowadays very faint. These changes are believed to be caused by a drift of dust aerosols raised by storms.

The atmosphere of Mars is very thin: according to the results obtained by the Mariner and Viking missions, the pressure at ground level is only about 2–10 millibars, the atmospheric constituents being 95.3% carbon dioxide, 3.0% nitrogen, 0.1% oxygen, and some other ingredients. Only 0.03% of water vapour was found. The atmospheric temperature is low: a range of from −80°C to −30°C was found at the Viking Lander 2 site in Utopia Planitia (225.7°W, 47.7°N). If the atmospheric pressure and temperature are both low, then even a small amount of water vapour can exist in a saturated state, and this can be why water as well as carbon dioxide hazes or mists are to be seen covering some markings. In addition, a mass of supersaturated vapour could condense into cloud or haze if it encountered a heat source.

The so-called 'great dust storms' are sometimes observable as bright yellow clouds in the late summer of the southern hemisphere. These have been observed since the time of Schiaparelli, being recorded at almost all perihelic oppositions. For example, at the 1956 perihelic apparition the beginnings of the great dust storm at Noachis was detected by Japanese observers, and its development was well followed; a similar storm, beginning in the same Noachis region, was caught and followed by American observers in 1971. The reason why observers in particular parts of the Earth had these privileged views is that the rotation period of Mars (24 h 37 m 23 s) almost coincides with our own, so that the longitude of the central meridian advances by less than 10° at the same time on successive nights. Consequently, from one particular site on the Earth's surface a given region is continuously observable, partly observable, and then continuously unobservable over a cycle of some 5 weeks.

The 1956 storm began at $L_s = 245°$, and the 1971 storm at $L_s = 260°$. During the 1973 apparition the beginning of another dust storm was recorded at $L_s = 300°$ by observers in the Pacific areas of the Earth, just after the southern hemisphere's summer solstice (see Fig. 4.6). The Viking orbiters in 1976 also imaged some evidence of dust storms in late summer.

Observations of the occurrence and development of any yellow clouds are very welcome, since the data will give concrete evidence of the movement of the Martian atmosphere. Furthermore, the airborne dust works as a most important component of the air (just like water vapour in the terrestrial atmosphere), and hence observations of storms are vital to the study of Martian meteorology.

TELESCOPIC EQUIPMENT

The intriguing nature of the planet's naked-eye features is matched by its fascination as a telescopic object. It is always exciting to make out markings on the attractive ochre disk as it slowly rotates in the telescopic field, the markings moving gradually from right to left for observers located in the northern hemisphere and

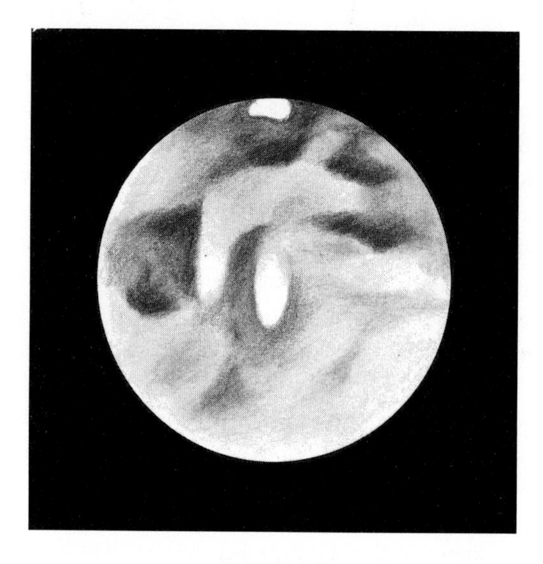

FIGURE 4.6
The yellow cloud 1973: 15 October, 11.55 GMT; $\omega=79°$, $\phi=17°S$, $L_s=301°$, $\delta=21.4''$; 150mm refractor, ×400. In 1973, a brilliant cloud appeared in the Solis Lacus region and was well observed in Japan. This shows the cloud two days after the occurrence; it was brightest on this day. (M. Minami, Fukui.)

using inverting telescopes. Even beginners can easily detect such famous dark areas as the wedge-shaped Syrtis Major when they pass across the central meridian. In addition, a deposit of water and carbon dioxide snow is almost always noticeable as a whitish bright cap at one of the poles. Through an inverting telescope the south polar cap is seen at the top of the disk at a perihelic opposition, and the north cap is visible at the bottom of the disk at an aphelic opposition. It is also possible to catch sight of several whitish or yellowish clouds, the latter becoming evident when they blot out parts of the familiar dark markings.

A good-quality telescope with an aperture of 100 mm will give a pleasing general impression of the planet when it is near opposition, but it should be noted that Mars does not remain near the Earth for very long, and to keep it under observation over as long a period as possible, in order to observe some seasonal changes on its surface, a reflector or refractor of at least 150 mm is needed. A telescope between 150 mm and 250 mm in aperture is easy to handle, and is large enough to cover Mars over most of an apparition. Telescopes between 300 mm and 400 mm in aperture will show more surface detail provided that the seeing conditions are sufficiently good to provide regular crisp images, but such large telescopes must be mounted on a permanent equatorial stand under an observatory of some kind. Although a simple altazimuth mounting looks awkward, it has the advantage of cheapness and portability, and it is not true that a large, motor-driven equatorial telescope is necessary for Martian observation — on the contrary, a well-collimated 200-mm f/8 reflector on an altazimuth mount can be most effective for regular work, as is proved by the amount of data produced by careful observers with instruments of this kind. Newcomers are, therefore, recommended to start work with a 150–200-mm Newtonian reflector on a simple mounting. It is far more important to have excellent optical quality and a site with good seeing conditions.

The high surface brightness of Mars allows a considerable magnification to be used — telescopes between 150 mm and 250 mm in aperture will permit powers of ×400 or ×500. Orthoscopic or Plössl eyepieces are recommended for all planetary work.

OBSERVING THE 'MODERN MARS'

The reason for encouraging long-term observation using moderate, easy-to-handle instruments is that there is not the same necessity nowadays to try to detect minute markings or fine structure on the Martian surface. Ever since the successful Mariner and Viking missions, even though their coverage was of short duration, there has been little point in making strictly topographical surveys from the Earth. On the other hand, climatological or meteorological surveys can be carried out by the amateur with a modest telescope, and these are of real scientific value.

It has already been pointed out that the disk diameter of Mars, even at opposition, varies greatly from one apparition to the next, since it depends on the orbital point at which the Earth overtakes Mars. It is tempting to assume that only perihelic oppositions, when the planet appears largest, are worth observing, but this is incorrect for two reasons.

The first is that Mars at a perihelic opposition is located in Sagittarius, and therefore appears at the lowest possible altitude as observed from northern latitudes, especially in Europe and the northern USA. Even though the disk is large, therefore, the poor seeing will tend to obscure the extra detail. Conversely, when Mars is near aphelion at opposition it is much higher in the sky, and the crisp images may reveal an unexpected amount of detail on the much smaller disk. In England, from the time of Hooke until the 1930s and such observers as the Reverend T.E.R. Phillips, aphelic apparitions were always well observed.

The second reason for observing Mars even when the disk is small is that to observe the course of the seasons properly it must be observed at all points on its orbit. Therefore, the complete 15-year opposition cycle must be covered in order to monitor its seasonal changes as comprehensively as possible — the perihelic apparition reveals only the spring-to-summer season of the Martian southern hemisphere.

As a general principle, Mars should be observed at all possible times when its diameter is over 10″ arc. Experienced observers usually try to observe it when it is even smaller — down to 5–6″ arc.

OBSERVING TECHNIQUES

It is now time to describe how to go about observing the planet, as well as how to record the details seen. First of all, every part of the disk must be examined attentively: casual observers will easily detect several of the dark markings as well as bright regions such as the polar cap, but it is very important to notice the positions, sizes, and intensities of these dark and bright regions. To establish these properly will take some minutes, but the maximum time spent making the observation should be about 20 minutes, mainly because the planet is rotating quite quickly on its axis: if too long is spent surveying the details, the appearance of the disk will have changed noticeably.

Familiarity with the surface features, which can only come with frequent observation, is necessary before this speed can be achieved. A good map showing the main features and their names is also necessary. One standard map was published by the IAU in 1957: this was reprinted in the *BAA Journal* [1], where another map by the Japanese observer S. Ebisawa may also be found. Another useful map, by de Vaucouleurs, was published in *Sky & Telescope* [2]. Reference to a map will lead the observer to identifying less obvious but still important markings, with the result that familiarity with the surface features will improve rapidly.

Once the details have been caught and identified, the next task is to record them either visually or by photography. Chapter 14 discusses the techniques of planetary photography, so in this chapter I concentrate on visual observation of Mars.

As already stated, it is necessary to complete the observation in not more than 20 minutes, since even in this time the planet's axial rotation increases the longitude of the central meridian by 5°. The diameter of the blank disk, which may be drawn in the notebook using a pair of compasses, is usually fixed at 50 mm, but it could also depend upon the apparent diameter of Mars at the time, and also upon the seeing conditions. When Mars is far from opposition, a diameter of 30–40 mm may be employed, but it can be increased to 60–70 mm when the apparent disk is large and seeing conditions are good.

An experienced observer will spend about half the observing session surveying every point on the planet's surface, and the rest of the time making a sketch. It may not necessarily be completed at the telescope: the polishing and finishing can come immediately afterwards. Every observer must be able to memorize at least some details for later recording.

Freehand drawing could proceed as follows. First, let the planet drift through the stationary field so as to establish the north and south points on the disc, align these points with the top and bottom of the blank disk, and, if the planet shows a defect of illumination, draw in the terminator at the correct orientation. Then draw in the bright polar cap, followed by the larger and darker markings. Check also the outlines of the bright regions — for example, circumpolar hoods and morning or evening hazes. Sometimes there are several bright patches on the disk which have to be represented with dotted outlines.

If you feel confident that all the outlines are correctly rendered, write down the time as the actual time of observation. Then start filling in finer details on the drawing, starting from the eastern (afternoon) side, since the markings here are being rotated out of view at the eastern limb. When drawing details in the dark markings, pay careful attention to the intensities of different parts, representing these by appropriate pencil shading. Soft pencils are not necessarily the most suitable, as they easily make the drawing dirty; the choice of hardness is also influenced by the quality and texture of the paper used.

As the drawing nears completion, you will be aware of several features that cannot be represented directly, such as the colours of some markings or some singular phenomenon. These must be recorded in writing as 'Observing Notes' by the side of the drawing. These remarks serve as a memorandum, and can be very useful when reporting the observation to other observers.

The observation must be accompanied by the date and time, the seeing, telescope details, and the magnification used. The following numerical data should also be included: longitude of the central meridian (ω), latitude of the disk centre, or inclination of the north pole towards the Earth (ϕ), the planet's orbital longitude (L_s), its diameter (δ) and its phase angle, or the angle subtended by the Sun and the Earth at Mars (ι). The phase can also be expressed as a fraction, given by $(1 + \cos \iota)/2$.

The numerical value of δ can be written to one decimal place, but it is sufficient to quote the other ephemeris values to the nearest whole number, and the time of observation need only be given to the nearest 5 minutes — 21.47 GMT may be rounded to 21.50, for example.

The date and time are given in GMT (which is numerically identical to UTC, although in practice the two are interchangeable). In the United Kingdom, where the date changes at 24 h/0 h GMT, it is advisable to write the date as, for example, July 7/8 to avoid any possible confusion. In the Japanese islands, which precedes Greenwich by 9 hours, the time of observation usually ranges from 12 h to 20 h GMT when Mars is at opposition, complementing the observations in Europe (about 18 h–06 h) and America (about 0 h–14 h). These times may be longer or shorter depending on the season of the year when Mars is visible, and also on its declination.

Various scales of seeing are in use, although one can simply write 'very good', 'good', 'moderate', 'poor', or 'very poor'.

After making the 'finished' sketch, another drawing can be made showing the features in less detail, marking it with intensity estimates of the main markings. The scale to be used ranges from 0.0 (assigned to the brightness of the polar cap) to 10.0 (the darkness of the night sky), with 2.0 representing the normal intensity of the equatorial deserts. The second drawing should also be labelled with the time

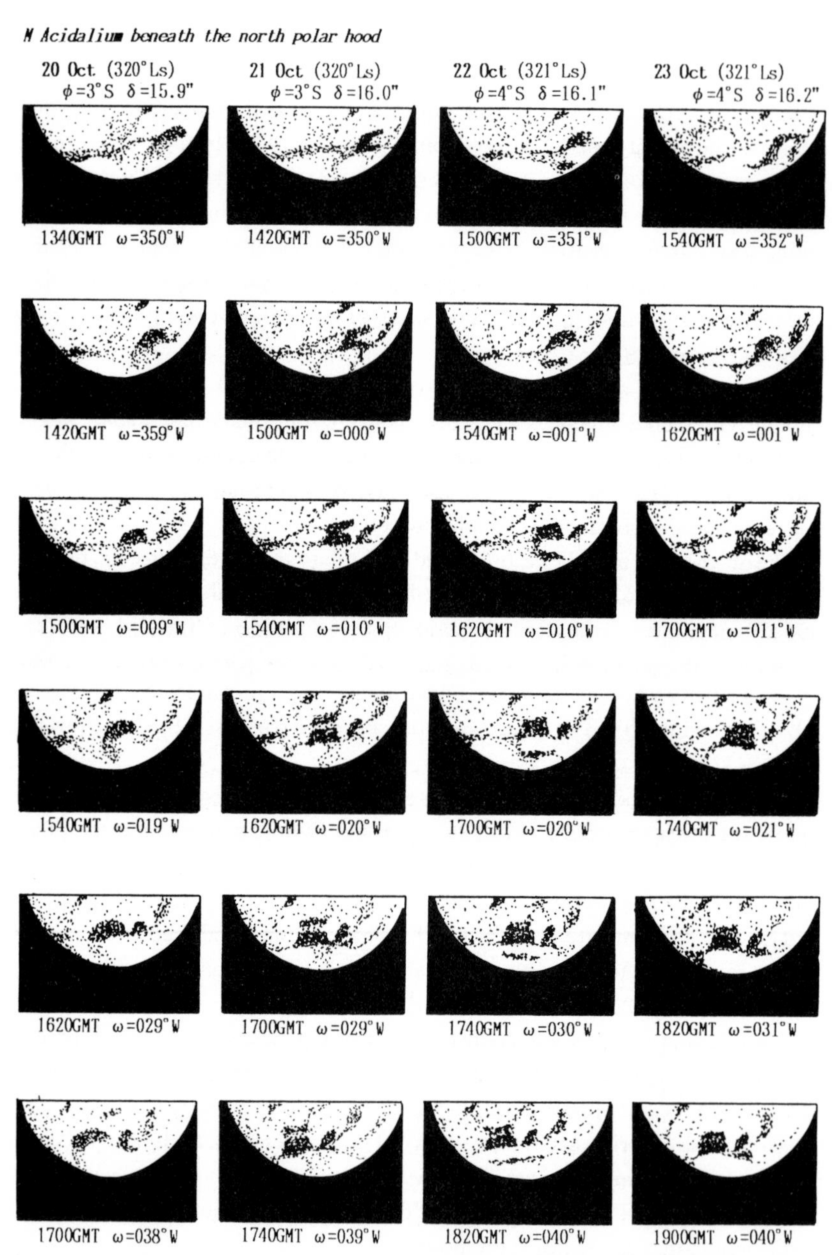

FIGURE 4.7

Mare Acidalium beneath the north polar hood in October 1990. An example of observations made every 40 minutes. This method is effective in detecting daily and hourly changes of the clouds and markings. In 1990 the north polar region was observed heavily covered by a polar hood (in late winter of the northern hemisphere) which was, however, shown to be very changeable every day as the dark marking called Mare Acidalium was seen through it. (M. Minami, 200mm refractor, Fukui.)

and the value of ω. It is also useful for owners of medium-sized telescopes to make observations through coloured filters — any data obtained through a blue or violet filter is particularly valuable, since this suppresses the reddish tint of the deserts and enhances the white and blue-white areas. Such an observation should be recorded on a third sketch. It is not easy to see much through a violet filter, since the eye's sensitivity to this colour is poor, and photography using TP 2415 film is recommended for this branch of Martian observation (see Chapter 14).

Having described how a routine set of observations should be made, we should now like to persuade the observer to try to repeat these every 40 minutes! During this time the longitude of the central meridian increases by about 10°, so the result will be a series of drawings showing views of the surface at this interval of longitude (Fig. 4.7). Furthermore, since the day on Mars is about 40 minutes longer than our own, if observations are made at the same times on successive nights the second drawing made on the following night will show the same view as the first drawing on the first night, and so on.

It is a good thing for new observers of the planet, as well as experienced amateurs, to undertake these repeated views. In the first place, making sketches of the same presentation will train the observer's visual skill by drawing attention to subjective differences from one night to the next, and secondly the series may reveal steady changes in the true appearance of particular features as time passes.

Those observers with some experience are advised to communicate with specialized groups. Among the international organizations which welcome observations of Mars may be mentioned the BAA (recently the Mars section was restored), the ALPO (the Mars Recorder), and the Oriental Astronomical Association (OAA), whose Mars Section has a history of observation extending back over 60 years. It should be pointed out that BAA amateurs have followed Mars at each opposition since 1892.

THE PRINCIPAL MARTIAN FEATURES

The following notes give brief verbal descriptions of some of the major features.

The starting point is Sinus Meridiani at 0°W, 0°N, which lies at the westernmost part of the bar-shaped Sinus Sabaeus. To the west of the prime meridian is seen a large dark area composed of Mare Erythraeum, Margaritifer Sinus and Aurorae Sinus. In the northern hemisphere there lies a big dark marking called Mare Acidalium, together with Niliacus Lacus, and they are easily detected at aphelic apparitions. The desert region between Margaritifer S. and M. Acidalium is called Chryse.

The large eye-shaped marking Solis Lacus (90°W, 25°S) is also conspicuous. Olympus Mons or Nix Olympica (133°W, 18°N) and its preceding Tharsis ridges are visible in some seasons.

The next marking to be caught is Mare Sirenum, which is, however, rather difficult to resolve and nowadays looks shorter than formerly. On the other hand,

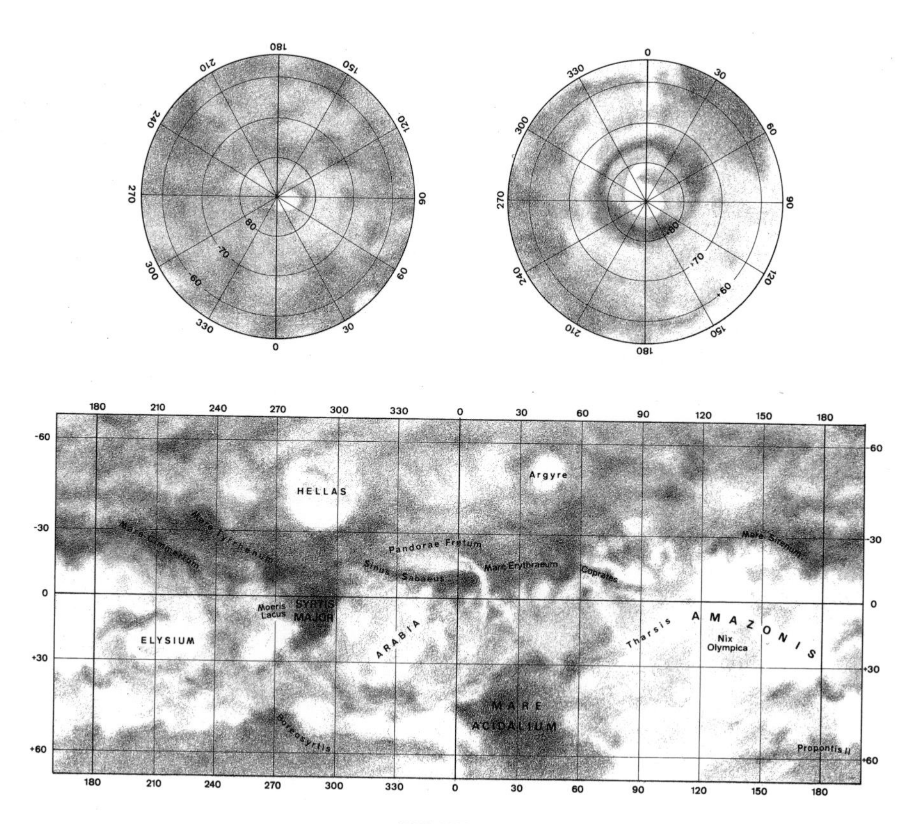

FIGURE 4.8
Map of Mars showing the average appearance of the features as recorded over several apparitions (J.B. Murray).

the succeeding Mare Cimmerium is always evident, with some complex structure. Hesperia is a light region, but rather hard to see at aphelic apparitions — at these times, however, a bright roundish area in the northern hemisphere, called Elysium and partly bounded by shadowy Cerberus and Trivium Charontis, will face the observer.

Mare Tyrrhenum and Syrtis Major (290°W, 10°N) are outstanding landmarks in any opposition, and to the south of them there lies the roundish basin called Hellas (295°W, 50°S) which is yellowish at perihelic apparitions and bright white at aphelic ones. If bright Hellas and dark Syrtis Major cross the central meridian, then Sinus Sabaeus will appear from the west. The easternmost part of Sinus Sabaeus underwent a change throughout the 1980s, but in 1988 the region recovered its usual intensity. This area is situated near the important place where the great dust storms frequently originate. The adjacent regions to the south of Sinus Sabaeus are called Deucalionis Region and Noachis.

RECENT OBSERVATIONS

To conclude, here are some recent results based on the latest observations.

The apparitions of 1986, 1988 and 1990/91 were perihelic, and the south polar cap was observed to regress regularly. It was expected that as the south polar cap thawed, dust storms might occur in the southern hemisphere. It turned out, however, that no great dust storms were observed during these recent perihelia.

In 1986 there were fewer cloud disturbances. However, a one-day dust storm near Solis Lacus was observed on 3 August at $L_s = 217°$ and a strong whitish veil over Chryse was noted from 15 to 19 August at $L_s = 225°$. It was also reported that on 8 November a local storm was observed near Hellas by an Italian observer.

In 1988 two local storms were observed. The first was on 15 June, when a dust disturbance was reported to be found at Noachis, Hellas and Hesperia at $L_s = 214°$ by American observers. This, however, did not expand globally, and faded rapidly when the region came into view from the Far East. On the other hand, it was observed from the Far East that Argyre was in resonance and full of dust during that time.

Secondly, in the latter part of the 1988 apparition, on 27 November at Taipei and two days later at Okinawa, we observed that the desert area to the south-east of Mare Sirenum was covered by a dust cloud; we received news from the United States that the cloud was originally found at an area near Solis Lacus on 25 November at $L_s = 315°$.

In 1990 there also occurred two local dust storms. The first, at Chryse at the beginning of October ($L_s = 310°$), was mainly observed in the United States, but soon subsided. The other storm was discovered on 2 November by European observers and also caught in a CCD image in England. The storm remained in the region from Chryse to Solis Lacus, and was well followed from Europe and America. The cloud did not, however, survive long, and when the region became visible from Japan (on 11 November), it had almost dispersed. By 15 November the region appeared to be free of obscuration.

As already stated in 1988 Mare Serpentis recovered the darkness that was lost in mid-1980. The recovery was possibly caused by the above-mentioned dust storm in June 1988. Similarly, a canal called Phasis reappeared in 1986 (or perhaps before) and was still detectable in 1988 and 1990. This canal was recorded once before, in 1877, by Flammarion, Green, Schiaparelli, Terby, Trouvelot and others, but subsequently disappeared. The area which includes Phasis is called Daedalia–Claritas, and this appeared largely darkened in 1973, after which it faded. This new Phasis seems to be a remnant of the 1973 dark patch in Daedalia.

Conversely, a dark region including a patch called Nodus Laocontis (found in 1948 by Japanese observers and reaching maximum development in the 1950s) has now become obscure, and was not recorded at the latest major apparitions up to 1990. Similarly, Moeris Lacus and the canal called Thoth–Nepenthes are currently faint. These lie east of the Syrtis Major, which is on the rising slope from east to west. Syrtis Major itself appears changed in shape compared with formerly. On the south of Syrtis Major a dark spot inside the Huygens crater, which has not always been visible, is easily seen at the moment.

The north-west end of Mare Sirenum is also changeable — for instance, the part known as Titanum Sinus appeared enlarged in 1954, while it has now faded, and Mare Sirenum appeared much shorter than usual in 1986. Sinus Gomer, on the north side of Mare Cimmerium, is nowadays minimal, although it was prominent in the 1940s and 1950s.

About the author

Masatsugu Minami, a Japanese amateur astronomer, was born in 1939 at Fukui City. He graduated in physics at Kyoto University in 1961, and received a PhD (Kyoto University) in Quantum Field Theory in 1970. During 1980–81 he worked at the Blackett Laboratory, Imperial College, London.

He began observing Mars in 1954 using a 150 mm refractor, and observed the great Martian dust storms of 1956, 1971 and 1973. In 1986 and 1988 he stayed at the Taipei Observatory, Taiwan (Republic of China) to observe these two very favourable apparitions. Taiwan is 10° south of Japan, and so the planet shone 10° higher in the sky. During the four most recent apparitions up to 1990 he secured 808, 998, 838, and 888 drawings of the planet.

He is now Director of the Mars Section of the Oriental Astronomical Association (OAA), established in Japan in 1920.

Contact address – Research Institute for Mathematical Sciences, Kyoto University, Kitashirakawa, Sakyo-ku, Kyoto 606, Japan.

References

1. See volume 96, No. 3, of the *BAA Journal* (1986).
2. De Vaucouleurs' map appeared in *Sky & Telescope* (May 1971).

Notes and comments

Forthcoming apparitions – Observers may find it useful to have a list of dates of the forthcoming perihelia and aphelia of Mars and its midsummer and midwinter dates, as well as details of future apparitions.

	Pehihelion	S. midsummer	Aphelion	N. midsummer
1993			26 Apr	6 Jun
1994	4 Apr	3 May		
1995			14 Mar	24 Apr
1996	21 Feb	20 Mar		
1997			29 Jan	12 Mar
1998	8 Jan	5 Feb	18 Dec	
1999	25 Nov	24 Dec		27 Jan
2000			3 Nov	14 Dec

Opposition of 1993, 7 January – Occurring in Gemini, maximum disk diameter 14.9″ arc, exceeding 10″ arc between the middle of October 1992 and the end of February 1993. Equatorial presentation at opposition.

Opposition of 1995, 12 February – Occurring in Leo, maximum disk diameter 13.8″ arc, exceeding 10″ arc between the middle of December 1994 and the end of March 1995. North pole fairly well presented at opposition.

Opposition of 1997, 17 March – Occurring in Virgo, maximum disk diameter 14.2″ arc, exceeding 10″ arc between the middle of January and early May. Maximum presentation of north pole at opposition.

Opposition of 1999, 24 April – Occurring in Virgo, maximum disk diameter 16.1″ arc, exceeding 10″ arc between the beginning of March and early July. North pole fairly well presented at opposition. – Ed.

CHAPTER FIVE · *Observing Minor Planets*

FREDERICK PILCHER
ILLINOIS COLLEGE, JACKSONVILLE, ILLINOIS, USA

LIKE THE MAJOR PLANETS, THE MINOR PLANETS OR ASTEROIDS MOVE in elliptical orbits. Except for a very few which may approach the Earth more closely than any of the major planets, their positions as viewed from Earth describe retrograde loops near opposition. At this location they are generally at their closest to Earth and their brightest, and most minor planet observing is done within a few weeks of opposition. Many minor planet orbits are more highly inclined to the ecliptic plane than those of the major planets, and minor planets are commonly observed up to 20° and farther from the ecliptic.

IDENTIFICATION

Minor planets have entirely star-like appearances in the telescope. To distinguish a minor planet from true stars in the field requires the following procedure. First the ephemeris, giving right ascension and declination at regular intervals (usually of 10 days), must be obtained. For fast-moving asteroids approaching the Earth more closely than any of the major planets, a shorter ephemeris interval often is used. Ephemerides on eight dates at 10-day intervals bracketing the dates of opposition when the planets are brightest are published annually for all numbered minor planets in the *Ephemerides of Minor Planets* (henceforth abbreviated EMP) by the Institute for Theoretical Astronomy, St Petersburg, Russia. Ephemerides of many brighter planets are published monthly by the *Minor Planet Observer*. These and other sources are given in full at the end of the chapter.

The ephemeris positions are marked against a star atlas showing all stars to a somewhat fainter magnitude than the minor planet. The position at the date of observation may be interpolated between these marks. The telescope is turned to the corresponding position in the sky, and the asteroid should stand out as the starlike object not on the map which is seen in the field near its predicted position.

Some star configurations may be confusing, and in any case it is absolutely essential that movement of the suspected object in the predicted direction and at the predicted rate against the background stars be clearly observed before claiming identification of the asteroid. This author recalls many personal misidentifications, particularly for objects near the magnitude limit of the telescope or star map, or in crowded star fields, which were revealed when the alleged object did not move and showed itself to be an interloping star. If several stars are nearby, movement may be apparent within one or two hours. Even in sparse star fields the displacement should be obvious by the following night..

The best star atlases for identifying minor planets which have been published at the time of this writing are those by Dr Hans Vehrenberg. The *Atlas Falkau*

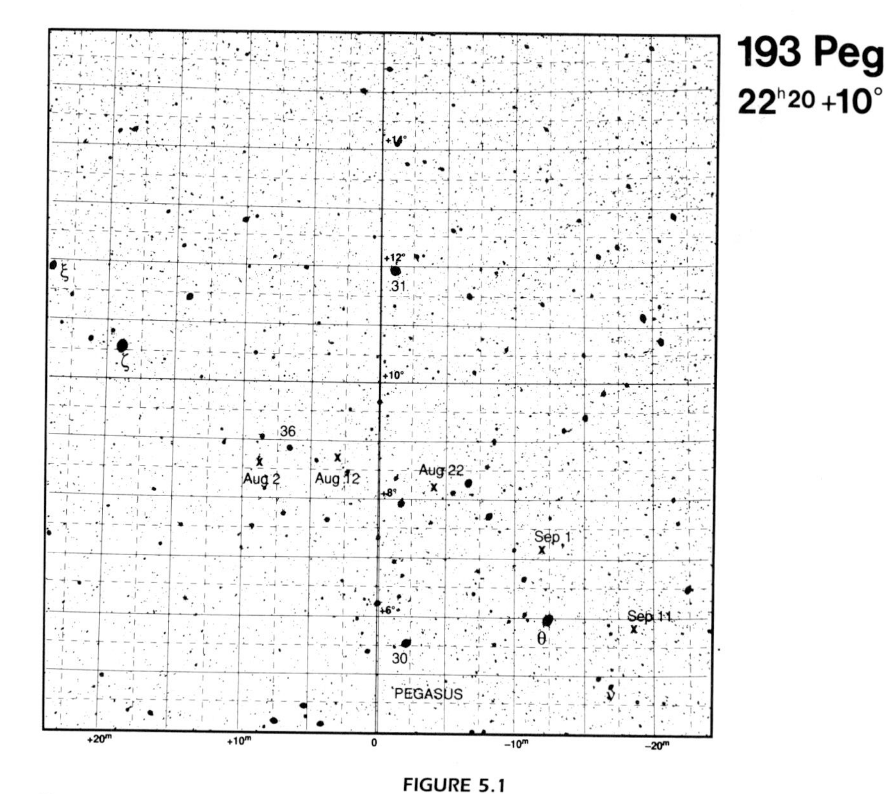

FIGURE 5.1
The 1989 opposition track of minor planet 12 Victoria plotted on the Vehrenberg Atlas Falkau chart 193, centred at 22 h 20 m, ×10°, with coordinate grid superposed and star designations added. (Reproduced by permission of Dr Hans Vehrenberg.)

shows stars to visual magnitude 12–13, and the larger *Atlas Stellarum* to about visual magnitude 14. The Vehrenberg atlases have coordinates in epoch 1950.0.

An illustrated example of plotting the observing positions of an asteroid is shown in Fig. 5.1. The ephemeris of minor planet 12 Victoria, coordinates 1950.0, is obtained from EMP 1989 (Table 5.1). The columns are, respectively, the 1989 data at 0 h UT, the right ascension, declination, distance from Earth in AU (Δ), distance from the Sun in AU (r), phase angle (β), and visual magnitude.

The Vehrenberg atlas chart covering the minor planet's path and the coordinate grid are aligned, and the positions at 10-day intervals are plotted with X's. Star designations are drawn, or for observers using setting circles the minor planet coordinates at 10-day intervals are printed beside the chart. To avoid permanently marking the atlas chart, a photocopy may be employed. Alternatively, a blank paper may be placed over the aligned Atlas chart and transparent coordinate grid, and all marks made on this paper. Centre grids of blank page with X's and star designations and of the Vehrenberg atlas photograph are then aligned on a clipboard for use at the telescope.

The telescope is turned to the position of the minor planet interpolated for the time of observation, and the position of the interloping minor planet is plotted at each observation on the photocopy or attached page. The UT of the observation must also be recorded. When possible, two or more positions on the same night are noted. This author uses a small sketch of the immediate field, with the asteroid positions shown relative to background stars for different times of observation.

A copy of this author's observation sheet for 12 Victoria is shown in Fig. 5.2. The plotted positions relative to field stars, measurements off the coordinate grid to $0.^{m}1$ time in right ascension and $1'$ arc in declination, and small sketches to illustrate movement on a single night, are all included.

TABLE 5.1

EPHEMERIS POSITIONS OF MINOR PLANET 12 VICTORIA FOR 1989

	RA	Dec	Δ	r	β	Mag.
Jul 23	$22^h\,31^m.41$	$+7°\,51'.4$	0.972	1.835	$23°.2$	9.5
Aug 2	$22^h\,28^m.68$	$+8°\,36'.1$	0.926	1.843	$19°.2$	9.3
12	$22^h\,23^m.15$	$+8°\,44'.5$	0.896	1.852	$14°.9$	9.1
22	$22^h\,15^m.85$	$+8°\,15'.5$	0.883	1.864	$11°.0$	8.9
Sep 1	$22^h\,8^m.14$	$+7°\,13'.3$	0.890	1.877	$9°.3$	8.9
11	$22^h\,1^m.57$	$+5°\,48'.1$	0.918	1.892	$11°.1$	9.0
21	$21^h\,57^m.40$	$+4°\,13'.7$	0.966	1.908	$14°.7$	9.2
Oct 1	$21^h\,56^m.33$	$+2°\,42'.7$	1.032	1.926	$18°.5$	9.6

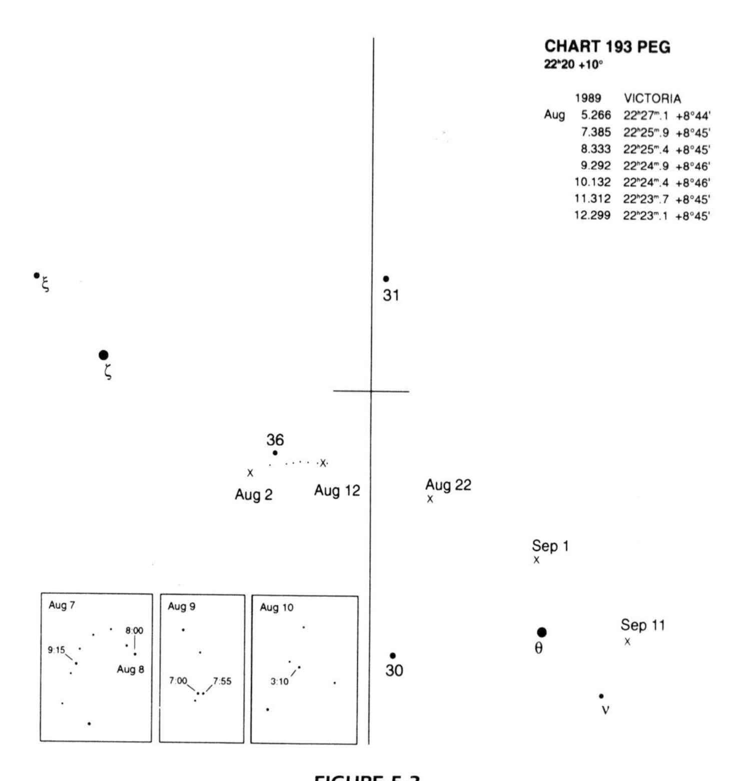

FIGURE 5.2
The author's 1989 observations of minor planet 12 Victoria are plotted on a sheet superposed on the Vehrenberg Atlas Falkau chart 193, with nightly sketches and measured coordinates included. The use of this sheet is discussed in the text.

To use Fig. 5.2, the reader is invited to trace the printed coordinate grid axes and dots marking observed positions on to a blank page, and align this page with Fig. 5.1. The appearance of the minor planet in the field of nearby stars can be seen, and the reader can measure the plotted positions and compare these with the author's measurements.

Tracking a minor planet as it moves through the sky can be both a challenge and a satisfaction for even a casual observer. No observer can claim to have 'done everything' until he has tracked a few minor planets and plotted their successive positions. Tracking minor planets is one of the best techniques in existence to weld the observer, telescope and charts into an efficiently functioning team.

Minor planet observing also requires opportunism. With any aperture, of all the planets within reach at a near-perihelion opposition, more than half will be fainter than the limit of this same equipment at a more distant near-aphelion opposition. One has at most two or three weeks in which to observe a given planet, and then many years may elapse before it is once again bright enough to view.

ASTROMETRY

The positions obtained by use of the Vehrenberg atlas grids are not sufficiently accurate to be useful in improving the orbital elements. Accuracy better than 0.1 second in right ascension and 1″ arc in declination, at UT measured with National Time service radio signals correct to the nearest second of time or 0.00001 day, is necessary to obtain publication in the prestigious *Minor Planet Circulars*. Visually this may be done with careful use of a filar micrometer to measure offsets of the minor planet from nearby stars listed in the Smithsonian Astrophysical Observatory Catalogue (SAOC) of star positions.

Photography is usually the most suitable method of obtaining precise minor planet positions, although some amateurs are beginning to use CCDs. A wide-angle photograph is preferred when equipment is available, because more comparison stars are available against which the minor planet's position may be measured to the required accuracy in both right ascension and declination. The time of mid-exposure to 0.00001 day is again required. The positions on the emulsion of the minor planet and a number of comparison stars are measured with a measuring engine, and the position of the asteroid is reduced by the standard method of dependencies, or, preferably, plate constants. As the position of an asteroid, even at a fixed time, may vary by several seconds of arc from different positions of the Earth, the longitude, latitude, and altitude of the observatory from which the photograph was made must also be reported to compensate for this Earth-centred parallax.

The astrometric positions of a given planet during a given apparition should be collected and forwarded to the Minor Planet Center, as a set rather than individually, as soon as possible after the series of observations has been completed. [See Chapters 16 and 17, respectively, for further discussion of astrometric work and the use of CCDs by amateurs. — Ed.]

DISCOVERY PROGRAMMES

Currently, most new discoveries of minor planets are being made around the 16th visual magnitude. An increasing number of amateur observers are now photographing to this limit, and thus acquiring the ability to discover new planets: an observer seriously interested in this work should first of all acquire the equipment to photograph a field several degrees wide, because of the larger number of planets which can appear on a single exposure.

The Minor Planet Center will give a provisional designation to an unidentified minor planet observed on only two nights. Positions on several nights over at least a week or two are necessary to enable the Minor Planet Center to compute a reliable preliminary orbit, from which the identifications at other oppositions required to assign a permanent number may be made, and the observer credited as official discoverer. The observer with a sufficiently comprehensive set of observations, not necessarily the one with the first observation in time, is accorded credit for the discovery and the right to assign the name.

With more than 5000 asteroids currently catalogued, some go unobserved for many years. Recovering these, especially following ten or more years of non-observation, may lack the glamour of a new discovery, but is of great scientific value.

Guiding a long time-exposure at the sidereal rate spreads the light of an already faint asteroid into an elongated trail. Fainter asteroids can be photographed when the telescope is guided to follow the movement of an average asteroid retrograding parallel to the ecliptic. This causes all stars to form trails, but concentrates the light of asteroids into points. However, asteroids with orbits inclined substantially to the ecliptic will still form appreciable trails, not parallel to the star trails, by this procedure. This is why high-inclination asteroids currently being discovered are mostly brighter than new low-inclination discoveries.

Minor planets which approach the Earth more closely than any of the major planets are of special interest. The former distinctions between Amor-type (perihelion between the orbits of Earth and Mars), Apollo-type (perihelion inside the orbit of Earth), and Aten-type (semimajor axis less than that of Earth), are nowadays being seen as less significant. Several of these, appearing for a brief time brighter than magnitude 15 and including both newly-discovered and returning objects, flash past Earth each year. At their closest and brightest, they may move up to several degrees of arc across the sky each day. Most discoveries are made by professionals with large Schmidt cameras, but timely follow-up photographs necessary to compute a reliable orbit can be made by amateurs, and are among the most scientifically useful observations an amateur can obtain. The IAU Central Bureau for Astronomical Telegrams of the Smithsonian Astrophysical Observatory provides rapid dissemination of information on newly discovered or recovered objects.

No minor planet with aphelion inside the Earth's orbit has yet been found. Always close to the Sun in the sky, and never coming to opposition, such hypothetical objects would be very difficult to observe. Richard Hodgson, Dordt College, Sioux Center, Iowa, USA, has proposed that scans of the region between horizon and celestial pole in late spring and early summer by mid-latitude observers have the greatest potential for discovery of minor planets inferior to the Earth. The probability of success of such a venture is small, although discovery of comets, novae, or supernovae in an otherwise seldom-observed part of the sky would constitute a useful by-product of the programme. But the importance of a single such asteroidal discovery would be enormous.

PHOTOMETRY

Photoelectric photometry can obtain brightnesses of minor planets correct to ± 0.01 magnitude. It has been extensively used by professionals and amateurs alike in the past 20 years, and much valuable work remains to be done even with modest

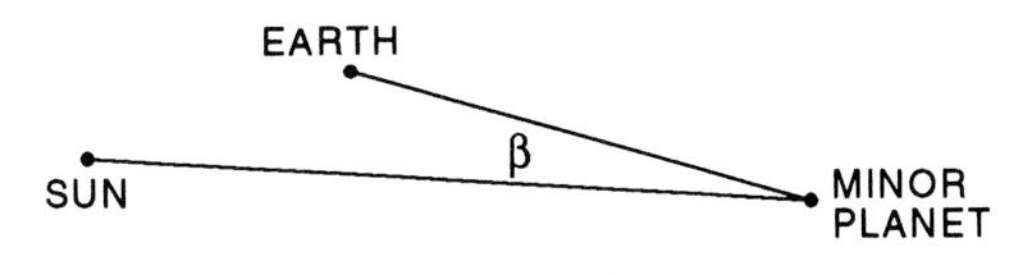

FIGURE 5.3
Illustrating the phase angle β of a minor planet.

apertures. Variations in minor planet magnitudes are caused by many different factors, all of which provide different and scientifically useful information about the physical nature of the minor planet.

Distance – The most striking variation in minor planet magnitude is caused by its varying geocentric (Earth) and heliocentric (Sun) distances. The geocentric distance factor (Δ) typically causes a minor planet to be fainter by about 0.1 magnitude each additional week preceding or following opposition, and is the principal reason why most minor planet observing is done within a few weeks of opposition.

Phase angle – It is well known that the observed, or apparent, magnitude of the Moon is brightest at Full Moon and becomes rapidly fainter with greater phase angle — its angle from the point in the sky opposite the Sun. An analogous effect occurs for all minor planets, in which even after correction for changing geocentric and heliocentric distances the magnitude becomes rapidly fainter with increasing phase angle (Fig. 5.3).

Absolute magnitude – The absolute visual magnitude H of a minor planet is defined as the magnitude it would appear at if observed at 1 AU from both Earth and Sun, and zero phase angle, a configuration which can never actually occur. It depends both upon the size of the asteroid, or more precisely upon its projected area, and the albedo. The slope parameter G determines the variation of magnitude with phase angle. The relationship between observed visual magnitude V, absolute magnitude H, slope parameter G, heliocentric distance r in AU, geocentric distance Δ in AU, and phase angle β, is given by the rather complicated equation:

$$V = H + 5\log(r \cdot \Delta) - 2.5\log[(1 - G)\phi_1 + G \cdot \phi_2]$$

ϕ_1 and ϕ_2 are functions of the phase angle β expressed as follows:

$$\phi_1 = \exp\{-A_1\,[\tan(\beta/2)]^{B_1}\}$$

$$\phi_2 = \exp\{-A_2[\tan(\beta/2)]^{B_2}\}$$

where $A_1 = 3.33$, $A_2 = 1.87$, $B_1 = 0.63$, $B_2 = 1.22$.

Computation of H and G for a given minor planet requires a series of photoelectric measurements of minor planet magnitudes over a range of phase angles.

Rotation – Owing to small objects like minor planets being significantly non-spherical in shape, or to a lesser extent having spots of different albedos on their surfaces, their brightness changes as they rotate. In the most common case in which the light curve is due to rotation of an elongated object, there are two maxima and minima per rotation. For some small-amplitude objects one or three maxima per rotation are suspected. Additional high-quality light curves of these enigmatic objects are especially desired at different aspects in their orbits around the Sun to resolve these ambiguities. Typical rotational light curves have amplitudes from 0.1 to 0.5 magnitude, with occasional larger values being found, and are observable with the techniques of photoelectric photometry.

In the past twenty years there has been an enormous increase in the number of photoelectric light curves of minor planets being made. This is another field in which amateur observers can make scientifically valuable observations, because any good-quality photoelectric light curve of any minor planet extending over an interval of several hours is eminently publishable. A small number of minor planets now have extremely well-established rotational properties, but many others can benefit greatly from additional observations. The tabulation in *Asteroids II* is a useful guide to selecting which minor planets an observer may wish to study.

Differential photometry alone between a minor planet and a nearby star can provide the amplitude of a light curve and the times of rotational maxima and minima from which the rotation period can be deduced. In general a longer light curve, lasting several hours or all night, is more useful than a shorter one. A long series of light curves of a single object, extending over several weeks around opposition, allows a more accurate rotation period to be derived. It may also permit several additional properties explained below to be deduced.

An accurate rotation period may possibly allow the determination of the exact number of unobserved rotation cycles during the long interval of unobservability between one opposition and the next, leading to a further refinement in the period. Subtle changes in the shape of the light curve as the phase angle changes provide additional constraints on modelling the shape of the asteroid. Small changes in the observed Earth-centred, or synodic, period during a single apparition may provide otherwise unobtainable information on the true sidereal rotation period, and also provide a means to distinguish direct from retrograde rotation. Absolute photometry with measurement of standard stars and calculation of extinction coefficients enables, in addition, the magnitude at mean light to be found. Over an extended time this yields the variation of visual magnitude with phase angle and hence provides for the calculation of H and G.

An example of a fine set of photoelectric photometric observations of planet 751 Faīna was obtained by British amateur Richard Miles and published in *Minor Planet Bulletin* [1]. These photoelectric light curves, shown in Fig. 5.4, constitute the first ever made of this planet. Observations on 11 nights distributed over 73 days have been combined to include the entire light curve with no significant gaps.

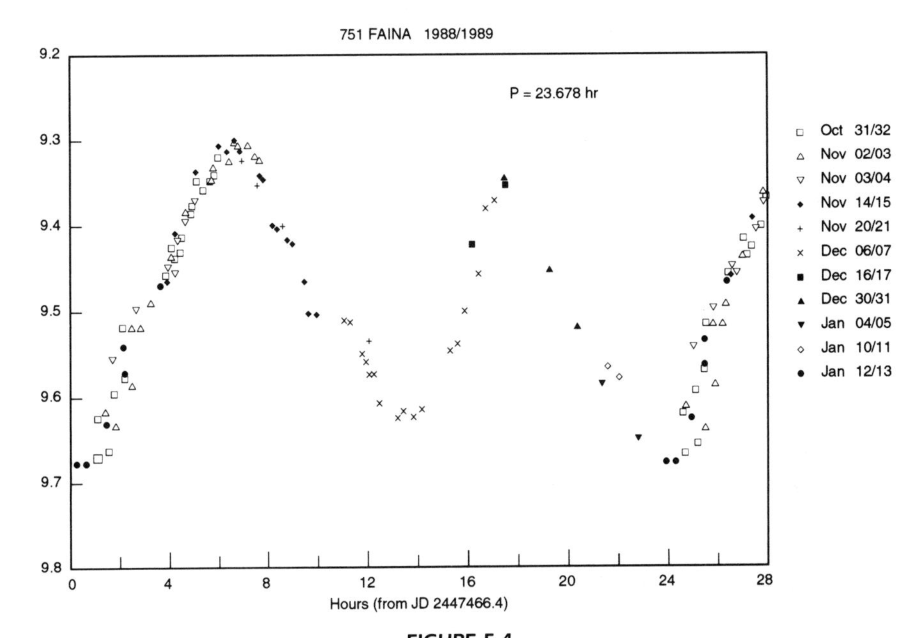

FIGURE 5.4
Composite rotational light curve of minor planet 751 Faīna in 1988–89. (Reproduced by permission of Richard Miles.)

A reliable rotation period of 23.678 hours is obtained. Reduction of observations over a wide range of phase angles to standard stars has yielded the absolute magnitude H and slope parameter G, shown in Fig. 5.5.

For the small number of planets whose rotational amplitude exceeds 0.5 magnitude, sharp-eyed visual observers, and those with variable star experience, can detect brightness variations with the eye alone. This can be done in any field in which the asteroid may be located by establishing a series of brightness steps among the field stars and assessing the asteroid's brightness against this series of steps every few minutes for a time interval of several hours. Times of maxima and minima can often be established with an accuracy of 10–15 minutes. But unless the light curve is followed for 6 hours or longer, an ambiguity in the number of unobserved cycles between one night's observations and the next may prevent the determination of a reliable rotation period. [See the notes on visual photometry at the end of the chapter. — Ed.]

A small body in space naturally rotates about an axis through its smallest dimension, and when viewed from a pole-on orientation the largest cross-sectional area of the object is presented. At this viewing the rotational amplitude is zero. Alternatively, at equatorial viewing the rotational amplitude is a maximum, and mean light is faintest. From one opposition to the next the angle between the rotational pole and the line of sight to Earth changes, and with it the absolute magnitude and amplitude. Photoelectric photometry performed at different aspects

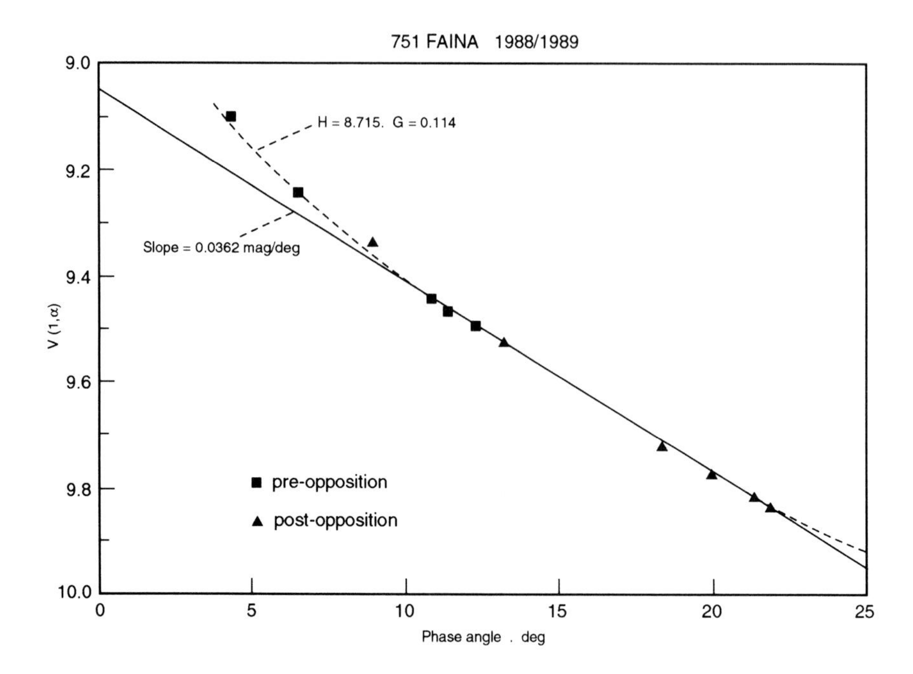

FIGURE 5.5
The phase relation of minor planet 751 Faīna. (Reproduced by permission of Richard Miles.)

around the orbit can yield information on both the relative dimensions of the different shape axes and the orientation of the rotational pole. A solution for the polar orientation usually yields two alternative locations far removed in the sky, making it very difficult or impossible to establish one definitive location.

OCCULTATIONS

One type of observation which often requires only a very small telescope and in which amateurs are making outstanding contributions of great scientific value is the occultation of stars by minor planets. When a minor planet happens to pass between the Earth and a star, its 'shadow' of the star sweeps across the Earth's surface at the speed of the minor planet relative to the Earth. As this speed is known to great accuracy from the known orbit, the time interval of occultation enables the length of the occulting chord of the asteroid cross-section to be found. If many such occultation chords are observed by different people in the shadow path, the two-dimensional cross-section can be constructed. A mean diameter accurate to about 2%, and in sufficiently well observed cases any ellipticity of profile and even limb irregularities, can be deduced.

The occultation may be observed either visually or with a photoelectric chart record. As the event approaches, the asteroid appears to merge with the star, and their combined light is observed. At the occultation ingress, the light decreases abruptly to that of the asteroid alone, and at egress the light is equally abruptly restored to that of both star and asteroid combined. A dimming of a magnitude or more, with no passing clouds to produce spurious events, is normally necessary to observe an occultation visually. This requires the asteroid to occult a somewhat brighter star. A much smaller dimming of a star fainter than the occultating asteroid can be recorded photoelectrically.

The times of ingress and egress must be determined within a few tenths of a second or closer. For visual observations the requirement is for a National Time Service radio and tape recorder on which the one-second signals are recorded together with the observer's commentary giving the instants of dimming and brightening. Photoelectrically, accurate time as well as brightness must be recorded on the chart.

Dr David W. Dunham of the International Occultation Timing Association (IOTA) coordinates the prediction of occultations and reduction and publication of minor planet profiles from actual observation. For most minor planets, the uncertainty in their orbits may cause the actual occultation track to be several hundred kilometres north or south of prediction. Similarly, the actual time of occurrence may be as much as two or three minutes earlier or later than predicted. Observers well outside the predicted occultation track may have nearly as great a chance of observing an event as those within it.

A photoelectric light curve should be made near the time of occultation to relate the limb profile observed to its position in the light curve.

As an example of this work,. Dr R.L. Millis and his colleagues published the mean diameter and limb profile of a well-observed occultation by 88 Thisbe of the star AO 187124 on 7 October 1981 in the *Astronomical Journal* [2]. Figure 5.6 shows the limb profile from positive observations and a constraint based upon a negative observation, and Fig. 5.7 the actual shadow ground track.

Bibliography

A list of a few of the most significant publications for each of several aspects of minor planet studies follows. It is not intended to be complete, but devotees of that particular aspect should have the related works on their bookshelves.

Books

Cunningham, C.J., *Introduction to Asteroids*. Willmann-Bell, Richmond Va, 1988, 208 pp. This is probably the most comprehensive current exposition of minor planets at the amateur level on the market. In an appendix it includes a comprehensive list of additional reference materials.

Gehrels, T., Matthews, M.S. and Binzel, R.P. (eds.), *Asteroids II*. University of Arizona Press, Tucson, Ariz, 1989, 1258 pp. This exhaustive compilation by many authors will be for many years to come the one standard professional-level book

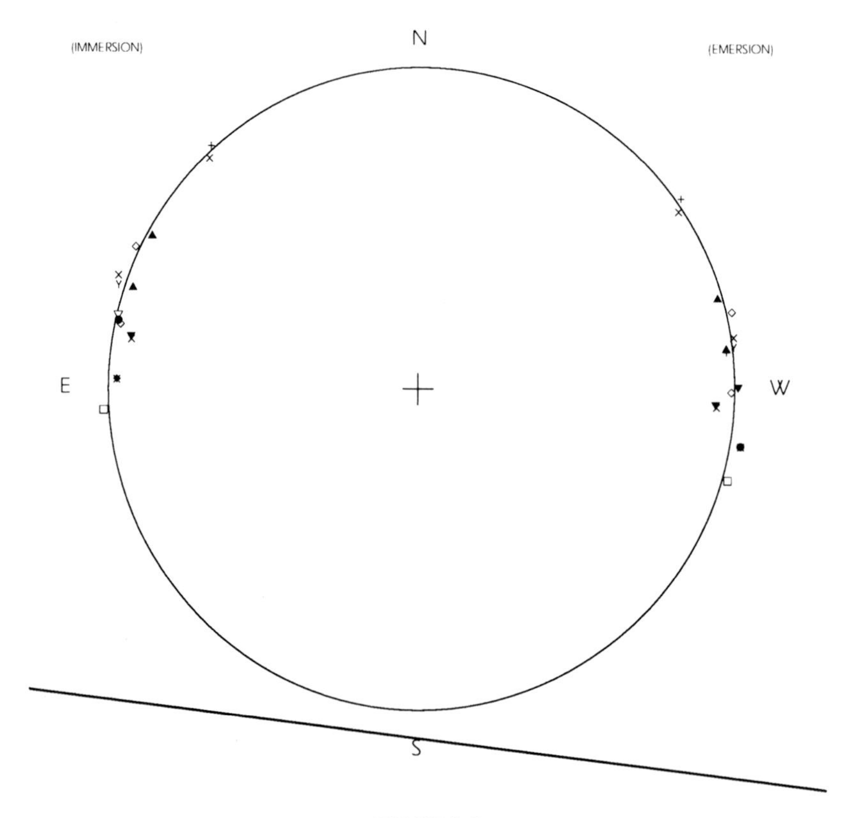

FIGURE 5.6
A circular limb profile fitted by the method of least squares to an occultation by minor planet 88 Thisbe on 7 October 1981. This solution yields a diameter for the face of This be seen at the time of occultation of 221.8 ± 1.4km. The response times of visual observers were included as the free parameters in the least-squares solution. The diagonal line near the bottom of the figure represents the constraint placed on the solution by negative observations at Erwin Fick Observatory. (Reprinted from the *Astronomical Journal*, February 1983, by permission of Dr R.L. Millis.)

on asteroids. While some of the chapters are highly technical, many others contain descriptive material comprehensible to advanced amateurs.

Kowal, C.T., *Asteroids: Their Nature and Utilization.* Ellis Horwood, Chichester, UK; Halsted Press, New York, 1988, 152 pp. Kowal's text covers much of the material in Cunningham's, though more briefly and tersely. An interesting chapter is on the utilization of asteroid raw materials by twenty-first-century spacecraft. The short book concludes with 60 pages of tables providing, for more than 3000 asteroids, the radius, semimajor axis, eccentricity, and inclination, and when known the taxonomic class and rotation period.

Tattersfield, D., *Orbits for Amateurs with a Microcomputer.* Stanley Thornes (Publishers), Ltd., Cheltenham, England, GL53 0DN, published in the USA by

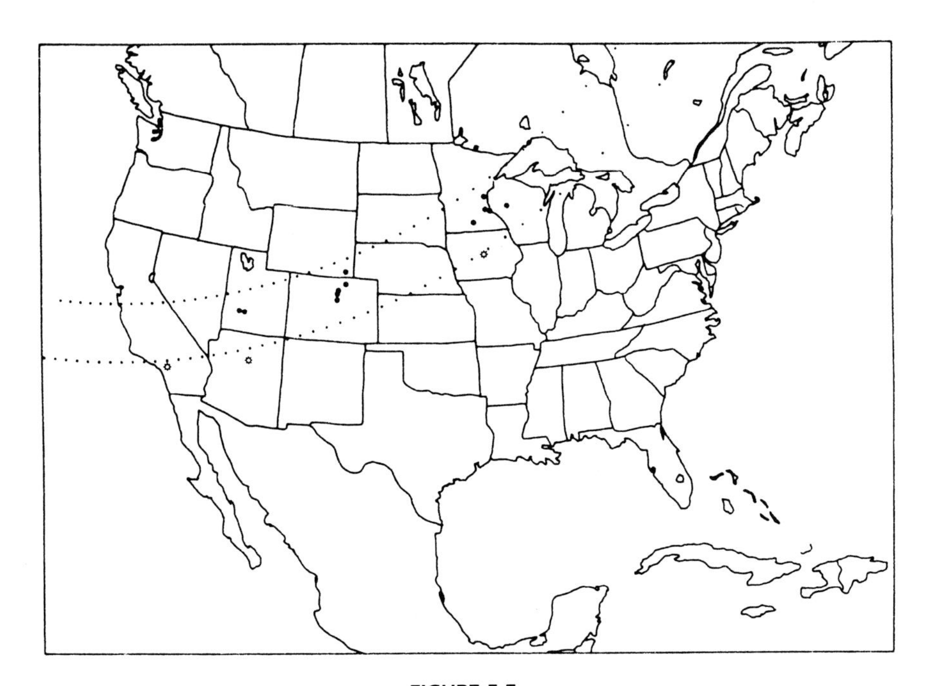

FIGURE 5.7
The actual ground track of the 7 October 1981 occultation by minor planet 88 Thisbe, determined from the least-squares solution shown in Fig. 5.6. Closed circles denote sites where the occultation was observed. The other symbols indicate sites where photoelectric observations showed no occultation. (Reprinted from the *Astronomical Journal*, February 1983, by permission of Dr R.L. Millis.)

Halsted Press, 1984, 171 pp. This book contains sample computer programs for computing an ephemeris from a given set of elements, determining the elements from a set of observations, and perturbing the elements due to the gravitational attraction of the major planets. Some mathematical competence is required, but it is not necessary to take several or even any graduate-level courses in the complex theorems of celestial mechanics that formerly made orbit calculation the exclusive domain of a handful of professionals.

Periodicals
Ephemerides of Minor Planets for the Year 19— (abbreviated EMP). Prepared each year by the Institute for Theoretical Astronomy, St Petersburg, Russia. This is the official annual list of orbital elements and ephemerides for all numbered minor planets. Other sources of ephemerides are copied from this volume. It may be obtained by serious amateur as well as professional observers of minor planets from Mr Dennis Bowman, White Nights Trading Company, 520 N.E. 83rd St., Seattle, WA 98115 USA.

The *Minor Planet Circulars* (abbreviated MPC) are a true astronomical monthly, published at the time of Full Moon when asteroid observation is at a minimum. They contain precise astrometric positions of minor planets and comets, assign new permanent numbers to planets with newly obtained definitive orbits, publish provisional orbits of objects not yet sufficiently well observed to merit numbering, explain the meanings of newly-assigned names, and publish ephemerides of comets and of minor planets not listed in the EMP. A subscription to the MPC may be purchased from the Minor Planet Center, Smithsonian Astrophysical Observatory, Cambridge, Massachusetts 02138, USA.

The Central Bureau for Astronomical Telegrams, Smithsonian Astrophysical Observatory, Cambridge, Massachusetts 02138, USA, provides rapid dissemination, including electronic mail, of information about newly-discovered minor planets, comets, novae, and so forth. In the UK and Europe a similar service is available from *The Astronomer* magazine (Guy M. Hurst, 16 Westminster Close, Kempshott Rise, Basingstoke, Hants RG22 4PP, England).

The *Minor Planet Bulletin* (abbreviated MPB) is the quarterly publication of the Minor Planets Section of the ALPO. It contains ephemerides of a few brighter planets soon to come to opposition, predicts other upcoming minor planet phenomena worthy of observational attention, and reports on observational results of members. Annual subscription may be placed through Derald D. Nye, 10385 East Observatory Drive, Tucson, AZ 85747, USA.

The *Minor Planet Observer* (abbreviated MPO) is published monthly by Bdw Publishing, Box 818, Florissant, CO 80816, USA. This newsletter is a guide to minor planet observing, describing almost all events related to minor planets of 13th visual magnitude or brighter which occur in the month following publication. In it are found opposition finder charts and ephemerides, appulses of minor planets to deep-sky objects or bright stars, occultation paths and finder charts, The Photometry Page, and letters from observers about real observing experiences. Contained in its pages is everything a beginning minor planet observer needs to get started, and to sustain his program for a long time afterwards!

Occultation Newsletter is the publication of the International Occultation Timing Association (IOTA). It presents predictions of minor planet as well as lunar occultations, and reports results of observations. Order through David W. Dunham, IOTA, 1177 Collins Ave., SW., Topeka, Kansas 66604, USA.

Star atlases

Atlas Stellarum and *Atlas Falkau*, by Hans Vehrenberg. These atlases show stars to approximately magnitudes 14 and 13 respectively, with coordinate grids to epoch 1950, the same epoch as standard ephemerides. Hundreds of asteroids become brighter than the faintest stars on these atlases, and hence may be identified readily by reference to them. Each contains 303 maps covering the entire sky from the north celestial pole to -25° declination. Southern hemisphere supplements are available. European customers should order directly from Dr Hans Vehrenberg, Schillerstrasse 17, 4000 Düsseldorf 1, Germany. The North American distributor is Sky Publishing Corporation, PO Box 9111, Belmont, MA 02178-9111, USA.

Acknowledgments

The author wishes to thank Dr Richard P. Binzel and Dr Brian G. Marsden for many helpful comments during the preparation of this manuscript.

About the Author

Frederick Pilcher was born in 1939 in Calgary, Alberta, Canada, of United States parents. He earned a BS from Washburn University and MS from the University of Kansas, both in physics, and has been Chairman of the Physics Department of Illinois College since 1962. In 1968 he began observing and writing about observing asteroids. He independently developed the technique of utilizing the Vehrenberg star atlas to identify asteroids at the telescope, and demonstrated the falsity of the popular earlier myth that asteroids were very difficult to find and identify.

In 1973, with Jean Meeus, he authored *Tables of Minor Planets*, privately published but widely distributed. In the same year he become a charter member of the ALPO Minor Planets Section. In 1981 he become Recorder of the Minor Planets Section of the ALPO, in which position he continues.

He prepared the Discovery Circumstances section of *Asteroids* (1979) and *Asteroids II* (1990), published by the University of Arizona Press.

Throughout the years he has steadily pursued his hobbies of visually observing faint minor planets, reading widely the world's professional literature on minor planets, and popularizing the subject in amateur-level publications and activities.

Contact address – Illinois College, Jacksonville, Illinois 62650, USA.

References

1. Miles, R., *Minor Planet Bulletin*, **16**(3), 25–27 (1989).
2. Millis, R.L. *et al.*, *Astronomical Journal*, **88**, 229–235 (1983).

Notes and Comments

Visual photometry – Andrew Hollis (Director, BAA Asteroids and Remote Planets Section) has added the following comments. — Ed.

> In small telescopes, asteroids always appear as single points of light. Indeed the name 'asteroid' means 'star-like', and was coined by William Herschel in 1802 shortly after the earliest discoveries. It should also be noted that Uranus and Neptune also appear star-like when observed using low

magnification. In this respect, observation of these bodies has many similarities with stellar astronomy.

Unlike most stars, however, the rotation of solar-system bodies can readily be determined from the observation of periodic variations in the intensity of the irregular shape of the smaller members of the solar system and as a result of variegated patterns of light and dark covering the surface. Although the amplitude of any variation is usually quite small (typically 0.2 magnitude or less), some planets can exhibit much greater ranges, as for example 216 Kleopatra, which varies by up to 1.6 magnitudes. Even though for any particular opposition no apparent variation can be detected, it does not mean that there will be none at a future opposition — the pole of the rotational axis can at times be directed towards the Earth, along the line of sight of the observer, when no light variation would be expected.

Several types of photometric study are possible visually. Brightness estimates made every 10 to 15 minutes can be plotted on a 'rotational' light curve which, if the amplitude is sufficient, can be used to determine the axial rotation period of the object.

Observations taken over a period of several months will show the change in brightness as the distance from the Sun and Earth varies, and also as the phase angle alters. This latter curve allows an estimate of the surface physical characteristics of the planet to be made.

A chart should be prepared which gives suitable comparison stars for estimating brightness. Because the planets move relative to the background stars, a number of comparison stars strategically placed along the path are selected for this purpose. Whenever possible, comparison stars are picked which have been previously checked photoelectrically so that their magnitudes are accurately known. When few such comparison stars can be found, additional stars can be selected from the AAVSO *Star Atlas* or the BAA *Variable Star Fields*. Least reliable are stellar magnitudes taken from the SAO *Star Catalog*.

Several factors must be borne in mind:

1. A reliable sequence of comparison stars must be identified.
2. Observations should be continued over several nights, which need not be consecutive.
3. If light changes are detected on a given night, then observations should continue for several further nights so that a complete rotation of the planet is seen — normally made up of two maxima and two minima in the visual light curve. If this is not done, several different rotation periods may, in fact, fit the observations.

If the regular change has an amplitude greater than 0.3 magnitude, then the double period (taking in two maxima and two minima) is most probably the true rotation period of the object. For the asteroids, typical periods lie between 5 and 20 hours.

Magnitude estimates — extending the limit – Attempting to determine, visually, rotation periods from the modest fluctuations exhibited by most minor planets, makes any refinements in technique worth pursuing. Andrew Hollis points out how the Argelander step method (see Chapter 10) can be refined to match the limits of the eye's discriminatory powers. It takes account of the fact that successive magnitude estimates made with the eye are likely to be inconsistent for reasons outside the observer's control, and that an 'average' reading from a number of mental estimates needs to be made. — Ed.

In the Argelander method, allowance is made for the unsteadiness of the atmosphere and the imprecision of the human eye. Indeed it could be said that the method relies on these two factors to decide upon brightness steps. The following technique is used:

1. If, after prolonged viewing, one object appears brighter than the other for the same amount of time as it appears fainter, then the two objects are assigned the same magnitude.
2. If one appears brighter more often, but on a few occasions it is fainter, then it is noted down as being one 'step' brighter.
3. If one appears brighter most of the time, but on occasion is equal to the other, then it is two 'steps' brighter.
4. If one is always brighter, but on occasions only just so, then the difference is three 'steps'.

Further 'steps' can also be recorded, but are less reliable to estimate. Several comparison stars should be used to achieve maximum accuracy. Note that no attempt is made to estimate the size of an individual observer's 'step', which is different for different observers owing to physiological factors. Typical 'steps' fall in the range 0.06–0.09 magnitude.

Discovering new minor planets – A most interesting article by Robert McNaught appeared in *The Astronomer*, September 1989. Entitled 'Amateur searches for minor planets', he points out that in a period of almost ten years (1978–87) 197 new minor planets were discovered photographically by amateurs, of which 160 were found from Japan and the remaining 37 from Italy. The discoveries were all made using wide-field instruments (typically f/3) with apertures ranging from 160 mm to 460 mm.

These 'discoveries' are not necessarily the first time the object in question has ever been recorded, but are usually recoveries of objects that were never sufficiently well observed for accurate predictions to be published. They are checked at the

Minor Planet Center and, if definite identification with earlier observations can be made, the object is given a permanent identification and the discoverer is given credit.

A minor planet must be observed during at least two apparitions, and if no pre-discovery positions can be traced, confirmation awaits its recovery when it returns to a subsequent opposition — if it can be found.

McNaught comments that his experience with using the UK Schmidt Telescope at the Anglo-Australian Observatory suggests that one new object of about magnitude 16 is likely to be discovered in every 80 square degrees photographed near the ecliptic. If one can reach magnitude 17, the density rises to about one new object every 5 square degrees, and there are about 50–100 new objects of magnitude 21 per square degree! Searching for new objects in the magnitude 16–17 range is certainly a feasible project for amateurs — Brian Manning (Chapter 16) has even succeeding in making discoveries from England. — Ed.

CHAPTER SIX · *Observing Jupiter*

JOHN B. MURRAY
DEPARTMENT OF EARTH SCIENCES,
THE OPEN UNIVERSITY, UK

OF ALL THE PLANETS, JUPITER IS THE EASIEST AND THE MOST rewarding to observe, and the study of its markings is one of the few remaining areas of astronomy where the amateur observer can contribute in a substantial way to scientific knowledge. For most of the time, it has by far the largest apparent diameter of all the planets as seen from the Earth, usually being two to three times larger than Mars or Saturn. Even at its farthest, it is still larger than either planet at their closest. Only Venus can occasionally exceed it in apparent size for a few weeks every $1\frac{1}{2}$ years, but when Venus is at its closest, it always presents its dark side to us. Jupiter is also well placed for observation for much of the time; during favourable apparitions it is observable from temperate latitudes for up to ten months of the year. In addition to these great advantages, Jupiter has prominent markings which are constantly changing, and when its atmosphere is in turmoil observations of the same region only a day apart will show substantial changes. Yet despite this exciting and dynamic nature, there are surprisingly few dedicated Jupiter observers, and it has to be said that any amateur who can afford the time to make regular observations of its surface changes over an extended period will find himself a member of an élite which comprises only a handful of observers worldwide at any one time.

HISTORY OF TELESCOPIC OBSERVATION

Seventeenth century – Ever since the discovery by Robert Hooke of a prominent dark spot in Jupiter's southern hemisphere on 9 May 1664, the observation of the changing position of markings has been the principal work of amateur observers. Intensive observation of this spot by J.D. Cassini in the following year meant that Jupiter's rotation period of just under 10 hours was the first to be discovered of all the planets. By 1690, however, Cassini had accumulated sufficient observations of the rotation periods of various spots to realize that those near Jupiter's equator had rotation periods around 9 h 50 m 30 s, about 5 minutes shorter than those at higher latitudes, which average about 9 h 55 m 40 s. This is due to the fact that all observed spots are cloud patterns in Jupiter's upper atmosphere, and differential wind currents produce the observed differences in rotation period.

Eighteenth century – Most features in Jupiter's atmosphere are short lived, often appearing and disappearing over a few days or weeks, rarely surviving for more than a year, but Hooke's spot in the southern hemisphere of Jupiter was a particularly long lived feature, and was observed intermittently for 50 years, being last recorded by J.P. Maraldi in 1713. Whether it actually disappeared after this date, or whether no one bothered to record its presence, is a matter for considerable debate. The following hundred years was a low period in the observation of the planet; astronomers had moved on to more fundamental problems, and though here and there individuals such as Sir William Herschel and J.H. Schröter continued to observe it, only a very incomplete record of its activity survives.

Nineteenth century – The nineteenth century saw a steady revival of interest in Jupiter, almost exclusively by amateur astronomers, and observers such as H. Schwabe in Germany (who also discovered the sunspot cycle after long and painstaking observation) and the Reverend W.R. Dawes and later W.F. Denning in England published important observations of the planet. Then in 1878, a feature which had been observed in the southern hemisphere for a number of years became much more conspicuous, being larger and darker than any previously observed spot, and taking on a striking red colour. It became known as the Great Red Spot, and it has survived to the present day, which is quite remarkable when one remembers that it is only a cloud configuration in the atmosphere. Some consider that it may be the same feature as Hooke's spot, but it seems odd that it was not reported between 1713 and 1830.

The striking appearance of the Great Red Spot from 1879 to 1882 caused great interest and drew worldwide attention to the planet. Measurements of its rotation period were made by several observers, and many maintained their interest in the planet in the following years. However, although increasing numbers of observations were made, individuals published their results in isolation, and things might have remained this way had a group of amateur astronomers not combined together to form the British Astronomical Association (BAA) in 1890. This body was formed so that observers with small telescopes could combine their results and thus have a better chance of contributing something useful to science. A Jupiter Section was formed, and reports of observations were published for each apparition of the planet, so that a regular record of the planet is available from this time onwards.

At first, the reports consist mainly of verbose descriptions and drawings, but after the publication of Stanley Williams' classic paper of 1896, in which he announced the discovery of no less than nine persistent wind currents at different latitudes, based on his own central meridian transit timings, the wide ranging possibilities of his methods became evident, and were gradually followed by other observers on a regular basis. Prominent among these were Captain P.B. Molesworth and the Reverend T.E.R. Phillips, who became Director of the Section in 1901. By the time the South Tropical Disturbance appeared in the same year, therefore, there was an enthusiastic group of observers regularly observing the planet, so that the birth and early development of this extraordinary feature that was to dominate the planet for the next 39 years was recorded in minute detail.

Twentieth century and spacecraft observations – Phillips was a keen and methodical personality with a great gift for inspiring enthusiasm in others, and he gathered round him a talented young group of observers, notably F.J. Hargreaves and B.M. Peek, who succeeded Phillips as Director in 1934. They made some of the most exciting discoveries in the inter-war years, including the outbursts of the South Equatorial Belt in 1919, 1928 and 1943, circulating currents of 1919–20 and 1931–34, and the oscillating spots of 1940–42. After the war there was a steady increase in amateur observation in the United States, and the Association of Lunar and Planetary Observers, with such skilled observers as E.J. Reese, began publishing important work.

The advent of space exploration in the 1960s meant that there was a sudden revival of interest in the planets by professional astronomers, and the New Mexico State Observatory began a regular programme of high-quality photography of Jupiter, which resulted in the discovery of the vortex motion of the Great Red Spot and its 90-day oscillation, and also the identification of an unusually long-lived equatorial plume that lasted from 1963 until 1977. In 1969 Lowell Observatory began a continuous photographic patrol of the planets, from six observatories spaced in longitude and latitude around the Earth, so that virtually continuous photographic coverage was possible, and in one year more photographs of the planets were taken than in the previous hundred or so years since planetary photography began. The Lunar and Planetary Laboratory, University of Arizona, also began publishing reports of their photographic observations of Jupiter in the 1970s, in a build-up to the visits by NASA spacecraft Pioneers 10 and 11 to the planet in 1973 and 1974, and Voyagers 1 and 2 in 1979.

These space missions provided spectacular close-up pictures of the planet, which revealed the detailed structure of the belts and zones, the Red Spot and many smaller features. Successive images during the brief encounters were also measured for cloud movements, which showed vortices in many spots, and also demonstrated that some surface currents, hitherto thought to be intermittent and related to outbursts at particular latitudes, were always present in features too small to be seen from Earth. More important were the spacecraft discoveries about the composition, temperature and structure of the atmosphere, the interior and magnetic field, and the satellites, which are outside the scope of a book on observation.

This upstaging by spacecraft and well-funded professional observatories put amateur work in the shade during the 1960s and 1970s, but professional funding declined after the Pioneer and Voyager visits, and many professional planetary programmes were abandoned. Since 1980 the amateur has again found himself as the main source of information on the activity of Jupiter, and groups of observers from many countries, such as the ALPO in the USA, the Société Astronomique de France and other societies in Europe and Japan are now publishing regular reports of activity on Jupiter. The BAA Jupiter Section is continuing to carry on its pioneering work, recently on the latest series of South Tropical Disturbances, under its new Director, J.H. Rogers.

METHODS OF OBSERVATION

Starting to observe Jupiter – The most skilled and experienced observers have on occasions been known to make useful observations of Jupiter with telescopes as small as 80 mm aperture, but the beginner is unlikely to see anything worth recording with less than 200 mm. If this is beyond your means, then find out where your nearest local astronomical society is from *Sky & Telescope* or the Federation of Astronomical Societies, as many such societies have their own telescope for members' use. The best instrument for observing Jupiter (or any planet) is a long-focus refractor of the widest aperture practicable, with the object glass as high as possible off the ground, away from the air turbulence just above ground level. Such instruments are extremely rare in amateurs' hands, however, most having to content themselves with the usual 200 mm Newtonian, but lifetimes of extremely valuable research can be carried out with this humble instrument, or with a professionally manufactured catadioptric.

Even when observing with an adequate instrument, it may take some time before a beginner's eyes get used to the delicate nature of planetary detail. What is described as a very dark spot by an experienced observer appears as an extremely faint smudge to the novice. Sometimes it is the contrast of dark sky and bright disk that renders the details difficult to see, and observation during bright twilight, when the planet first becomes visible to the naked eye, may show the surface features much more clearly than against a dark sky.

It will at once be evident that the disk is slightly flattened at the poles, and that a series of dark belts and bright zones cross the planet parallel to the equator. The position and intensity of these belts and zones may vary, and some may be absent altogether at times, or additional belts present at others, but an average

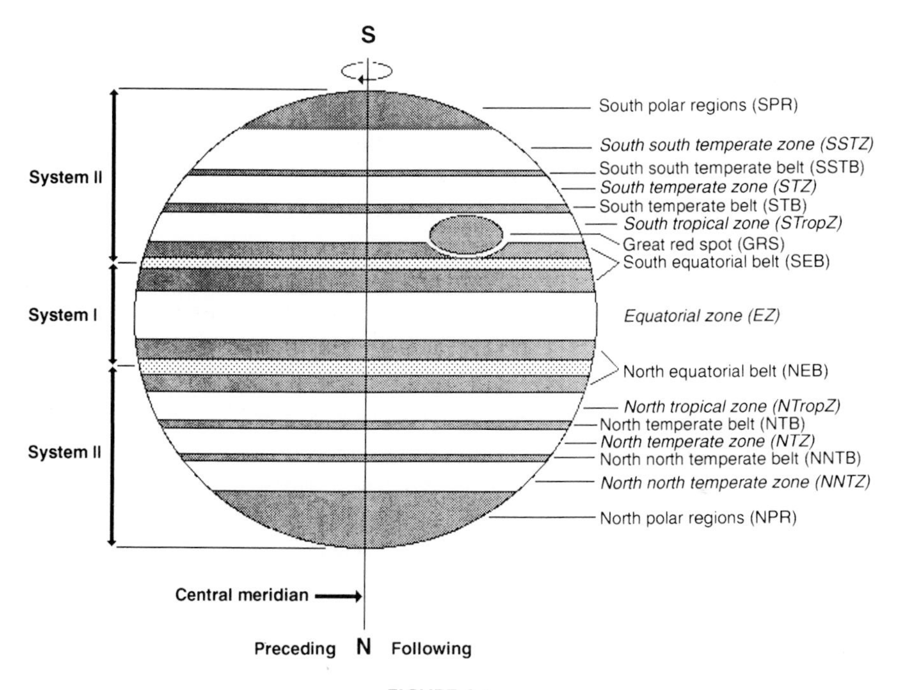

FIGURE 6.1
Nomenclature of the belts and zones of Jupiter.

appearance is shown in Fig. 6.1, with the standard nomenclature for each belt and zone indicated, together with their abbreviations which are universally used.

Seeing conditions – Part of the initial difficulty in seeing and measuring position of planetary detail is the distortion of the image by seeing conditions. The slightest variations in moisture content and temperature of the Earth's atmosphere produce differences in refractive index of the air, so that the planet will appear to distort and move around as you watch it. Sometimes it becomes totally blurred and looks like a ball of fire, in which case study is impossible, but usually it is rather like looking at a reflection in a lake. At first this is very confusing, but with experience the eye can follow the image around and accommodate the changing shape of the limb and the changing position of the spot as the image stretches and contracts. Every now and again a moment of steadiness occurs, and in these brief respites the eye and brain must work hard to extract the maximum information. Those who have persisted at Jupiter observation eventually become highly skilled, and scarcely notice the vagaries of the atmosphere that are so confusing to the uninitiated. However, every observer looks forward to those all too infrequent nights of excellent seeing, when Jupiter stands still and can be closely studied in all its glory.

Seeing conditions tend to be worst when the planet is low down in the sky, as the light from the planet is traversing a greater thickness of the Earth's atmosphere, and in summer seeing is often poor after a sunny day when buildings and the ground have heated up, and best in the early morning before dawn. Stable weather conditions in winter often produce excellent seeing, but there is no hard and fast rule, as appalling seeing and perfect seeing may be unexpectedly encountered at any time.

The best magnification for the observation of Jupiter will depend upon the preference of the observer. The author prefers to use a fairly moderate power eyepiece, around ×120 with a 200 mm, or ×200 with a 300 mm, as this gives a bright image and clearly defined colours and markings. However, many observers prefer to use a higher power, even though this makes the image duller and more blurred, because they find it easier to judge transit timings on a larger disk.

If the beginner is lucky enough to pick a night when the seeing is good and Jupiter is in a fairly active state, light or dark coloured spots may be seen, or other lumps, bays or irregularities in the belts. Chief of these is the Great Red Spot already described, though this may at times be very faint; otherwise the most prominent markings are usually the bright and dark spots along the edge of the North Equatorial Belt (NEB). After about ten minutes, it will become clear that any visible markings will have moved to the left (if you are observing with an inverting telescope in the northern hemisphere). This is because the rapid rotation of Jupiter has carried them 6° round the planet in that time.

POSITION MEASUREMENTS OF FEATURES

Central meridian transit timings – Any spot seen in the right half of the disk will eventually be carried by the planet's rotation to the central meridian, an imaginary straight line across the centre of the disk, joining the north and south poles (see Fig. 6.1). If the time is noted when the spot crosses the central meridian, then the longitude of the spot can be found by looking up the longitude of the central meridian of the illuminated disk of Jupiter in tables published annually in the BAA *Handbook*, the *Astronomical Almanac*, or similar publications in other countries. It is the determination of the longitude of markings in this simple way that has formed the basis of most of the important discoveries described earlier, and unless you have access to a CCD camera (Chapter 17), central meridian transit timings are probably the most important contribution you can make to the knowledge of Jupiter.

In practice, and particularly at first, it is difficult to judge the position of the central meridian, as well as the exact moment when the spot crosses it. When taking your first transit timings it is best to make detailed notes minute by minute, as the spot may seem to be on the meridian for five or ten minutes at a time, or alternatively may seem to wander backwards and forwards from one moment to the next. If the time when it first and last appears to be on the central meridian is noted, then the best estimate of the transit time is halfway between the two. It is absolutely vital that a reliable watch is used, and that it is checked before and after the observation period. The major source of error in positions derived by transit timings is the use of a watch or clock that is a few minutes out. It is easiest to use a digital watch, as extremely reasonably priced accurate watches can now be bought with 24-hour display, but these may lose or gain 30 seconds a month so should always be checked. Time should be recorded at least to the nearest minute in UT, which corresponds with British time in winter.

After a few observing sessions, you will get used to the technique, and in the end you should be able to judge transit times to within a minute or two, i.e. with an uncertainty of just over one degree, which is sufficiently accurate for useful analysis. You will continue to improve, however, in both the amount of detail that you can see and the accuracy with which you time transits, over a period of several years. In the author's case, it was three to five years before he reached his full potential, though the biggest improvement took place in the first year or two.

Every observer will have a small personal equation in his transit timings, that is he or she will consistently tend to estimate transits slightly too early or too late. This does not matter in the slightest, because it is usually very small, and can in any case be allowed for. However, observers should never attempt to try to correct their personal equation at the telescope. More importantly, they should always observe the planet with the orientation of the image the same way up. Suddenly switching from an inverting eyepiece to an erect image can introduce large changes in one's personal equation, which are unknown because a personal equation is a product of both the perception of the centre of the disk and the anticipation of the planet's rotation, and the relative contribution of the two effects differs between observers. This is particularly important for those observing near the equator, where the planet may be conveniently viewed from either orientation, especially for some types of telescope such as the Newtonian.

Finally there is a slight error in transits which occurs near quadrature, where the phase effect causes the terminator to be visible down one side of Jupiter. The uneven illumination of the planet shifts the perception of the central meridian slightly to one side of the centre of the illuminated disk, but this phenomenon, known as the Phillips effect, is about the same for all observers and can again be allowed for.

Recording observations – Transit observations are usually written out in an abbreviated format, with a serial number for each transit in an apparition, followed by a description of the feature, the time of transit, and its longitude in either System I or System II. System I refers to the more rapidly rotating spots near the equator (see Fig. 6.1) and System II to the rest of the planet. If in doubt, both longitude systems should be given. For greater accuracy, both edges of a spot can be timed as they cross the meridian, as well as the centre. The first edge to cross the meridian is called the preceding (p.) edge, and the second the following (f.) edge. Table 6.1 shows an extract from the author's observing notebook for 1978.

The abbreviated descriptions in the table translate as follows:

794	Centre of tiny projection into the white spot south of the north component of the South Equatorial Belt.
795	Preceding end of dark bridge joining the north component of the South Equatorial Belt and the Equatorial Band.
796	Centre of faint white spot between the two components of the south component of the South Equatorial Belt.
797	Centre of large white spot south of the south component of the North Equatorial Belt.
798	Following end of the south component of the South South Temperate Belt.
799	Following end of a darker section of the north component of the North North Temperate Belt. This feature is ill defined.
800	Centre of hump projecting southwards from the south component of the South Equatorial Belt. This transit was not actually observed, owing to passing cloud, but the time when it was on the meridian has been estimated from observations just before and just after the actual transit.

The long-winded descriptions above will make it clear why abbreviated descriptions are preferable in tables of transit timings.

In Table 6.1, despite the relatively poor seeing, features are crossing the central meridian at the rate of about 30 per hour. However, these observations were made by an experienced observer with a large instrument, a 460-mm refractor, beyond the means of most amateurs. A beginner with a 200-mm telescope may only see two or three features cross the meridian per hour, depending on the state of the planet and the seeing conditions. An experienced observer with the same telescope, however, will record something like 8–12 transits per hour.

TABLE 6.1

1978 March 12 d. Seeing poor — detail easy, but image jumping about.
46 cm O.G. × 300.

	UT	λ_1	λ_2
794. Tiny proj. into w. sp. S of NEBs:	21 h 29 m	286°.3	–
795. P. end bridge SEBn/EB:	21 h 29 m	286°.3	–
796. Fnt. w. sp. twixt 2 comps. SEBs:	21 h 31 m	287°.5	193°.9
797. Large w. sp. S. of NEBs:	21 h 38 m	291°.8	–
798. F. end SSTBs:	21 h 39 m	–	198°.8
799. F. end dk. section NNTBn (ill-def.):	21 h 41 m	–	200°.0
800. Hump S. of SEBs (cloud; est.):	21 h 42 m	–	200°.6

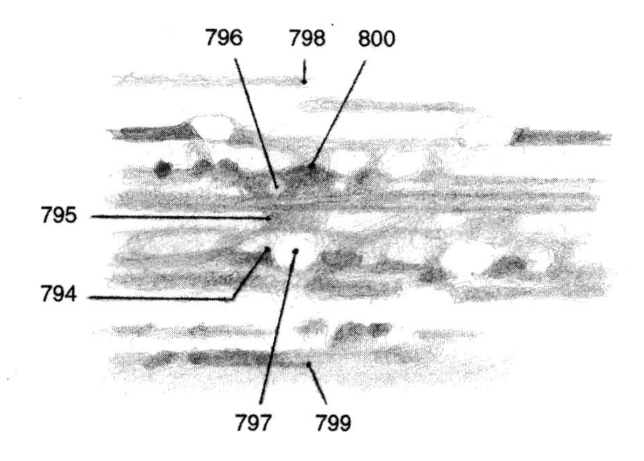

FIGURE 6.2
Strip sketch of Jupiter drawn on 12 March 1978, including features described in Table 6.1.

Strip sketches – An important adjunct to transit observations is a sketch of the features observed. Different observers may use different abbreviations, and belts appearing at unusual latitudes may be labelled differently, so to avoid possible confusion, as well as to give an accurate representation of the appearance of the feature described, a sketch of some kind, however rudimentary, should always accompany a list of transit observations. Since Jupiter rotates so rapidly, more than one drawing of the disk will be necessary if observation is continued for more than an hour, so a strip sketch is more commonly used, in which the features are drawn as a kind of map, with a wide range of longitude represented on the same drawing. Figure 6.2 shows a strip sketch which accompanies the series of transits from which Table 6.1 is taken.

Disk drawings – A series of drawings of Jupiter's disk may also be used to accompany a series of transit observations, and most astronomical associations such as the BAA supply Jupiter blanks. These are printed blank disks of a standard diameter on a black background with the polar flattening of the planet accurately represented, on to which sketches of Jupiter may be drawn. It is best to draw in the positions of the belts first, then draw the main spots and irregularities as quickly as possible before the planet rotates, and then fill in the details afterwards, starting on the preceding (left) side of the disk. Though more time-consuming than strip sketches, artistically gifted observers may prefer the aesthetic appeal of a series of disk drawings, but any transits that occur whilst drawing the planet should always be recorded.

Measuring drawings – The Société Astronomique de France use disk drawings to measure longitudes of markings, and although the accuracy of this method is about four or five times poorer than with transit timings, necessity may force some observers to use the method. It is particularly useful when a lot of interesting detail has just passed the central meridian at the start of observation, or is coming up to transit as the planet is setting, or when transits are likely to be missed because of cloud interruptions. The method comes into its own at the beginning and end of the apparition, when the planet may be visible for only 20 minutes in the twilight. It should never take precedence over transit timings, however, and whilst making such drawings any transits of features should be timed as usual.

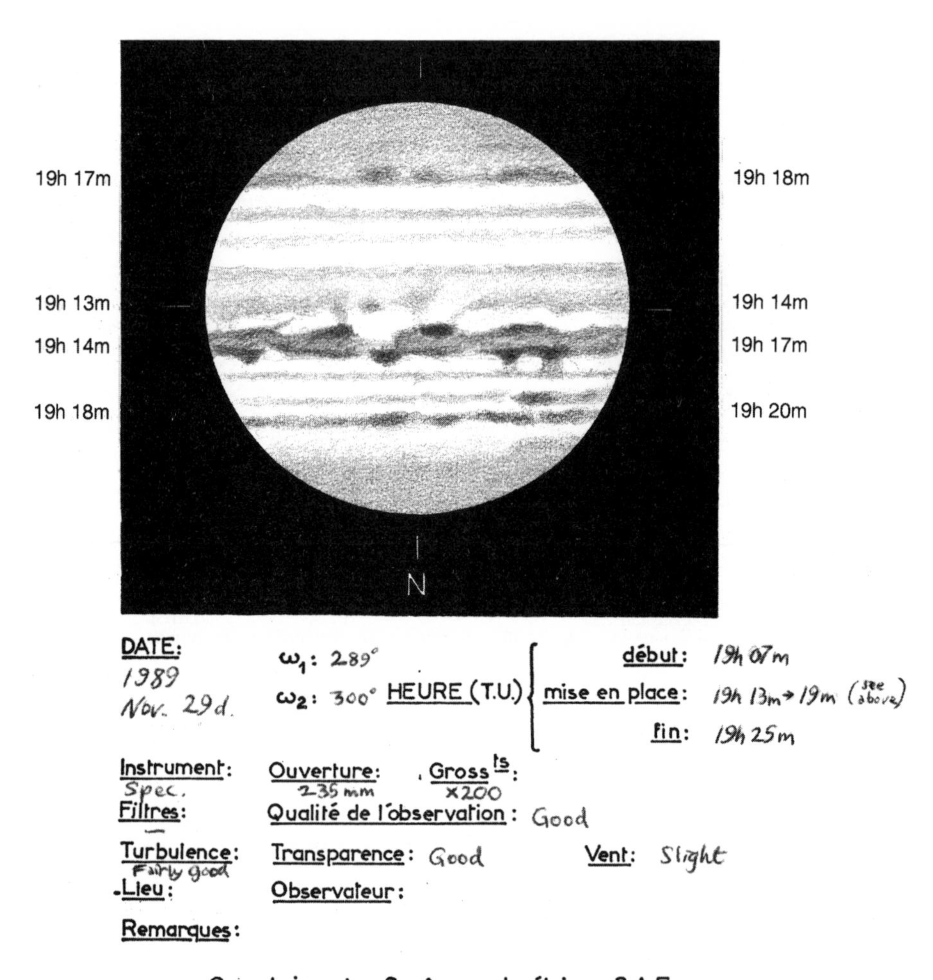

DATE: 1989 Nov. 29d.

ω1: 289°
ω2: 300°

HEURE (T.U.) début: 19h 07m
mise en place: 19h 13m → 19m (see above)
fin: 19h 25m

Instrument: Spec. Ouverture: 235 mm Gross^ts: ×200
Filtres: — Qualité de l'observation: Good
Turbulence: Fairly good Transparence: Good Vent: Slight
Lieu: Observateur:
Remarques:

Commission des Surfaces planétaires S.A.F.

FIGURE 6.3
(a) Sketch of Jupiter intended for subsequent measurement, drawn on a blank issued by the Société Astronomique de France. See text for details.

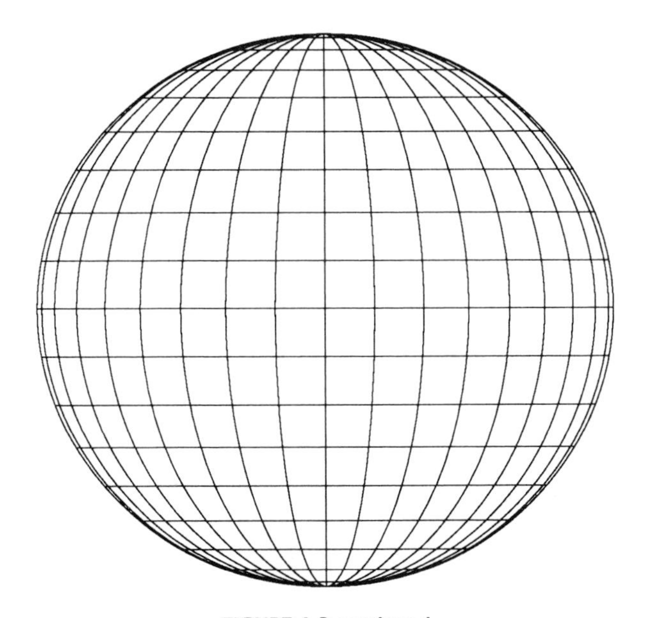

FIGURE 6.3 continued
(b) Latitude'longitude grid at 10° intervals, for measurement of positions on drawings. Inclination of the equator to the observer is 0°in this example.

When making disk drawings for subsequent measurement, the greatest care must be taken to draw in the positions of features accurately. The belt positions should be drawn in first, checking that the latitude is correctly represented in each case. The spots and markings should then be drawn in taking *one belt at a time,* starting at the preceding (left) side of the disk. The time of starting to add details to each belt should be marked on the left side of that belt, and the finishing time on the right. Accurate timing is essential here, as the spots move one degree in less than 2 minutes. Again it should be emphasized that the greatest care must be taken in positioning the features, each one being carefully ckecked before proceeding to the next, but at the same time the rapid rotation make speed of work important. Figure 6.3a shows a drawing made for subsequent measurement.

If time allows, make two or even three such sketches of the planet, as this will give two or three independent values of position for each marking, whose average will be a more reliable final value, and there will also be some indication of its accuracy. The resulting picture is not a true representation of the planet, as the entire drawing may take 20 minutes or more to complete, and each belt has been drawn with a different central meridian longitude, but the method gives more accurate results than trying to draw the entire disk at one go.

To derive positions from such a sketch, the time that each spot was drawn must first be estimated from the starting and finishing times noted to the right and left of the belt concerned. For example, if two spots are drawn on the South Temperate

Belt (STB), and the STB spots were sketched in between 21 h 04 m and 21 h 08 m, one may estimate that the preceding spot was drawn at 21 h 05 m and the following one at 21 h 07 m. The longitude of the central meridian at the time the spot was drawn may then be looked up in tables, and the longitude of the spot is given with sufficient accuracy by

$$\lambda = \omega + \arcsin\left(\frac{d}{D}\right)$$

where λ is the longitude of the spot, ω is the longitude of the central meridian at the time it was drawn, and D is the distance from the central meridian to the limb measured (with a ruler on the drawing) along a line normal to the central meridian at the same latitude as the spot; d is the measured distance of the spot from the central meridian, being positive on the following side and negative on the preceding side. Rather than laboriously measure and calculate the longitudes of each marking in this way, however, it is simpler to trace the latitude/longitude grid in Fig. 6.3b, and use it as an overlay from which the longitude differences from the central meridian can be read off.

Micrometer observations of latitude – So far, only measurements of longitude have been considered and, generally speaking, these are more important than measurements of latitude simply because most of a given spot's motion is in longitude. Unfortunately, this has meant that the observation of latitudes of features on Jupiter has been greatly neglected in the past, usually being restricted to one or two series of measurements of the latitudes of the belts per year, though some professional observatories made more regular latitude measurements of features in the 1970s, with fascinating results. It is certain that belts and spots vary in latitude, sometimes considerably, from month to month as well as from year to year. Great variations in latitude of belts from one part of the disk to another are often quite obvious, which makes it clear that measurement of latitudes only once or twice per year is inadequate. The amateur who has access to a micrometer may carve out a field of research for himself, simply by regularly measuring latitudes of spots and belts on Jupiter's surface. Such a programme of research need not be incompatible with taking transits of features at the same time, especially when a double-image micrometer is being used.

The use of a filar micrometer is briefly dealt with in Chapter 11. Although it may sound an ideal instrument to use for position measurement of markings on Jupiter, in practice its use is limited to latitude measurements only. Longitudes *can* be measured, but a single set of measurements will take 2 or 3 minutes, even for an experienced observer, and the elimination of systematic instrumental errors requires that at least four sets of measurements be made of each feature, so that only a few features could be measured per hour, and a series of longitude determinations such as that listed in Table 6.1 would be impossible. The second problem is contrast, for at high magnification even the thin webs of a filar micrometer appear as thick black bars that actually hide the delicately contrasted features of Jupiter which one is attempting to measure.

Latitudes of the belts and spots is another story, however, for latitude does not change during the observing session, so the time factor is less critical. A good equatorial mount and a faultless clock drive are essential for filar micrometer measurements, however. To measure the latitude of a belt on Jupiter, the two parallel movable webs A and B of the filar micrometer are adjusted to coincide, preferably where A gives a zero reading. The adjusting screw of web B is then left untouched while the measurements are made with web A alone, taking care always to move the web against the compression of the spring to avoid backlash. The sequence of operations in the measurement of the latitude of a belt are shown in Fig. 6.4a. The webs are rotated so that they are oriented parallel with the belt to be measured and, using the telescope slow motions, the image of Jupiter is moved until the north limb of the planet coincides with web B. Web A is then moved until it lies upon the belt to be measured, and the reading on the micrometer screw A is noted. The measurement is repeated at least twice more, taking care that web B still lies exactly on the limb while A is upon the belt. The telescope is then moved until web B lies over the belt to be measured, while A is moved to coincide with the south limb. After repeating this measurement the same number of times as in the first case, the telescope is moved until web B lies on the south limb of Jupiter, and A upon the belt. Finally, the image of Jupiter is moved until B lies over the belt, and web A is moved onto the north limb. Each of these last two readings is repeated as were the first two configurations.

This measurement of the belt against both the north and the south limb and from two different directions is necessary because, when moving the web A, the observer is likely to stop systematically too soon or too late upon the limb or the belt, so that provided the object to be measured is always approached against the compression of the spring, the webs will be moved apart as one measurement is made, and together for the other, so any systematic error will be cancelled out. The fact that web A makes measurements on alternate sides of B also cancels out any zero error in the setting of A and B to coincidence before measurement begins.

In practice the measurement of latitudes with a filar micrometer has other problems. Even with a clock drive that is satisfactory for any other purpose, the tiniest error in motion, such as periodicity in the driving worm, will cause the reference web B to move off the limb or the belt being measured, and even if the drive is absolutely perfect, atmospheric refraction and particularly changing seeing conditions may cause the image to move around constantly. This is why it is important that each measurement should be repeated at least three times.

Measurements of latitude with a double-image micrometer – A double-image micrometer, particularly the type described by Dollfus [1], has distinct advantages

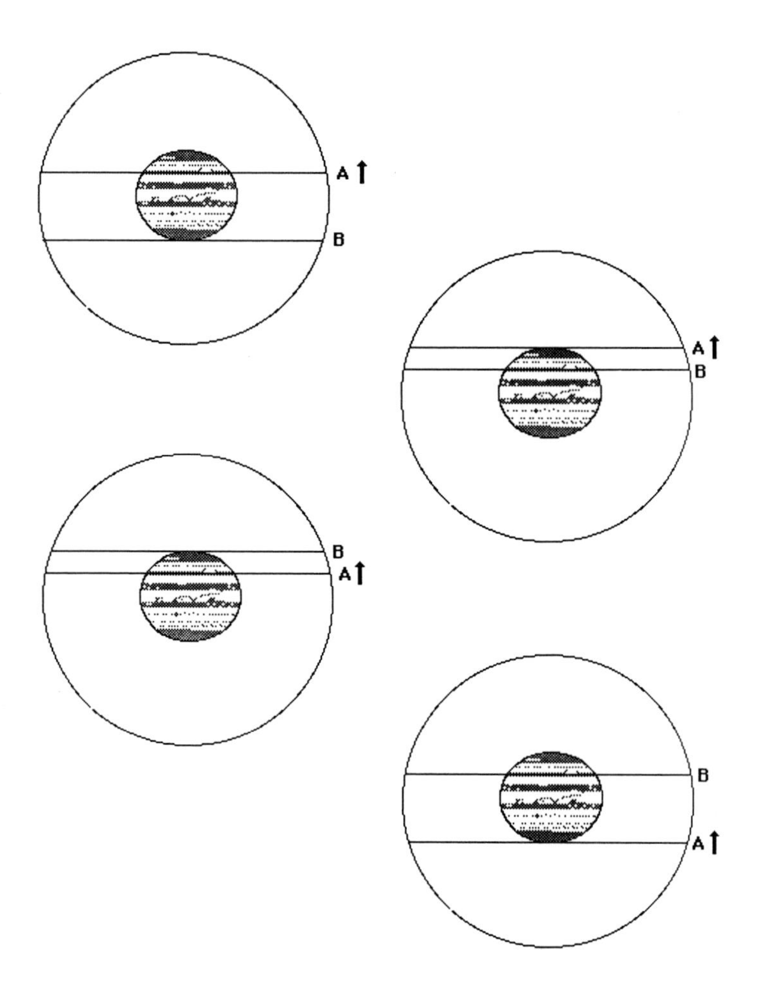

FIGURE 6.4
(a) Sequence of positions of webs of a filar micrometer during measurement of the latitude of Jupiter's STB. See text for details.

over a filar micrometer for the measurement of belt latitudes. Instead of using webs to measure separation, two images of the planet are seen, and the limb of one image is placed against the other as a measuring reference. This means that faulty clock drives and unsteady seeing no longer cause problems, for the two images will wander around together, with virtually no differential movement. Most observers also find it much easier to place the limb of one image over a marking on the other, as the marking can still be seen through the edge of the planet; limb darkening helps this visibility. Finally, the design of the double-image micrometer described by Dollfus has no slack, so the problems of backlash are avoided altogether.

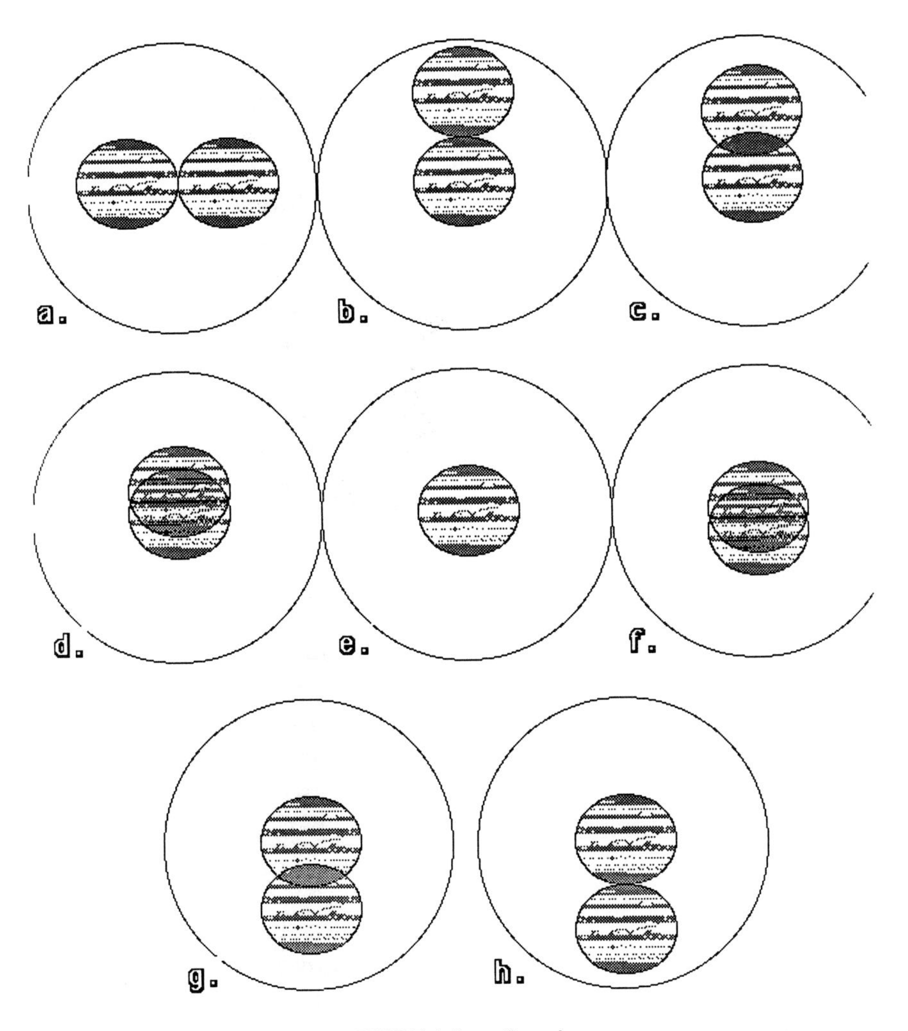

FIGURE 6.4 continued
(b) As (a), but for a double-image micrometer.

The method of measuring a belt latitude with a double-image micrometer is illustrated in Fig. 6.4b. First align the two images side by side so that the belts on the two images merge. Then rotate the micrometer through 90°, and adjust the separation of the two images so that the limbs of the two images touch at the poles. Note this value (position (b)). Then turn the image separation dial to bring the two images together, noting the values at the following configurations:

(c) the north limb of image 1 upon the belt to be measured on image 2.
(d) the south limb of image 2 upon the belt on image 1.
(e) coincidence of the two images.

Continue to turn the dial in the same direction so that the images move apart again, noting

(f) the south limb of image 1 upon the belt in image 2.
(g) the north limb of image 2 upon the belt in image 1.
(h) the south limb of image 1 touching the north limb of image 2.

The above series give four separate measurements of the distance from the south limb to the belt, and from the belt to the north limb. Because of the greater accuracy of this instrument, only one reading need be taken at each configuration, so that a complete series can easily be completed in 5–8 minutes, two or three times quicker than with a filar micrometer. It will be evident, however, that only latitudes of objects on the central meridian can be measured with a double-image micrometer, but this presents no problem for belt latitudes.

Reduction of latitude measurements – Table 6.2 shows two series of measurements of the latitude of the South Temperate Belt carried out with a double-image micrometer by the author on 13 June 1973. The original values read on the particular type of micrometer used have been converted to simple linear distances for clarity; the units are arbitrary. These are equivalent to the values that would be read on a filar micrometer.

It has become standard practice to list zenographical latitudes rather than zenocentric. The difference between the two systems need not concern the observer, but the distinction is important, as in places they differ by 4°. To reduce values read at the telescope, such as those in Table 6.2, to zenographical latitudes, first the inclination of Jupiter's equator with respect to the Earth, or D_E, must be looked up in the *Astronomical Almanac* or the BAA *Handbook* for the date in question. Then $\sin\theta$ is derived from

$$\sin\theta = \frac{s - n}{s + n}$$

TABLE 6.2

1973 June 13 d

	STB centre to S. limb	STB centre to N. limb
Direct:	243.5	684.3
	243.1	683.9
Indirect:	247.9	678.7
	247.4	678.0
	981.9	2724.9

where s is the measured distance from the belt to the south limb, and n the corresponding distance to the north limb. If $E = 1.0714$ and $\beta = \theta + E$, then the zenographical latitude L is derived from

$$\tan L = 1.0714 \tan \beta$$

Using these formulae on the measurements shown in Table 6.2, D_E is found from the *Astronomical Almanac* for 1973 to be $-0^\circ.34$ on June 13.

$$\sin \theta = \frac{s - n}{s + n} = \frac{981.9 - 2724.9}{981.9 + 2724.9} = -0.4702$$

therefore,

$$\theta = -28^\circ.05$$

$$E = 1.0714\, D_E = 1.0714 \times -0.34 = -0^\circ.3643$$

$$\beta = \theta + E = -28^\circ.05 + -0^\circ.3643 = -28^\circ.41$$

$$\tan L = 1.0714 \tan \beta = 1.0714 \times \tan(-28^\circ.41) = -0.5796$$

therefore L, the zenographical latitude, is $-30^\circ.10$.

Although tedious to calculate by hand in this way, particularly for a long series of measurements, the availability of cheap programmable calculators today means that a simple programme can be written which will allow a complete series of latitude measurements of all Jupiter's belts to be reduced in a few minutes.

Photography – An outline of methods of planetary photography is given in Chapter 14, and will not be further detailed here. At first sight, photography may seem to be the ideal method of studying Jupiter, since a photograph, once taken, can be studied at length thereafter. Unfortunately, there are many factors that seem to do their best to make sure that the amateur never gets a photograph worth studying in the first place. Yet, despite the difficulties, some amateurs have persisted, and somehow manage to get usable pictures of planets against all the odds. The great asset of amateurs is their inexhaustible fund of determination and enthusiasm, but they should be warned that unless they have an exceptionally good observing site, the fact that Jupiter requires several second's exposure means that nights when the seeing will allow them to take satisfactory pictures of Jupiter may be limited to only a few each year.

Even if an excellent photograph of Jupiter is finally obtained, it loses 90% of its value if the time when it was taken is not recorded accurately. This should be noted at least to the nearest minute, but if the photograph is really good, recording to the nearest 10 seconds will improve the accuracy of longitudes that can be measured from it. Jupiter photographers, even professional ones, have failed to realize the importance of this in the past; for example, the vast number of photographs taken at the Pic du Midi from 1940 to 1970 were normally taken in batches of 20 images on one plate in moments of good seeing over a period of 5 or 10 minutes.

Unfortunately, only the times of the first and last exposure of each batch were noted, so that longitudes of markings on most of the photographs can only be obtained to the nearest 3° — less accurate than an ordinary transit timing. Some camera bodies allow a digital readout of the time to the nearest second to be printed on the film alongside each picture. This saves a lot of trouble, but otherwise the systematic noting down of the time of each exposure is perfectly adequate.

Many photographers will find the achievement of producing good quality photographs of Jupiter an end in itself, and indeed the Director of the BAA Jupiter section, or his equivalent in other countries, will always be delighted to receive copies of such photographs. However, the main scientific value of a photograph will be from its subsequent analysis, and a good photograph of Jupiter can be measured as described earlier for the measurement of drawings. This is much more easily done if the photograph can be printed to a standard size, so that the grid overlay in Fig. 6.3b can be used to read off longitudes of a large number of features. A photograph can also be used to measure belt latitudes. This is best done by measuring distances to the north and south limbs as described earlier, but using a ruler instead of a micrometer.

The problem of measuring details on photographs of Jupiter is the limb darkening, which makes the limb invisible on undodged prints of the planet (see Fig. 6.5a). Professionals measure the negatives with a plate-measuring machine, but for the amateur it is simplest to lay a piece of card with an elliptical hole in it over the image during printing. The centre of the image can then be exposed to show the detail, and the card is then removed to expose the limb clearly (Fig. 6.5b).

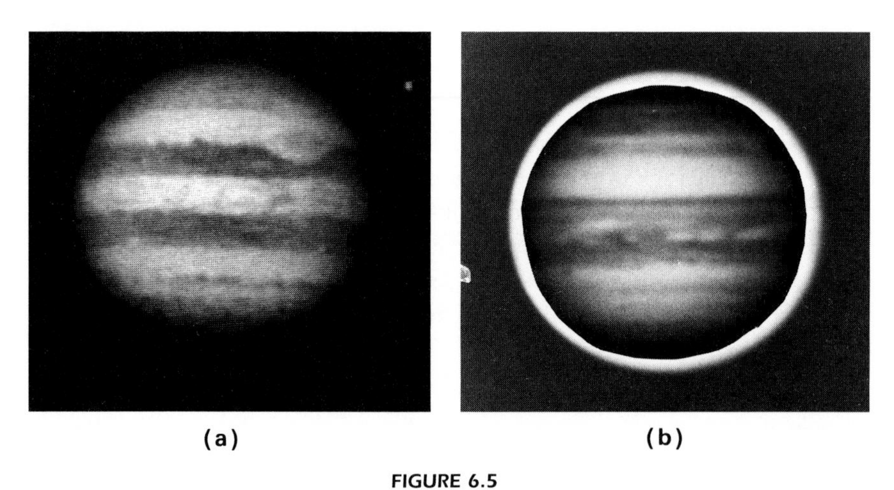

FIGURE 6.5
(a) A CCD image of Jupiter taken by Terry Platt at 22h 10m UT on 28 December 1988 with a 318mm telescope. Note the strong limb darkening. (b) Print of a photograph taken by J.B. Murray with the 1070-mm reflector at the Pic du Midi observatory on 23 September 1973 at 20h 35m 57s with a red filter and an exposure of 4 seconds. The print has been exposed twice as described in the text, once to show details on the planet, and once to show the limb clearly for subsequent measurement.

CCD cameras – At the moment, it looks as if the CCD camera may well revolutionize amateur planetary observation of the future, and allow the gap between amateur and professional planetary work to be dramatically reduced. Details on the use of CCD cameras are described in Chapter 17, but their particular advantages in the observation of Jupiter are very short exposure times, consistent linear response, and precise measuring capability. This means that they combine the advantages of photography and visual observation and eliminate some of the disadvantages.

Exposure times can be reduced to around 1/10 of the value required for photography, so that fleeting moments of good seeing can be captured, with precise times recorded for each. If an image is not of sufficient resolution, it can immediately be discarded and another taken, as the disks on which the images are stored can simply be erased and re-used again and again. It is even possible to design software which will select good images out of thousands, thus behaving rather like the selection process of the human eye and brain. This means that the resolution of a CCD camera used in this way can be as good as that seen through the telescope by an experienced observer.

With the right software, precise measurements of features on Jupiter are possible, and even the creation of maps of large parts of the planet from a series of images. This means that the position of atmospheric features can be recorded by taking good images every hour or two, rather than spending the same time constantly at the telescope making transit timings.

The linear response of a CCD camera means that if one feature on the image is ten times brighter than another, then ten times the amount of light has fallen upon it. The same is not true for photographic film, which suffers from reciprocity failure, inconsistency due to changes in temperature, humidity, age of the film and the particular batch number of the film. None of these inconsistencies effects CCD cameras, so that precise measurements of intensity and, with colour filters of different wavelengths, measurements of colour are possible as described in the next section. It is also much more difficult to saturate part of the frame by overexposure, as CCD cameras have a very wide range of sensitivity to different light levels, again more akin to the human eye than a photograph.

The potential of CCD cameras is enormous, and there has been a great increase in the use of CCD cameras by amateur astronomers in the last few years, as prices have come down to affordable levels. It is now possible for an amateur to take pictures with a 300 mm telescope which are equivalent to good photographs from the Pic du Midi observatory, and useful frames may be obtained with a 200 mm instrument, or even smaller. Compare, for example, the CCD images by Terry Platt in Figs. 6.5 and 6.6 with the Pic du Midi photographs in Fig 6.7.

Whether this enormous potential is realized remains to be seen. As with photography, the taking of a few images per apparition will be extremely valuable for observations of belt latitude, intensity and colour, but will give next to no information on the changing position of markings and atmospheric currents,

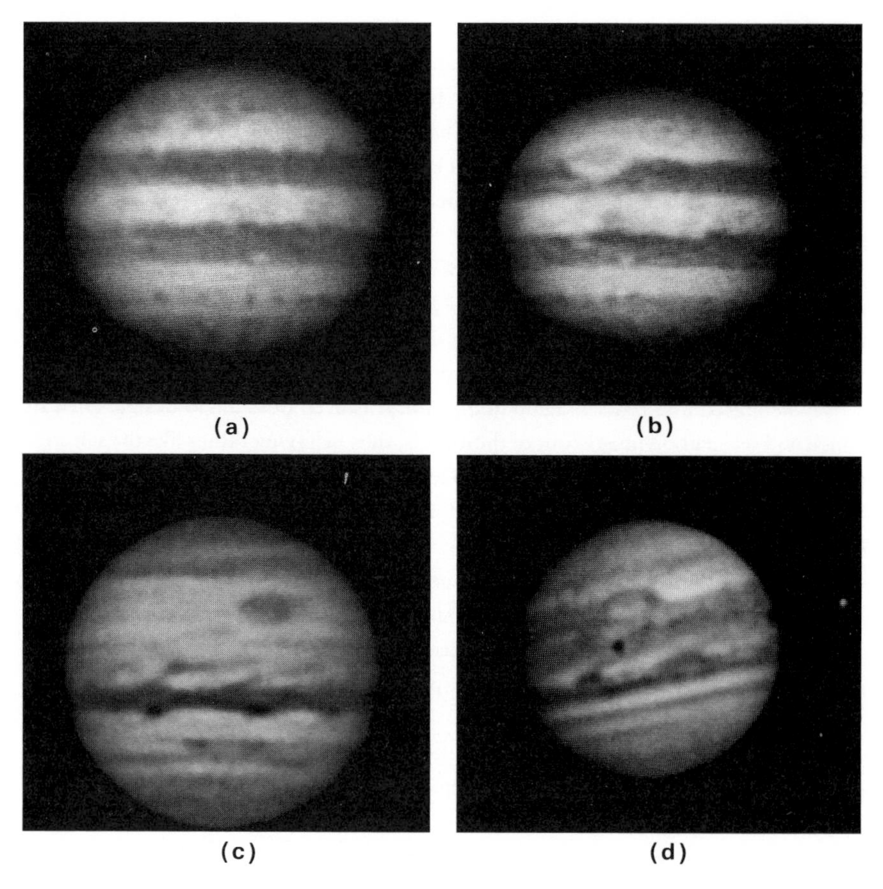

FIGURE 6.6
Four CCD images taken by Terry Platt with his homemade 318-mm Tri-Schiefspiegler off-axis reflector, using Sony ICX021L or ICX027L CCD devices. They were taken on (a) 28 December 1988 at 20h 52m; (b) 22 January 1989 at 19h 14m; (c) 11 January 1990 at 21h 32m; (d) 1 January 1991 at 01h 43m. The changes in Jupiter's atmosphere are well illustrated in these pictures, (a) and (b) showing a comparatively calm atmosphere, with only small spots visible and the belts more or less as in Fig. 6.1, whereas (c) shows the SEB almost missing, with larger spots and festoons visible, and (d) shows the atmosphere unusually turbulent, with major disruptions to the belts and zones. The Great Red Spot is visible in (b), (c) and (d), and the shadows of satellites Europa (left) and Io in (d), together with Europa itself on the following side of the planet. The images have been subsequently processed with a sharpness-enhancing filter and a noise-smoothing filter to improve contrast of fine details. Exposures were about 0.5 seconds at between f30 and f50.

which forms the bulk of serious amateur study. Unless owners of such cameras are able to take pictures regularly throughout the apparition, following the guidelines of an observing programme set out later, or unless such cameras become as widely used as telescopes, then the bulk of amateur work on Jupiter will continue to be undertaken by visual observers and their transit timings.

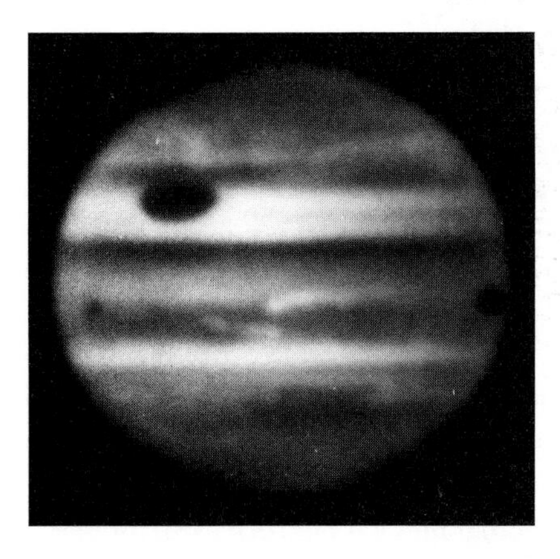

FIGURE 6.7
Photos of Jupiter taken by J.B. Murray with the 1070-mm reflector at the Pic du Midi observatory on 1 September 1973 at 22h 54m 54s and 22h 55m 51s, with blue (left) and yellow (right) filters, and exposures of 4 and 2 seconds, respectively. The satellite Ganymede is close to the preceding limb, and its shadow on the following limb. The prints have been dodged by hand to show the limb clearly.

INTENSITY AND COLOUR MEASUREMENT

Although the word 'measurement' is used above, true measurements of colour and intensity have until now been carried out only by professional astronomers. Such measurements have in the past been made on photographs, which have been calibrated using a lamp of known spectral response to expose a series of patches of known relative density on to each film or plate. The negatives are then scanned with a microdensitometer; and the use of colour filters while photographing Jupiter that isolate well defined regions of the spectrum means that the intensity of the different belts and zones at different wavelengths can be measured, taking into account the spectral response of the film or plate. This means that the colour of each belt and zone can effectively be measured. CCD cameras bypass many of these operations.

Intensity – Such rigorous analysis is beyond the means (or inclination!) of most amateurs, but nevertheless useful information on the intensity of a belt or zone, and particularly on any changes in relative intensity over the years, can be made simply by estimating the darkness of a feature on the planet. This is normally done on a scale of 0 to 10, with 0 being the brightest visible zone and 10 the black sky.

There are many problems in estimating intensity in this way. Firstly, the brightest visible zone, used as a reference zero, may itself vary considerably in brightness over the years, or even between one part of the planet and another. This will artificially alter all the other intensities measured. Secondly, apparent intensity of small features will vary both with the seeing and with the size of telescope used, since a very thin dark belt of intensity 5, for example, may appear as a faint fuzzy belt of intensity

2 under poor seeing or small aperture. Finally, the intensity values observed bear little resemblance to measured values derived from microdensitometry; the eye and brain together compensate for the enormous differences in intensity between black sky and even the darkest features, and exaggerate the differences between different features on the planet.

Despite these problems, observers show a remarkable amount of agreement in their estimates of intensity, and it is rare for two experienced observers to differ by more than one point in their estimates of the same feature. As long as the above-mentioned problems are borne in mind, intensity estimates can provide useful information on the state of Jupiter's atmosphere, and are an important adjunct to measurements of position.

The estimation of intensity by eye on photographs has not been used, for unless the photographs are properly calibrated there is uncertainty in the response of the original film and the paper on which the photograph has been printed. However, the information in a CCD frame is linear and accurate, and provided the image is of high enough resolution then true relative intensities can be obtained using appropriate software without any necessity for calibration. This promises to be an extremely valuable method in the future.

Colour – Notes on the colour of the different regions on Jupiter can be taken at the telescope, but some colours seen may be artificially introduced by various disturbing effects. Refractors are generally unsuitable for observing colour in this way, as they are subject to chromatic aberration. Similarly, the chromatic effects of the atmosphere itself will cause spurious colours in any telescope when the planet is low down in the atmosphere, making the southern edge of belts appear blue and the north side red when observed from the northern hemisphere, though this effect may be corrected with a Ramsden eyepiece. The appearance of colours on Jupiter is also affected by contrast effects, particularly in small apertures. A reddish belt, for example, may make the surrounding features appear bluish or greenish simply by contrast with the red. Colours appear stronger in an instrument of large aperture; in fact, when the seeing is average this may be the only advantage of a larger instrument. Similarly, colour detection may often be improved by lowering the magnification, as this makes the image brighter. Colour photography is often helpful, though again the instrument needs to be large and the seeing unusually good for resolution to be high enough for colours to be seen unequivocally.

Notes on colour taken at the telescope inevitably involve a large element of subjectivity. 'Blue-grey', 'brown', 'slightly ruddy or neutral' and 'reddish brown' have all been used by different observers to describe the same feature at the same time, so any scientific analysis based solely on visual descriptions of colour could be of only the simplest kind, involving only the strongest changes of colour. A much more rigorous analysis can be undertaken by the use of colour filters.

Figure 6.7 shows two photographs of the same region of Jupiter, one taken with a blue filter and the other with a yellow filter. The Great Red Spot, which was orange-red at the time, appears as the darkest feature on the planet with the blue filter, because the spot is reflecting very little blue light, but with the yellow filter it is much brighter, as it is reflecting a good deal of yellow light. It appears even brighter with a red filter. Some parts of the south side of the North Equatorial Belt, however, appear darker with the yellow filter than they do with the blue, indicating that these are reflecting more blue light, and indeed the south edge of this belt often has a bluish tinge.

By observing with different colour filters, therefore, quantitative estimates of colour can be obtained simply by estimating intensities of belts and other features through filters with different wavelength ranges. A scale of 0 to 10 is used as described for ordinary intensity estimates above. Although it would be useful to use a range of filters, the use of just two filters spaced far apart in the spectrum, such as Wratten 25 (red) and 47 (blue), give a numerical value for redness when the blue value is subtracted from the red. Table 6.3 shows colour intensity estimates and the R – B index of redness by A.W. Heath for the zones and belts of Jupiter during the two apparitions of 1975–6 and 1982, together with visual descriptions of colour as seen in the telescope. The Great Red Spot was much less prominent in 1982, and far less red, some observers describing it as buff, but the R – B index shows that it is still the reddest feature on the planet. This is a technique which is more important than simple white-light intensity estimates, yet it is one that has been greatly neglected in the past. Among amateur astronomers, only A.W. Heath has specialized in this type of observation over many years, and has produced a valuable record of colour changes of which Table 6.3 is a sample.

Colour filter photographs of Jupiter, such as those in Fig. 6.7, can be extremely valuable as general indicators of changing colours of different features, but they should be properly calibrated as described earlier for any serious work. Once again, CCD cameras promise an important future in colour work, since the linear response of these cameras means that, using colour filters, only the transparency of the filter to different wavelengths need be known, and this is supplied by the manufacturer to sufficient standards of accuracy. Filters used during visual observation, such as the W25 (red) and W47 (blue) filters mentioned above, necessarily have a fairly broad bandpass which allows low light levels of other wavelengths besides red and blue so that sufficient light can get through to make accurate observation possible. The great sensitivity of CCD cameras means that filters with much more restricted and well-defined wavelengths can be used while still maintaining the exposure sufficiently short to catch the good moments of seeing. Certain types of CCD cameras have some infrared sensitivity, however, and many Wratten filters 'leak' in the infrared, so filters and cameras should be chosen so that unwanted wavelengths are not let through. Even colour-filtered CCD frames from one or two good nights per apparition could be enough to provide important precise information on long-term colour changes.

TABLE 6.3
MEAN ESTIMATED INTENSITIES OF THE BELTS AND ZONES OF JUPITER, WITH AND WITHOUT COLOUR FILTERS, FOR 1975 AND 1982, BY A.W. HEATH

	1975					1982				
	None	Red	Blue	R – B		None	Red	Blue	R – B	
SPR:	7.0	7.4	6.1	1.3	–	7.0	7.5	6.0	1.5	Grey
STZ:	9.5	9.6	6.7	2.9	White*	8.1	8.3	6.0	2.3	Yellowish
STB:	4.9	5.7	5.3	0.4	Warm grey*	5.5	6.1	5.4	0.7	Grey
Red Spot:	5.6	9.6	2.0	7.6	Dark red*	7.0	8.5	5.0	3.5	Pinkish
STropZ:	9.9	10.0	8.4	1.6	Bright white*	9.0	9.0	7.2	1.8	Yellowish
SEB(s):	5.0	5.9	4.9	1.0	Grey*	4.6	5.4	2.6	2.8	Brown
SEBZ:	8.8	9.4	7.3	2.1	–	–	–	–	–	–
SEB(n):	5.0	5.8	4.5	1.3	Slightly bluish*	5.0	5.5	2.8	2.7	Brown
EZ:	9.8	9.9	8.7	1.2	–	9.3	9.6	8.9	0.7	White/off white
NEB:	5.0	5.8	3.8	2.0	Reddish*	5.0	5.6	3.0	2.6	Brown
NTtopZ:	9.5	9.7	6.7	3.0	Yellowish*	8.3	8.4	6.4	2.0	Yellowish
NTB:	5.3	6.3	5.9	0.4	Grey*	–	–	–	–	–
NTZ:	8.5	8.7	6.3	2.4	–	–	–	–	–	–
NPR:	7.0	7.4	6.0	1.4	–	6.7	7.5	6.0	1.5	Grey

Listed values are for intensity estimated on a scale of 70 (as dark as the sky background) to 0 (brightest). Columns show intensity in the order: without filter, through red Wratten 25 filter, through blue Wratten 47 filter, and lastly the red minus the blue values, which gives an 'index of redness', the highest values being the reddest. Descriptions of colours as seen visually without a filter are then given; * those marked with an asterisk are not by A.W. Heath. Note the strong red colour of the great Red Spot in 1975, with a redness index of 7.6, and its fade to only 3.5 in 1982, though even in the latter year it is still the reddest feature on the planet. Note also that both components of the SEB have more than doubled the redness index in 1982, when warmer colours were seen visually too. Other features remain more or less the same colour, with redness indices differing by 1 or less. The figures are averages of observations made on 19 nights in 1975, and 18 nights in 1982, with a 300 mm Newtonian reflector, magnification × 190.

A PROGRAMME OF OBSERVATION

Any observations, however casual or routine, are always welcomed by observing organizations such as the BAA Jupiter Section, but anyone who wants to contribute useful observations on a regular basis will soon realize that some kind of general plan of observation is required. Most observers prefer to observe between 7 and 10 o'clock in the evening, so that, although Jupiter may be visible for ten months of the year, most observations are made in the two months after opposition in the second half of the apparition, when Jupiter is well placed for observation at a civilized hour. Accordingly, observations, particularly transit timings, carried out when Jupiter rises after midnight are therefore much more valuable simply for their scarcity value. Any observations made more than three months before opposition are like gold dust, so it is well worth setting your alarm for an hour before dawn in the early part of the apparition. The end of the apparition should not be forgotten either, and if Jupiter can be picked up in the twilight (daylight observations are often helped by a polarizing filter) as late as possible before conjunction, these observations will also be particularly valuable. The beginning and end of each apparition are especially important for transit observations, as a reduction of the gap in observations from one apparition to the next to as short an interval as possible will greatly aid the definite identification of the continued life of spots which survive the interval.

Transit timings, or equivalent measurements of the longitudes of markings using photographs or CCD frames, should form the basis of most amateurs' programmes of observation, though many may wish to specialize in intensity, colour or latitude measurements if they have a CCD camera or micrometer. A Jupiter specialist may consider himself among the true élite of observers when he first attains a thousand transits in an apparition. This magic figure, which is looked upon rather like a century in cricket, requires good weather and a favourable declination of the planet, preferably well above the equator. Just to illustrate what can be attained with dedication and the most favourable circumstances, P.B. Molesworth, observing from near the equator in the clear skies of Ceylon, recorded no less than 6758 transits in a single apparition between January and November 1900. This figure still stands as the all-time record for one apparition by a huge margin; but over a longer period the Reverend T.E.R. Phillips, observing in the cloudier skies of England, managed just over 30 000 transits between 1898 and his death in 1941, at an average rate of about 8 to 10 transits only per hour of observation. This lifetime achievement is also unsurpassed.

The dynamic atmosphere of Jupiter make its observation one of the most exciting branches of astronomy; nothing can beat a night at the telescope at the height of the Jupiter season with an SEB outbreak in full flow, and the sense of suspense whilst a critical feature among the unpredictably changing explosion of spots and belts is awaited with impatience as the globe slowly turns. But however enthusiastic and dedicated the observer, long hours at the telescope take their toll, and mental fatigue will inevitably set in without the constant exchange of information with other observers of the same specialization. All observers are therefore urged to join a local astronomical society, or to seek out and correspond with other observers both nationally and internationally. Such contact not only provides the inspiration necessary to carry out an exacting programme of observation, but also allows news of any sudden developments on the planet to be rapidly disseminated, and enables one observer to cover a period when another may be unable to observe.

It is also important to plot your transit observations or position measurements on a longitude chart as you go along, so that any unusual motions can be picked out at once, and dubious identifications confirmed.

It is likely that amateurs will continue to play an important part in the observation of Jupiter in the future. The interests of professional astronomers change, and they are in any case dependent upon research funds and the vagaries of applications for telescope time, and this applies to the Space Telescope as well. The great advantage of the amateur is that he is accountable to no-one and can observe when he likes, and the steady decrease in price of CCD cameras means that he has a new powerful tool at his disposal. Particularly important is the observation of the planet by amateurs during space missions to the planet, so that amateur work can be calibrated against a more detailed study; this will help in the evaluation of observation over the past 100 years. For full amateur potential to be realized, a degree of international amateur cooperation is essential, as an apparition with a high southerly declination is very unfavourable for northern hemisphere observers, but very favourable for those south of the equator. A good spread of observers in both longitude and latitude will mean that a much more satisfactory coverage of the planet can be maintained.

About the Author

John Murray, MA, MPhil, PhD, is a research scientist whose interests have been in the two fields of planetary astronomy and volcanology. After long amateur experience in planetary observation, he went to Magdalen College, Cambridge, in 1964 and joined the University of London Observatory in 1967 where he was involved in lunar work prior to the Apollo missions. He spent three years at Meudon Observatory, Paris, studying Mars and Jupiter, and observing at the Pic du Midi Observatory. He was a guest scientist at J.P.L. Pasadena during the Viking missions to Mars in 1976–77, and is presently working on interpretation of the topography of Mars. Since 1969 he has carried out field measurements of changes on Mt Etna twice a year, and has similar programmes at Colima Volcano, Mexico, and Poàs Volcano, Costa Rica.

Contact address – Department of Earth Sciences, The Open University, Walton Hall, Milton Keynes, MK7 6AA, UK.

Reference

1. Dollfus, A., *Compte Rendu de l'Academie des Sciences*, **238**, 1477 (1952).

Notes and comments

Observing projects – In an important article in *Sky & Telescope*, November 1988, the ALPO Jupiter Recorder Phillip Budine had something to say about current and future trends in Jupiter observation. These may be summarized as follows:

(1) SEB disturbances may be entering a new phase. In the 'classical' disturbance, the belt fades and is replaced by numerous small, dark spots and white ovals showing retrograde motion. In 1985 and 1987, however, the SEB darkened and developed one or more bright rifts.

(2) The Red Spot should be monitored continuously. Its length is now about 20° compared with a size of twice that suggested by early observations.

(3) Three long-enduring bright ovals in the STB have been observed since 1939. Designated BC, DE and FA, they are now quoted as 'shrinking into oblivion', although other short-lived white spots are being observed.

The overwhelming need, as emphasized in this chapter, is for transit timings. Amateur observers are the only source of this information, and monitoring of features and currents will supply important data for the Galileo orbiter when it starts work in 1995.

Spot a volcano? – Many years ago, in the first edition of *Guide to the Planets* (Eyre and Spottiswoode, 1955), Patrick Moore wrote about some observations he made of Io in 1951:

> 'During the transit of August 3 it was unusually bright, during the transit of August 24 unusually dark. This might be attributed to different parts of the surface being turned towards us, but such an idea does not fit in well with the strong probability that Io keeps the same hemisphere turned permanently towards Jupiter',

and elsewhere he recorded seeing it

> 'brighten up strikingly in the space of two hours for no apparent reason'.

Our present knowledge that Io is volcanically the most active body in the solar system makes it at least possible that Moore was recording changes in reflectivity caused by exceptionally violent eruptions on its surface. A systematic photometric monitoring programme might be worth considering, although Io's nearness to Jupiter, and its rapid change in angular distance, would make the task much more difficult than for an ordinary variable star. — Ed.

CHAPTER SEVEN · *Observing Saturn*

ALAN W. HEATH
DIRECTOR, BAA SATURN SECTION

'THE PLANET SATURN IS, PERHAPS, ONE OF THE MOST ENGAGING objects that astronomy offers to our view', wrote Sir William Herschel. Whilst there is no doubt that the planet is worthy of observation purely for aesthetic appreciation, it is far more meaningful if the observations made follow a carefully ordered plan and thus are of scientific value.

THE TELESCOPE

A 75-mm refractor with a magnification of around ×50 will show the planet and its ring system, but an aperture of not less than 125 mm is needed for the observations to be of value, as apertures less than this lack the resolving power to show significant detail. Ideally one should aim for at least 150 mm, and the larger the better.

It has been claimed that the best magnification for planetary detail is about equal to the diameter of the object glass or mirror in millimetres. To see the fine details of the belts and ring structure, a magnification of ×150 to ×300 is necessary, and therefore, according to the above rule, telescopes of 150 mm or more are required. Departure from this, by using a higher magnification, can only be achieved at the expense of contrast, which is important when seeking fine detail. There are now many telescope designs available to the discerning amateur, but in the writer's experience you cannot beat a good-quality, correctly collimated f/8 Newtonian reflector. A motor drive is not essential, but it adds considerably to the comfort and ease of observation, as the image remains in the field of view throughout the period of observation and time is not lost 'steering'. [Some further comments on the interesting question of 'optimum magnification' appear at the end of this chapter. — Ed.]

A valuable aid to observation is the use of colour filters, and some means of mounting such filters near the eyepiece is necessary. More will be said about this later, but, as a filter reduces the incoming light, the need for a sufficient aperture is again stressed.

OUTLINE OF DISK AND RINGS

Start with a prepared outline of the planet. Such outlines are available from the Saturn Sections of the BAA and the ALPO. The standard size outlines adopted by the organization concerned make analysis easier and should be used if possible: in the case of the BAA Saturn Section, the planet's equatorial diameter measures 45 mm, and the outer diameter of Ring A is 113 mm.

The apparent outlines of the disk and rings vary with *B*, the angle of presentation of the poles against the sky. Reference [1] gives instructions for hand-producing the outlines of disk and rings for any value of *B*, using an ingenious hand method. Since the original source is not easy to obtain, the method is reproduced here.

The first requirement is a table of minor axes of different components of the ring system, given a standard major axis, for different values of the inclination of the planet's axis with respect to the Earth (*B*). Table 7.1 gives these values for the outer and inner edges of Ring A, and the inner edge of Ring B, based on a value of 100 mm for the major axis of the outer edge of Ring A (inner edge of Ring A therefore being 88 mm, inner edge of Ring B being 66 mm). The method is illustrated

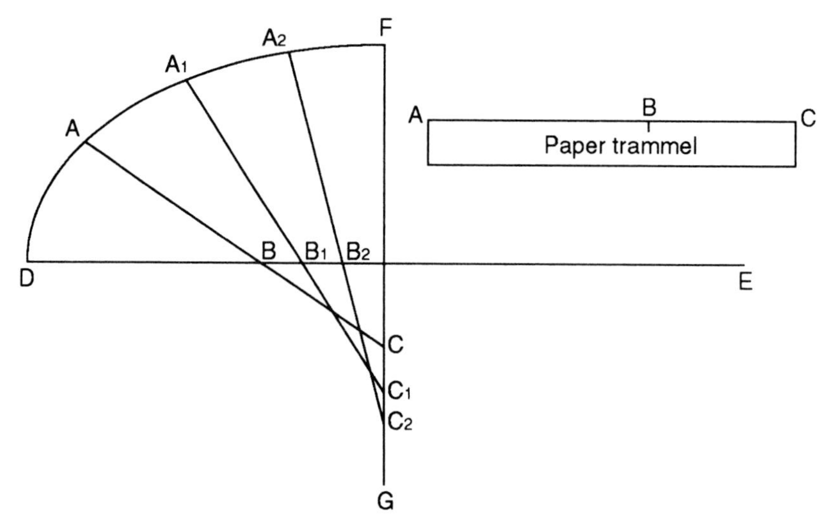

FIGURE 7.1
Method of drawing the ring outline for any value of B. (After M.B.B. Heath [1].)

TABLE 7.1

THE MINOR AXIS OF DIFFERENT COMPONENTS OF THE RING SYSTEM AT DIFFERENT INCLINATIONS OF SATURN'S AXIS WITH RESPECT TO THE EARTH (*B*). ALL DIMENSIONS IN MILLIMETRES. THE MAJOR AXIS OF THE OUTER EDGE OF RING A IS TAKEN AS 100 MM

B (°)	Outer edge Ring A	Inner edge Ring A	Inner edge Ring B
1	1.75	1.5	1.25
2	3.5	3.0	2.25
3	5.25	4.5	3.5
4	7.0	6.25	4.75
5	8.75	7.75	5.75
6	10.5	9.25	7.0
7	12.25	10.75	8.0
8	14.0	12.25	9.25
9	15.75	13.75	10.5
10	17.5	15.25	11.5
11	19.0	16.75	12.75
12	20.75	18.25	13.5
13	22.5	19.75	14.75
14	24.25	21.25	15.75
15	26.0	22.75	17.0
16	27.5	24.25	18.0
17	29.25	25.75	19.25
18	31.0	27.25	20.25
19	32.5	28.75	21.5
20	34.25	30.0	22.5
21	35.75	31.5	23.5
22	37.5	33.0	24.75
23	39.0	34.5	25.75
24	40.75	35.75	26.75
25	42.25	37.25	27.75
26	43.75	38.5	29.0
27	45.5	40.0	30.0
28	47.0	41.25	31.0

in Fig. 7.1, where DE and FG are two lines at right angles, DE representing the major axis and FG the minor axis of the desired ellipse.

To draw the complete ellipse, as many intermediate points as are required can be plotted using a 'trammel', a piece of paper or thin card whose edge is marked such that AB equals the semi-minor axis and AC equals the semi-major axis of the ellipse being drawn. Then, provided B is on DE and C is on FG, A will mark a point on the ellipse. The ellipse is completed by passing a line through the plotted points.

The following ratios may be found useful. Let equatorial diameter = E. Then the polar diameter = $0.89E$ and the major axes of the rings are:

Ring A (outer edge)	$2.24E$
Ring A (inner edge)	$1.97E$
Ring B (outer edge)	$1.93E$
Ring B (inner edge)	$1.48E$
Ring C (inner edge)	$1.23E$

DRAWING TECHNIQUES

The original sketch made while observing is important, for it should be a factual record of all that is seen at the time.

The time taken should not exceed 15 minutes, for the drift of features due to Saturn's rotation is fairly rapid. To promote speed, time should be spent observing the planet before starting to draw, and the method employed should be simple, rapid, and capable of giving an accurate record.

The observer's comfort is of prime importance: the hands and eye should be as free as possible to carry on with the sketch without the distraction of having to guide the telescope manually, although with practice it is possible to produce accurate drawings without any sort of drive.

Household steps form a simple support on which to lean or stand, and on which to support a drawing board. This could be fitted with a lamp holder to take a red or orange bulb, shielded from the eye. The light must not be so bright as to upset the eye's sensitivity — some observers have employed LEDs, possibly with a resistor to control the current. Paper of medium surface texture is best, and a range of pencils of different hardnesses is needed. Harder pencils are easier to control, but softer ones may be better for shading. A soft, chisel-shaped rubber is needed, and it should be kept clean by rubbing it frequently on a spare piece of paper. For the background sky, and shadows on the rings and globe, a matt black paint such as Process Black, available from art dealers, is ideal, although Cassini's division may be easier to draw with a sharp H or HB pencil.

When making a drawing, particular attention must be paid to the position and shape of the shadows both of the globe on the rings and of the rings on the globe. The correct placing and shaping of these shadows is a useful criterion of the quality and reliability of the observation. The North or South Equatorial Belt should be drawn next, with as much accuracy as eye-estimation will allow. A carefully-made drawing of even these meagre details is of value, since it may be used to compute a fairly reliable saturnicentric latitude for one or both edges of the belt.

Any other belts are then drawn as accurately as possible, and only after these things have been done should a search be made for delicate detail at or near the threshold of vision, remembering that the drawing should show only those features definitely and accurately seen.

Charcoal is useful for shading the limb of the planet, and it can be rubbed with clean cloth or paper tissue to show diffused regions and light shading, being reinforced with pencil if necessary. Be sure to check and re-check features already drawn while working on other parts of the planet, and finally check all regions again to make sure that nothing has been left out. Compare the general appearance of the drawing with the object itself, and see whether the drawing looks the same: if it does not, decide why not and correct the drawing!

If at all possible, the entire drawing should be finished at the telescope. If this cannot be done, it should be finished indoors *immediately afterwards* — it is useless to make notes or a hasty sketch at the telescope and then go indoors to make the drawing.

Line sketches in black ink often convey what is seen just as well as artistic and desirable shaded drawings. Often the observer can indicate the position of an object of interest, such as a dark or light spot, a shadow transit, or a satellite transit. Full notes should, of course, accompany all drawings.

To conclude this section, *never* include anything on a drawing that you did not see at the time. Do not ruin your reputation for accuracy by recording doubtful details.

Some examples of finished drawings by three different observers are shown in Fig. 7.2.

GENERAL OBSERVATION

A casual glance will determine whether seeing conditions are sufficiently steady for a useful observation to be made. A steady image is more important than the transparency of the sky for planetary observation. Indeed, conditions just preceding fog can produce very stable images.

Points to be noted in a general inspection of the planet and its field are:

1. Any changes in the shape or position of the belts on the globe.
2. Shape of shadows of rings on globe and globe on rings, and any irregularity in their outline.
3. Any star that seems likely to be occulted by the planet's rings or globe, although such events are likely to have been predicted beforehand.
4. Position of any bright or dark spots or patches on the globe or the rings, or any projections from a belt which appear to move with the planet's rotation.

(a)

(b)

(c)

(d)

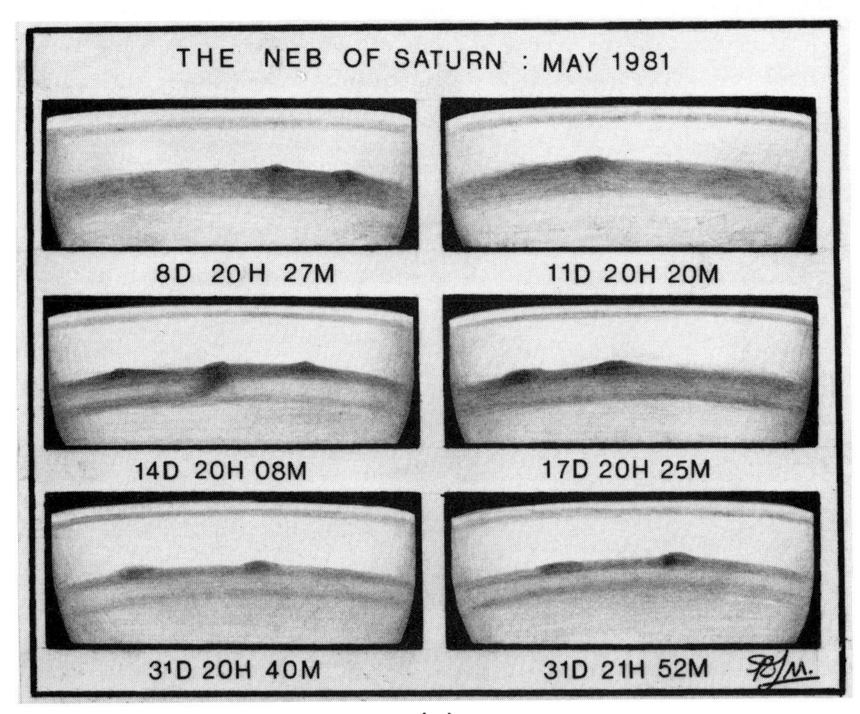

(e)

FIGURE 7.2
Four drawings of Saturn by UK observers. (a) 14 Nov 1966, 21 h 15 m UT, 210-mm reflector ×274; Titan off the disk, casting its shadow on the planet (P.D. Doherty). (b) 8 May 1981, 20 h 27 m UT, 320-mm refractor ×230, ×320 (R.J. McKim). (c) 23 April 1984, 00 h 20 m UT, 415-mm Dall–Kirkham reflector ×237, ×474 (D. Gray). (d) 10 May 1987, 01 h 40 m UT, 415-mm Dall–Kirkham reflector ×237 (D. Gray). (e) The same region of the North Equatorial Belt of Saturn, as observed in May 1981 with a 320-mm refractor (R.J. McKim).

5. Any changes in the visibility of irregularities in the outline of Ring C, both at the ansae and across the globe.
6. Any sign of a faint ring outside the bright rings.
7. Position of any dusky patches or small divisions (for example, Encke's division) on the ansae of the rings.

IDENTIFICATION OF FEATURES

The nomenclature of the belts and zones is the same as for Jupiter, but owing to the considerable change in *B* during the course of Saturn's year, and the apparent shift of presented latitude towards the north or south pole, identification of belts and zones is not always straightforward. The rings, when wide open, occult one hemisphere, and only when the rings are edge-on is the whole globe visible to us. Care must be exercised as not all belts or zones may be visible. A belt seen to the north of the NEB may not be the NTB — if far north it is likely to be the NNTB, with the NTB absent. Figure 7.3 shows the major features of the globe and rings.

Determination of the latitude of the belt or zone is perhaps the best way of being certain of identity and the following approximate latitudes are given as a guide:

Region	*Latitude range*
SPR	-60° to -90°
STB	-33° to -44°
SEB	-9° to -27°
NEB	$+12^\circ$ to $+30^\circ$
NTB	$+30^\circ$ to $+40^\circ$
NNTB	$+56^\circ$ to $+62^\circ$
NPR	$+60^\circ$ to $+90^\circ$

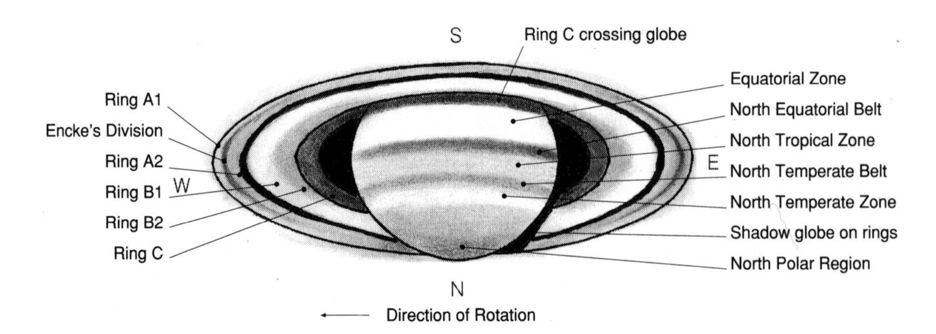

FIGURE 7.3
Nomenclature of the principal ring and globe features on Saturn. (The nomenclature of the belts and zones is as used for Jupiter.)

ESTIMATES OF RELATIVE INTENSITY

Estimates are made on a scale of 0 to 10, where 10 is the black background sky and 1 is the normal intensity of the brightest part of Ring B. Any feature brighter than this is assigned a value of 0.5, or 0 if exceptionally bright. This scale, introduced by the BAA Saturn Section in 1946, is used by European observers, but in the United States the scale is reversed, with 0 being the background sky and 8 the bright reference of Ring B.

The quickest and easiest method of recording intensities is to write the numbers down on an outline sketch of the globe and rings. A good plan is to make one series of estimates taking the features in order from the brightest to the darkest, and another series taking them in the reverse order, adopting means where any results differ. If different degrees of brightness are found in various parts of the same ring, belt, or zone, so that too many numbers have to be written on the sketch, larger-scale sketches can be made or explanatory notes added. The whole observation might be accurately done in about 20 minutes, or, with practice, less.

This estimation should be done as often as possible so as to gain experience, to avoid missing slight temporary variations, and to secure good averages, care being taken to avoid being biased by any previous estimates. Experienced observers show a certain amount of personal equation, but this can be allowed for if they are consistent. Some observers rate features too low, others too high, but the suspect observer is one who shows irregularities, sometimes high and sometimes low! Unsteady air can reduce the intensity of dark markings, and differences occur according to the aperture of the telescope being used. Intensity estimates are best made using the same telescope and magnification each time.

From numerous observations, results of great reliability can be derived, as differences between observers and apertures can largely be overcome by weighting and averaging. Particular attention should be given to recording the intensities of the polar areas, as they seem to change according to the tilt of the globe and rings.

While it is permissible to give estimates using halves or quarters (or their decimal equivalents), the use of decimal places such as 5.2 or 5.3 is highly suspect.

The method just described can also be used with colour filters (see below), first estimating with no filter, followed by red and then by blue. It is possible to reveal faint colours by this method, particularly those which are below the threshold of vision. The resulting figures are converted into 'light intensity' by subtracting the shading value from 10. As an example, an average shading value of the NEB of 5.8 in blue and 4.6 in red would give light intensity values of 4.2 in blue and 5.4 in red. 'Red minus blue' (5.4–4.2) is therefore 1.2, suggesting a slight brownish tinge. Here, it is permissible to use one decimal place in the averages. Visual colour notes of belts, zones, rings and shadows should also be made, especially by those using reflectors.

A most valuable study of the visual photometry of the planet from 1943 to 1981 was carried out by R. McKim and K. Blaxall [2]. The work of 126 observers, totalling over 80 000 estimates of intensity was used. The study showed that the short-term changes in colour and intensity are real, while long-term changes can frequently be correlated with the seasonably variable solar altitude at a particular global feature, showing that the meteorology of the planet is not merely controlled by released internal heat. All ring features exhibit intensity variations, many of which can be correlated with saturnicentric latitude of the Sun. This study demonstrates the value of cooperative study where large numbers of observations can be analysed by an organization, thus revealing information which no individual observer could achieve alone.

OPTICAL ILLUSIONS

From time to time, certain strange appearances are reported. The globe casts a shadow on the rings, and also the rings cast a shadow on the globe. Distortions of these shadows are reported at times, but most can be explained as optical illusions. Perhaps the best known is the Terby Spot, which takes the form of a bright spot adjoining the globe's shadow on the rings. It is generally considered to be a contrast effect, and sometimes it is more prominent than at others depending on the observing conditions and the planet's position in relation to the ring angle [3].

TRANSITS OF SPOTS AND OTHER MARKINGS – LONGITUDE ESTIMATES

Such phenomena are regarded as rare on Saturn, but many rather difficult dark spots, streaks, patches, humps and projections have been detected. The position and estimated time of transit of any such features over the central meridian of the globe should be recorded. Bright spots should also be looked for, but any which do not move with the rotation of the planet should be regarded as optical effects of the Terby 'white spot' variety.

The determination of the longitude of surface features is a very important area of study. Looking at Saturn through the telescope, it is easy to imagine the central meridian, the imaginary line which bisects the visible disk, perpendicular to the belts. Observations of the longitude of a spot consist of accurately recording the time when it transits (crosses) the central meridian. With practice, an observer can achieve an accuracy of one minute of time, and these timings can be reduced using longitude tables available in such publications as the *Journal of the Association of Lunar and Planetary Observers* (ALPO), which give the longitude of the central meridian at 0 h UT daily, with provision for calculating intervals.

System 1 (analogous to that on Jupiter) covers the whole of the two equatorial belts and the Equatorial Zone, which has a period of 10 h 14 m (rotation rate 844°.3 per day). System 2 includes the rest of the planet with a rotation rate of 812° per day. However, there is no sudden change of rotation period as with System 1 and System 2 of Jupiter: there is merely a gradual lengthening towards the poles. Botham's spot of 1960 showed a rotation period of about 10 h 40 m at latitude

60 N. This was in System 2; Hay's famous spot of 1933 was in the Equatorial Zone and had a rotation period of 10 h 14 m 15 s in System 1 [4].

Spots on Saturn are usually short-lived and nothing like as frequent as those observed on Jupiter. It is important that any spots which do occur are observed as fully as possible. Regular observations of the planet are necessary, and, though long periods of apparent inactivity may occur, the period when a feature appears should not be missed.

Outbreaks of dark spots on the NEB in the early 1980s produced sufficient transits for rotation periods to be obtained, a value of the mean being 10 h 13 m 58 s for 1980–81 and 10 h 13 m 49 s for 1981–82. Activity in the belt declined shortly afterwards.

THE WHITE SPOT OF 1990

In the autumn of 1990, a large and very bright white spot was discovered in the planet's Equatorial Zone by Stuart Wilber, an amateur astronomer of New Mexico, USA. Large white spots have been seen on Saturn in the past, the main ones occurring in 1876, 1903, 1933 and 1960. The dates of all the major spots are 27–30 years apart, so the large spot of 1990 was not entirely unexpected. The apparent 27–30-year periodicity must be treated with caution, however, as the spots of 1903 and 1960 were smaller, less spectacular, and not in the Equatorial Zone. The spots of 1876, 1933 and 1990 were all bright and located in this zone, and give a rough periodicity of 57 years. They have similar rotation periods and all have occurred in the ecliptical longitude range of 290–315°.

According to Tombaugh, the initial System 1 longitude of the spot was 334° on 28 September, and the centre of the spot remained more or less in this region until late October, after which it showed a steady motion in decreasing longitude. With the passage of time the preceding end became diffuse and the zone showed brightening ahead of it, thus making transits difficult to time with accuracy. This behaviour agrees with that of past spots, such as Hay's of 1933. One of the effects of the Great White Spot was to obscure the NEB at some longitudes, and the Hubble Space Telescope shows it overlapping parts of the belt's southern edge. The BAA obtained a period of 10 h 13 m 48 s for the spot, from 32 observations.

Spots at longitudes other than that of the Wilber spot appeared in the Equatorial Zone, but none of these was long-lived. Some smaller and inconspicuous spots appeared in the North Tropical Zone prior to the main outbreak, and some small spots in the Equatorial Zone were recorded in 1991. All of these were smaller and less bright than the main spot of 1990, but may have been associated with the initial event.

LATITUDE ESTIMATES

A study by Hollis of the observed latitudes of the belts for 1946–76 confirmed that latitude variations are real [5]. The equatorial belts vary in width: in the case of the NEB this was from a minimum of 6° to a maximum of 23°. The SEB showed a definite increase in latitude, amounting to about 4°, between 1946 and 1952, and also from 1963 to 1976. The value of long-term studies is clearly demonstrated.

Quite useful latitude estimates can be obtained by measuring careful drawings. Although the accuracy of a single drawing may not be high, a large number of results will give good averages, permitting comparison of the belt latitudes from one apparation to the next, as demonstrated by Hollis.

The measurements required are the polar radius r and the distance y of the feature (such as a belt edge) from the centre of the disk. The saturnicentric latitude of the feature, C, can then be worked out using the Crommelin formula, provided the saturnicentric latitude of the disk centre, B, is ascertained from the *Astronomical Ephemeris* or some other source. Then

$$B' = \arctan(1.12 \tan B)$$

$$\sin(b' - B') = y/r$$

$$\tan C = \tan(b'/1.12)$$

Latitude estimates are required for the edge of Ring C across the globe; the centre of the EZ; the N and S edges of the SEB and NEB; and for any other belts or zones that may be visible.

COLOUR FILTERS

The use of colour filters in visual observation is now common among observers, especially if larger-aperture telescopes are available to them. The most useful filters are red and blue, but other colours can be used also. It is very important that the transmission characteristics of the filters used are known, and it has been found that perhaps the best for general work are Wratten 25 (red) and Wratten 44a (blue). If the aperture is greater than 250 mm, then the Wratten 47 can be employed. Ideally, filters should be as monochromatic as possible, but such filters are often too dense for practical use visually: the filters suggested are a compromise between good light transmission and a bandwidth that is not too broad.

The cheapest way to obtain filters is in gelatine form from photographic dealers. Mounting the filters is a matter of choice, but a very convenient way is to cut two circles of convenient size from postcard, and then punch a hole in the centre of each — a loose-leaf hole cutter is ideal for this purpose, giving a hole of about the correct size, and with a neat edge. Carefully cut a piece of filter about 10 mm square (slightly larger than the hole), and sandwich it between the two pieces of card, which are then glued together. In the case of older-style eyepieces with a recessed face, the card circle can be cut to fit into the recess, with a short arc cut straight across so that it can be removed from the recess with a finger nail. Alternatively, the mount can be made in the form of a short tube which fits over the eyepiece.

With a little ingenuity a sliding or revolving filter holder can be incorporated into the eyepiece assembly, the filter being located in front of the eyepiece field lens. Filters should be stored in a suitable dustproof container when not in use, and needless to say the gelatine variety should not be touched by hand.

Before embarking on a programme involving filters, it is advisable to become familiar with the views given by them. It is best to select a suitable magnification and keep to this for all filters. First note the appearance of the planet without any filter. Follow this by using a red filter, and subtle differences of intensity will be noticed; follow this by looking with the blue filter, and again difference will be noted. A relative darkening of the EZ in blue is common.

Do not expect to see detail with filters at your first attempt. It is rather like having your first views of the planet and seeing only the most obvious features — finer detail comes later with patience and practice, and the same is true for observing through filters. The best magnification to use is probably the same one normally used without a filter.

All filters reduce the total amount of light-energy reaching the eye very considerably, and so the image seen with them will be much dimmer: this is most markedly the case with red and blue filters, which suppress that part of the spectrum to which the eye is most sensitive. However, given sufficient aperture in the first place, filters can be a very valuable aid in revealing detail, especially of weak colour.

THE RING SYSTEM

Large apertures are needed to reveal the details of the three major rings, known as A (the outer), B (the brightest) and C (the inner). This ring system is inclined at about 28° to the plane of the Earth's orbit, and during the $29\frac{1}{2}$ years Saturn takes to orbit the Sun the ring system is presented at all angles from edge-on to an opening of 28° north and south.

Edge-on Presentation

During the 'edge-on' phase, the Earth passes through the ring plane either once or three times. When these passages occur, the rings are, for short intervals, extremely difficult or even impossible to observe. The reason may be that the extremely thin ring system is presented exactly edge-on to the Earth, or the rings are exactly edge-on to the Sun with neither face being illuminated, or the unilluminated face is presented towards the Earth. At such a time, Cassini's division can be seen as two light spots in the almost edgewise invisible ring owing to sunlight passing through it.

The next edge-on presentation will occur in 1995–96, and the following stages will occur. All will be observable except, probably, the final N–S passage of the Earth through the ring plane, since Saturn will only five weeks before conjunction with the Sun.

1995 May 20: Earth passes from N to S of the ring plane (opposite faces presented to Earth and Sun).

1995 August 11: Earth passes from S to N of the ring plane (same face presented to Earth and Sun).

1995 November 19: The Sun passes from N to S of the ring plane (opposite faces presented to Earth and Sun)

1996 February 12: The Earth passes from N to S of the ring plane (same face presented to Earth and Sun).

It is important to record the appearance of the rings at this time. A hint of a faint ring outside the bright rings has been indicated at times when the rings are near edge-on, as the ring material is thus concentrated in the line of sight: Voyager confirmed the existence of rings outside Ring A, and this is perhaps what can be seen.

Although the edge-on rings are often invisible in small apertures, they can present a truly magnificent sight when the angle is a degree or so, with the satellites appearing like brightly-lit droplets on a spider's web. The rings take on average about 7 years 3 months to pass from being fully open to edge-on.

Ring Variations and Divisions

Variations in the brightness of the rings have been reported, and all three rings are variable. The outer Ring A is often of two different intensity levels, the two parts being separated by the elusive Encke's division. Other faint divisions can sometimes be seen, especially in Ring B, and any observer who detects such a division should, if without a micrometer, try to estimate its position as accurately as possible by eye and measurement of a drawing, and could express this as a fraction of the ring's width — for example, 2/5 from the outer edge. Reports should indicate whether or not a micrometer has been used, although during my tenure of office as BAA Saturn Section Director, which began in 1964, I have never received a single micrometer measurement!

A comparatively large aperture is very desirable for this work, and really good seeing conditions and a steady image are necessary to be sure of the objective reality of ring divisions other than Cassini's. Subjective impressions of divisions in the outer ring can be caused when small telescopes are used in poor seeing and the outer edge of the ring is not steady, sharp, or well seen. Differing brightness in the parts of a ring can also give the impression of a faint division.

The Rings Through Colour Filters

A curious effect known as the 'bi-coloured aspect' of the rings has been reported. It takes the form of a difference in the brightness of the east and west arms of the rings when viewed with coloured filters. Walter Haas in New Mexico first drew attention to this effect in 1949 when he noticed that the west arm of the rings was distinctly brighter than the east arm when seen through a Wratten 47 (blue) filter. Haas and others have seen this effect many times, it being visible with a variety of telescopes and magnifications: however, in many views the arms are equally bright [6].

The brightness difference for the visual observer has been chiefly remarked with blue filters, and often red and green filters used at the same time show no such effect, nor does a view without filters. Occasionally the appearance has changed within only one or two hours. It is important to note that the bi-coloured aspect was recorded on a series of photographs taken in blue light on 1 October 1963 by C.F. Capen with the 410-mm Cassegrain telescope at the Table Mountain Observatory, USA; this is a project which is now within the range of an amateur with CCD equipment and adequate aperture.

So far, no one has succeeded in correlating the bi-coloured aspect with such parameters as the axial tilt of Saturn towards the Earth, the Earth–Saturn–Sun phase angle, the position angle of the axis of Saturn in the sky, etc. No satisfactory explanation has yet been offered, but the bi-coloured aspect would appear to be worthy of a more systematic investigation than it has yet received. Walter Haas, who is Founder/Director Emeritus of the ALPO, recommends serious efforts for simultaneous visual and photographic observations of the effect, employing selected colour filters of known transmission. It is critical, of course, that the visual observers make their observations totally independently of each other.

SATELLITES

Five of Saturn's major satellites are easily visible in most amateur telescopes. The brightest is Titan at magnitude 8.3, with Iapetus, Rhea, Tethys and Dione falling between magnitudes 9.0 and 10.5. The fainter Enceladus requires larger apertures as do Mimas and Hyperion. Iapetus is particularly interesting in that it shows a considerable range of brightness, being some two magnitudes brighter at western elongation than when at eastern elongation: Voyager proved what had been assumed, that its hemispheres are strikingly different in reflective power.

Variability

All the satellites are variable in brightness to some extent, and colour variation has been noted for Titan, the only solar-system satellite with an appreciable atmosphere. Monitoring of its colour could therefore be valuable, particularly if carried out photoelectrically using standard filters, in revealing possible atmospheric variations. However, whilst the reality of variation in brightness of all the satellites has been established, this type of observation is very difficult to make visually by standard variable-star methods owing to their different and varying distances from Saturn, whose glare has an enormous effect on apparent brightness. Errors so introduced can be considerable, and it must be faced that the only reliable way to record variations of brightness is by photoelectric photometry. Astronomical publications such as the BAA *Handbook* provide charts and tables from which the satellites may be identified.

Transits and Occultations

The main satellites, apart from Iapetus, have orbits which are more or less in line with the rings and at times when the ring angle is very narrow, phenomena such as transits, eclipses, and occultations may occur. This applies not only to the satellites but to their shadows as well. The easiest to see is Titan and its shadow, and Titan when in transit is very dark, often being mistaken for the shadow. Transits of Rhea, Tethys, and Dione (and their shadows) can be observed only with telescopes of around 300-mm aperture or more, and then only when the seeing is good. The timing of transits and occultations is important, but the fainter satellites are difficult to see when near the globe owing to glare. Observers will find this much harder than observing similar events with the satellites of Jupiter.

The satellites of Jupiter do not always disappear or reappear exactly as predicted. The reasons for this are not yet understood, but differences of a few minutes have been noted in some cases. It is not known whether a similar effect applies to the satellites of Saturn, but a difference of 8 minutes was found by two British observers for an eclipse of Rhea in 1979 [7]. Owing to differences in orbital inclination, eclipse phenomena of Iapetus occur at different 'seasons' from those of other satellites, and it can also be eclipsed in the shadow cast by the rings. Observations of this type are rare, and every effort should be made to try to observe them when possible: predictions are to be found in the BAA *Handbook* and other annual publications.

Occultations of Stars

Events of this kind are very rare indeed, the most recent being the occultation of the star 28 Sgr, which is of magnitude 5.8, on 3 July 1989 [8]. Even rarer was the occultation of the same star by Titan a few hours later [9]. It has been calculated that Titan occults such a star less than once in 500–800 years, and the event was well observed in the British Isles, where the sky was clear over most of the country (another very rare event!). Occultations by the planet can reveal details of ring structure, and every fluctuation in brightness and colour of the star, with the time and star's exact position, should be noted during its passage behind the rings or as it disappears behind the limb of the globe. This can be done with much greater precision using the more sophisticated electronic methods now available to amateurs, which include photoelectric photometry and CCDs. Photoelectric photometry of the Titan occultation is described in Chapter 15, and the subjects of high-resolution planetary photography and the use of CCDs have their own chapters.

CONCLUSION

It will be seen that the planet Saturn has much to offer to the patient and dedicated observer. The added motivation of the work being scientifically useful comes with the careful organization of the observations and forwarding of these to an organization where analysis and subsequent publication of reports takes place. All observations should have the date given as year, month, day, hour and minute (and seconds where necessary) in UT. The telescope used and its magnification together with observing conditions needs to be included (the 5-point scale being suitable), and one must not forget the name and address of the observer! Organizations such as the BAA in Great Britain, the ALPO in America (Saturn recorder Julius L. Benton, Jr., Associates in Astronomy, 305 Surrey Road, Savannah, Georgia 31410, USA); and the Unione Astrofili Italiani (Universita degli Studi di Padova, Dipartmento di Astronomia, Vicolo dell'Osservatorio 5, 35122 Padova, Italy) are but a few.

Continuous observation of Saturn is of paramount importance. Some features are seen for only a short time and could easily be missed and go unrecorded. The role of the amateur observer is as important as ever it was, if not more so. The amazing results provided by the Voyager spacecraft were an inspiration to us all, but these observations covered only a matter of days. The amateur's role is to keep observations going and to follow up such pioneering work as that by Voyager. Variations in the intensities and latitudes of belts and zones over many years has been clearly demonstrated, and whilst occasional observations are of value, observation of Saturn is a long-term commitment which can be very rewarding, is of scientific use, and perhaps may even spring a few surprises.

About the Author

The author first became interested in astronomy during the time of the Second World War, and learned the main stars during the blackout. He has been actively involved since 1952, when observations were made with a 50-mm refractor, this telescope still being used for regular counts of Active Areas on the Sun. Having joined the Nottingham Astronomical Society at this time, he then joined the BAA in 1953 and became involved with many observing sections, using a home-made 200-mm Newtonian reflector. In 1963 a 300-mm Newtonian was obtained on loan from the BAA, and it was at this time that he was appointed Director of the Saturn Section, a position held to this day.

Planetary observation became the main astronomical interest owing to light-polluted skies, and the author has served as Assistant to Director of the Jupiter Section, is on the committee of the Mercury and Venus Section, has been Secretary of the Lunar Section, and has also served as Acting Director of the Solar Section.

The observatory is also used for educational purposes, including youth groups and adult education groups. A primary school teacher by profession, the author is a naturalist specializing in freshwater microscopy where he samples local ponds for the local Wildlife Trust records. Currently he is Chairman of the Long Eaton Natural History Society. His interest in photography (including home-processing) lends itself to both astronomy and natural history, and he is a short-wave radio ham.

Contact address – 136 Trowell Grove, Long Eaton, Nottingham NG10 4BB, UK.

References

1. Heath, M.B.B. Drawing the planet Saturn, *BAA Journal*, **63**, 342 (1953)
2. McKim, R. and Blaxall, K. Saturn 1943–1981, *BAA Journal*, **94**, 145, 211, 249 (1984).
3. Alexander, A.F.O'D. *The Planet Saturn*, pp. 225–226, 400. Faber & Faber, London, 1962.
4. Heath, A.W. The 50th anniversary of Hay's spot, *BAA Journal*, **93**, 259 (1983).
5. Hollis, A. The latitudes of Saturn's belts, *BAA Journal*, **91**, 41 (1980).
6. See, for example, articles in the *ALPO Journal*, August 1988 (p.204) and July 1989 (p. 110), and in the *BAA Journal*, **96**, 174 (1986) and **100**, 87 (1990).
7. Heath, A.W. Saturn, 1978–79, *BAA Journal*, **90**, 461 (1980).
8. See *Sky & Telescope*, October 1989, pp. 360, 364–365.
9. Hollis, A and Mitton, J. BAA observers capture Titan's spectacular occultation, *BAA Journal*, **99**, 205–207 (1990).

Notes and Comments

Magnification – The optimum magnification for planetary work has been the subject of many investigations. Among the most relevant parameters are the apparent diameter of the planet, the contrast of the markings, the atmospheric conditions, and the aperture being used. Thus, observation of Mars, with its small disk, often requires higher powers than work on Jupiter, while observers of the contrasty lunar surface habitually use rather higher powers than are employed for the vague shadings on Venus.

Having said that, it is widely accepted that smaller apertures will bear higher relative magnifications than large ones. This is because inevitable atmospheric tremors (and possibly mounting and driving defects) place a limit on maximum usable magnification, and for planetary work magnifications higher than ×500 are rarely feasible, regardless of aperture, while ×200 can be used with a 75-mm telescope, certainly for the Moon, Mars, and Jupiter.

Some examples of magnifications used by leading observers of Saturn during the last century include ×450 (Hepburn and Ainslie, 710-mm refractor), ×180–270 (Ainslie, 230-mm reflector), ×100–250 (Knight, 125-mm refractor), ×450–700 (Barnard, 1000-mm refractor), ×250 (Denning, 250-mm reflector), and up to ×620 (Dawes, 210-mm refractor). This and much other interesting information can be found in *The Planet Saturn* by A.F.O'D. Alexander [3]. — Ed.

CHAPTER 8 · *Meteor Observing*

PAUL ROGGEMANS
INTERNATIONAL METEOR ORGANIZATION

PEOPLE HAVE WATCHED 'SHOOTING STARS' SINCE HUMAN BEINGS first populated the Earth. Many natural phenomena, such as these swift trails in the starry sky, must have fascinated them, or even terrified them in the case of fireballs and meteorite impacts, since they had no available explanation for these phenomena. While many frequently observed natural phenomena were satisfactorily explained several centuries ago, meteors have only been studied properly during the last hundred years. Even today several questions are left unanswered, and scientific investigation has still not covered the subject adequately.

METEOR SHOWERS

Meteors are either *sporadic*, being single visitors from space, or else belong to so-called meteor *showers*. Meteor showers are visible during limited periods of time, when the Earth sweeps through a belt of meteoroid particles distributed along an elliptical orbit in the solar system. Each meteor stream intersects the Earth's orbit at a specific point. During a shower, all the meteoroids will appear to hit the Earth's atmosphere coming from a common point out in space, known as the *radiant*. A radiant is a small area in the sky defined in right ascension and declination: it is found when the meteors of a given stream are traced backwards on a gnomonic star map to the area where they all intersect.

Most meteor streams correspond closely to known cometary orbits. Comets often produce a great deal of dust at their perihelion passage, the particles of which distribute themselves along the orbit and give rise to a meteor stream. Careful observation of radiant positions and meteor orbits in the solar system enabled former meteor researchers to identify the principal meteor streams known today, but it is known that very rich meteor displays from streams that no longer cross the Earth's orbit have occurred in the past. Likewise, new radiants can be expected to start producing quite noticeable activity without warning. This dynamic complex necessitates a continuous effort by meteor observers in order to report these new appearances, while the known meteor streams need annual coverage to study their structure, evolution, build-up and gradual destruction. These aims need a lot of manpower: since professional astronomers have neither the time nor the labour to monitor meteor activity, the task falls entirely on the shoulders of amateur observers.

A list of major and some minor showers appears in Table 8.1. All regular showers are given a Zenithal Hourly Rate (ZHR), but this is a very variable value, with unexpected rates being a common experience in meteor work. The observing limits are defined from orbital elements associated with the shower, and in many cases these limits are very uncertain.

MINOR STREAMS

Some meteor streams are very abundant and have been observed very well, but the majority of meteoroids do not belong to these major showers, appearing at first sight to be independent particles. These bodies form the sporadic background, providing continuous meteoric activity, although they are mainly faint. However, careful study of the orbital elements of sporadic meteors does provide evidence for the existence of some minor meteor streams, although they may not be noticeable as such to the observer. Since the sporadic meteor population is believed to consist of once-compact streams that have disintegrated, some of the members are still likely to share comparable orbits, and these can be distinguished if a large number of orbits are studied, although the danger of including chance coincidences must be borne in mind.

Some of these minor meteor showers are also listed in Table 8.1, while statistical analysis has produced about 200 groupings of orbits that await confirmation — most of these are probably the result of random alignments.

PLANNING A METEOR WATCH

While some meteor streams are hard to detect against the sporadic background, a few others are still so closely packed into a section of their orbit that their meteors are only easily recognizable when the Earth and the particle cloud meet at the crossing of their orbits. A typical example of such a case is the November Leonid shower: about every 33 years these bring a spectacular display with thousands of meteors an hour. Events such as this are rare, but it is very likely that some of these brief displays pass unnoticed. More regular work by amateurs offers the only hope of recording them.

Most meteor observing sessions are aimed at studying a particular shower. The

TABLE 8.1

A WORKING LIST OF VISUAL METEOR SHOWERS. WHERE NO ZHR IS GIVEN, IT IS TYPICALLY 2–3 PER HOUR. SHOWERS MARKED (*) ARE PERIODIC AND PRODUCE THE DISPLAYED RATES ONLY IN CERTAIN YEARS. IN OTHER YEARS, ACTIVITY IS TYPICALLY LIMITED TO A FEW METEORS PER HOUR. THE SOLAR LONGITUDE (COLUMN 4) EFFECTIVELY MEASURES THE EARTH'S POSITION IN ITS ORBIT AT THE POINT OF CLOSEST APPROACH TO THE METEOR STREAM IN QUESTION. THE DAILY RADIANT DRIFT IS TYPICALLY ABOUT 1° OF RIGHT ASCENSION EASTWARDS

				Radiant			
Shower	Activity limits	Max.	Solar longitude	RA	Dec	Velocity (km/s)	ZHR
Quadrantids	Jan 01–Jan 05	Jan 03	282.7°	230.1°	+48.5°	41	100[a]
δ Cancrids	Jan 05–Jan 24	Jan 14	296	126	+20	28	5
α Crucids	Jan 06–Jan 28	Jan 19	299	192	−63	50	–
α Carinids	Jan 24–Feb 09	Feb 01	311	95	−54	25	–
θ Centaurids	Jan 23–Mar 12	–	312	210	−40	60	–
α Centaurids	Jan 28–Feb 21	Feb 08	318	210	−59	56	–
o Centaurids	Jan 31–Feb 19	Feb 12	322	177	−56	51	–
Virginids	Feb 01–May 30	Several	--	186	00	30	–
γ Normids	Feb 25–Mar 22	Mar 14	353	249	−51	56	–
δ Pavonids	Mar 30–Apr 16	Apr 06	16.5	308	−63	59	13
Lyrids	Apr 16–Apr 25	Apr 22	31.7	271.4	+33.6	49	20[b]
π Puppids*	Apr 15–Apr 28	Apr 23	32.6	110	−45	18	40
α Boötids	Apr 14–May 12	Apr 28	36	218	+19	20	3
η Aquarids	Apr 19–May 28	May 04	42.4	335.6	−01.9	66	50

TABLE 8.1

A WORKING LIST OF VISUAL METEOR SHOWERS. WHERE NO ZHR IS GIVEN, IT IS TYPICALLY 2–3 PER HOUR. SHOWERS MARKED (*) ARE PERIODIC AND PRODUCE THE DISPLAYED RATES ONLY IN CERTAIN YEARS. IN OTHER YEARS, ACTIVITY IS TYPICALLY LIMITED TO A FEW METEORS PER HOUR. THE SOLAR LONGITUDE (COLUMN 4) EFFECTIVELY MEASURES THE EARTH'S POSITION IN ITS ORBIT AT THE POINT OF CLOSEST APPROACH TO THE METEOR STREAM IN QUESTION. THE DAILY RADIANT DRIFT IS TYPICALLY ABOUT 1° OF RIGHT ASCENSION EASTWARDS

Shower	Activity limits	Max.	Solar longitude	Radiant RA	Radiant Dec	Velocity (km/s)	ZHR
Scorpiids/Sagittariids	Apr 15–Jul 25	Several	–	260	−30	30	10[c]
Lyrids (Jun)	Jun 11–Jun 21	Jun 16	84.5	278	+35	31	5
Boötids (Jun)	Jun 28–Jun 28	Jun 28	95.6	219	+49	14	2
Pegasids (Jul)	Jul 07–Jul 11	Jul 09	109	340	+15	70	8
Phoenicids (Jul)	Jun 03–Jul 18	Jul 14	112	031.1	−47.9	47	–
Piscis Austrinids	Jul 09–Aug 17	Jul 28	124	341	−30	35	–
α Capricornids	Jul 03–Aug 25	Jul 30	126	305	−10	23	8[d]
Aquarids	Jul 08–Sep 20	Several	–	–	–	40	–[e]
Perseids	Jul 17–Aug 24	Aug 12	139.2	046.2	+57.4	59	100
κ Cygnids	Aug 03–Aug 31	Aug 18	145	286	+59	25	5
π Eridanids	Aug 20–Sep 05	Aug 28	155	052	−15	59	–
α Aurigids	Aug 24–Sep 05	Sep 01	157.9	084.6	+42.0	66	15[f]
Piscids (S)	Aug 15–Oct 14	Sep 20	177	006	00	26	3
κ Aquarids	Sep 08–Sep 30	Sep 21	178	338	−05	16	3
Capricornids (Oct)	Sep 20–Oct 14	Oct 03	189	303	−10	15	3

σ Orionids	Sep 10–Oct 26	Oct 05	191	086	−03	65	3
Draconids*	Oct 06–Oct 10	Oct 09	194	262	+54	20	–
ε Geminids	Oct 14–Oct 27	Oct 19	206	104	+27	71	5
Orionids	Oct 02–Nov 07	Oct 21	207.7	094.5	+15.8	66	30
Taurids (S)	Sep 15–Nov 26	Nov 03	220	050.0	+13.6	27	12[g]
Taurids (N)	Sep 13–Dec 01	Nov 13	230	058.3	+22.3	29	8[h]
Leonids*	Nov 14–Nov21	Nov 17	234.4	152.3	+22.2	71	–[i]
Monocerotids (Nov)	Nov 15–Nov 25	Nov 18	235	117	−06	60	5
Phoenicids (Dec)	Nov 28–Dec 09	Dec 05	253.5	015	−55	18	–
Puppids/Velids	Oct 15–Jan 22	Several	–	120	−45	40	–
Monocerotids (Dec)	Nov 27–Dec 17	Dec 10	258	099.8	+14.0	42	5
χ Orionids (N)	Dec 04–Dec 15	Dec 10	258	084	+26	28	3
σ Hydrids	Dec 03–Dec 15	Dec 11	259	126.6	+01.6	58	5
Geminids	Dec 07–Dec 17	Dec 14	261.4	112.3	+32.5	35	100
Coma Berenicids	Dec 12–Jan 23	Jan 04	282	175	+25	65	5
Ursids*	Dec 17–Dec 26	Dec 22	270.2	217.0	+75.9	33	50[j]

[a]Sharp maximum.
[b]Strong displays in 1803, 1922, 1082.
[c]Fireballs.
[d]During the shower the radiant moves from 290°, −14° to 324°, −04°.
[e]A complex stream with S and N radiants, each with multiple maxima. The most active (ZHR 20) reaches a maximum on Jul 29, radiant at approx. 340°, −16°. Rich in telescopic meteors.
[f]Strong displays in 1935 and 1986.
[g]Fireballs.
[h]Fireballs.
[i]Storms in 1833, 1866, 1966.
[j]Strong displays in 1945 and 1986.

following conditions need to be borne in mind when setting out to observe a shower:

(1) The observations must be carried out during the activity period of the stream, and the radiant must be above the local horizon. The activity periods are derived from observations made during previous years; isolated particles can be observed before and after these dates, but such activity will be very low.

(2) The catalogue position of the radiant is strictly valid for the date of maximum activity only. On other dates its position moves by about 1° per day as a result of the Earth's orbital motion changing the configuration with the orbit of the stream, making the radiant appear to move in an easterly direction. Precise data for several radiant positions are published in reference (1).

(3) Visual, photographic and telescopic observation can be done only during darkness on cloudless and moonless nights. Radio observations can be made during the daylight hours — some meteor streams have a radiant so close to the direction of the Sun that they can only be observed by radio.

(4) The name of the radiant is chosen according to the name of the brightest star close to its position in the sky. Since other authorities may use other stars to mark the radiant, the position in right ascension and declination, taking drift into account as well as the geocentric velocity, should be used to establish the likely identification of shower meteors. In amateur circles, observers often ignore the apparent velocity of meteors when associating them with radiants, but it is really an important characteristic in identifying showers.

(5) The position of the radiant, a means of calculating its altitude, awareness of the phase of the Moon and its time of rising and setting, are all required when a meteor watch is being planned.

NAKED-EYE OBSERVATION

This is the simplest method, and also a powerful one, since the eye has a large field of view and can detect stars and meteors over a wide range of brightness.

There are three ways in which naked-eye observation can contribute to meteor astronomy. The easiest is simply to count the number of shower meteors seen every hour. With more skill and experience, the magnitude of these meteors can also be estimated. This is useful because the magnitude of a meteor depends upon its mass, and the magnitude distribution teaches us something about the mass distribution and particle population in the meteor stream. Most observers record both frequency and magnitude, and this method of observation is strongly encouraged by the IMO because it is not too difficult and reveals the composition of the meteor stream.

The third technique is to plot the meteor trails on to a star map, and, although the IMO supplies a gnomonic star atlas for this purpose, it must be admitted that recording meteor positions requires an exceptionally good knowledge of the constellations, which is possessed by few people. Furthermore, when rates rise above 10–15 meteors per hour the observation is interrupted so many times for taking notes that some meteors are missed, spoiling the coverage of the shower for particle density analysis. Since the positional accuracy is an order of magnitude better when photographic or telescopic techniques are used, the method is useful chiefly for practice purposes, or when there is no other way of securing positional data.

The best time to make a start at naked-eye meteor observation is just before the maximum of a major shower, although observers with unpolluted, dark, transparent skies will be able to detect sufficient meteors for practice purposes on any night. In addition to an ordinary star atlas for identifying the constellations, some or all of the following items will be needed:

- A deck chair, sun-bed, or air-bed.
- A sleeping bag or blanket to keep out the cold (and mosquitoes).
- A waterproof cover to keep off dew.
- Pillows to help the head incline at the right angle.
- Warm clothes, food, and hot drink.
- A watch with a clear display, giving time to the second.
- Gnomonic star maps in case a particular meteor needs to be plotted (Fig. 8.1).
- Several pens or pencils (not red), clipboard, paper, and a red torch.
- A portable tape recorder is useful.

Even in group work, each observer is required to observe independently, and no data should be combined or mixed. This is because the final derived ZHR is calculated as the rate which would be seen by a *single observer*, under ideal conditions, and with the radiant in the zenith. However, groups of observers working singly can help alleviate boredom during slack periods.

The results of a meteor watch in which frequency and magnitude are being recorded should contain the following information:

1. Date and time of the beginning and end of the watch.
2. The location and geographical coordinates of the observing site.
3. The sky conditions, recording any clouds or obstructions. When clouds appear within the field of observation, estimate the percentage of sky covered at different times.
4. The limiting naked-eye magnitude has an important influence on the observed rates. Since there is always some doubt about the faintest star visible with the naked eye, this is not easy. One way is to use IMO charts of small areas of sky and to count the number of stars visible within the area: conversion tables indicate the limiting magnitude represented by this number. Another way is to obtain an appropriate variable-star chart, showing a sequence of stars of descending brightness, accurate to 0.1 magnitude. The limiting magnitude is a most significant factor in determining the ZHR, so it should be determined as accurately as possible, together with any changes that occur during the watch.

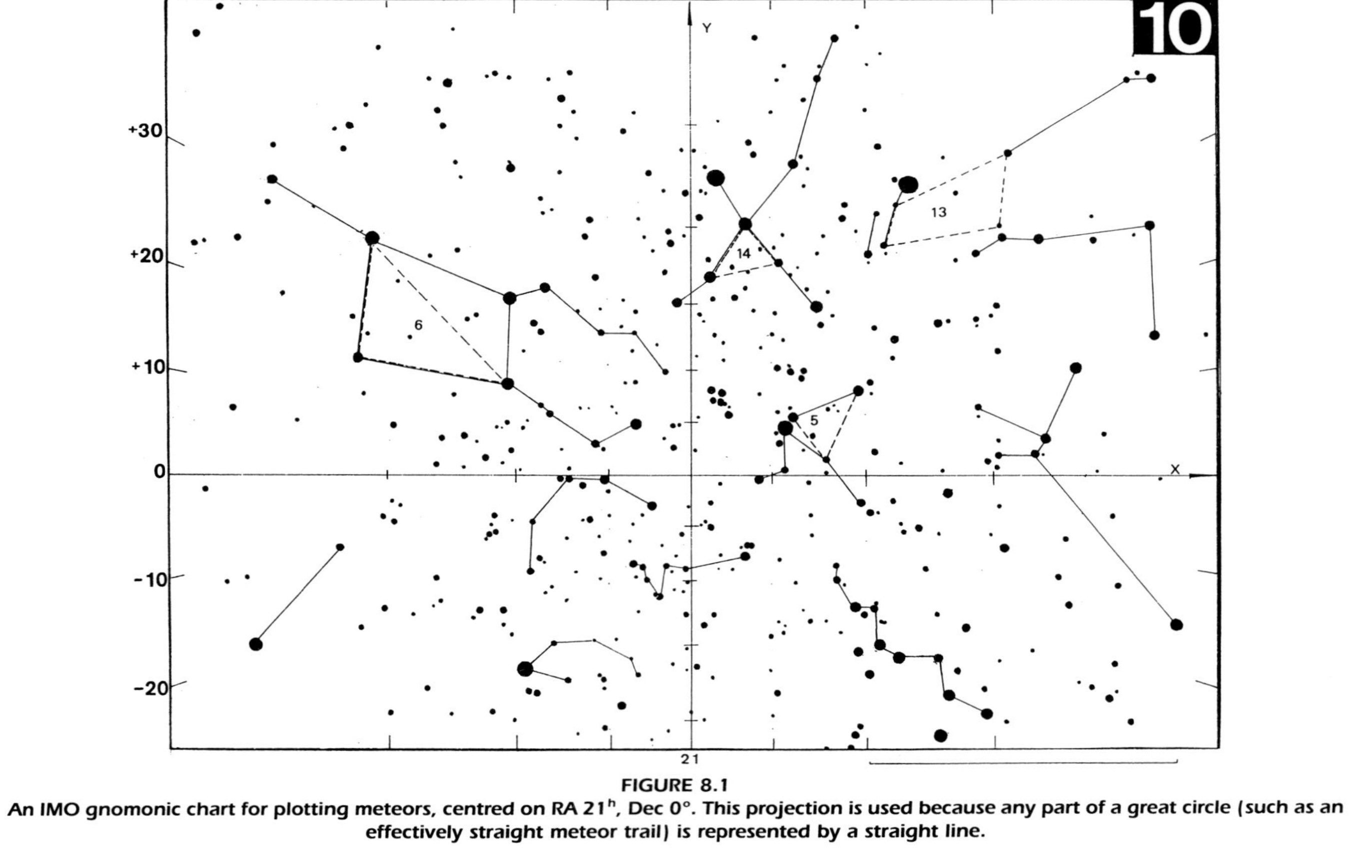

FIGURE 8.1
An IMO gnomonic chart for plotting meteors, centred on RA 21^h, Dec 0°. This projection is used because any part of a great circle (such as an effectively straight meteor trail) is represented by a straight line.

5. Breaks should be noted in order to know the exact net effective observing time.
6. A portable tape recorder will allow observations to be recorded while keeping an uninterrupted watch. A clock making an audible signal every hour is also useful, and the time when each observing session begins must also be recorded. Recording the time regularly while observing is recommended, the strict minimum being at every new hour interval.
7. Note the RA and Dec of the centre of the field of view. Any meteor that attracts the observer's notice while attention is being concentrated on this field centre must be recorded. The field of view is not confined to a certain area but includes as much as the observer can cover when looking around near a chosen direction in the sky.

The record of each meteor can include its time of appearance to the nearest minute (although for two-station visual work, or photographic observations, timings to the nearest second should be the aim), magnitude, and probable shower association. To establish the latter, some preparation is needed: using a gnomonic star map, plot the direction in which meteors from the given radiant will cross each constellation, and mentally reconstruct these while observing. In addition, the typical velocity of shower meteors should be noted. A Perseid meteor enters the atmosphere at 60 km/s, which is fast — Alpha Capricornids (25 km/s) are slow by comparison. It is useful to make a velocity estimate for each meteor using the following scale:

1 = very slow	(satellite re-entry)
2 = slow	(Capricornids)
3 = medium	(Quadrantids)
4 = fast	(Perseids)
5 = very fast	(Leonids)

A very efficient way of obtaining velocity estimates is in degrees/second. This may sound difficult, but with training it is feasible. It is necessary to be familiar with a one-second time unit (typically the time needed to say 'twenty-one'), as well as angular measure against the sky. A meteor typically is visible for less than a second, so when one is seen it is mentally continued forward in order to estimate how long the path would have been had it lasted for one second. For instance, a meteor that lasted for 1/3 second over 5° would receive a velocity estimate of 15°/s (fast). This method is preferred by most experienced observers who plot meteors, and it is an effective tool for more accurate shower association.

If path plotting is undertaken, the information detailed above is still necessary. It should be stressed that a few good-quality paths are far more useful than a large number of questionable reliability.

Meteor counts enable the activity profile of a meteor stream to be derived. To do this, all reported hourly counts are reduced to the ZHR – the standard hourly rate to be expected for a single observer with the radiant at the zenith, an unobstructed sky, and a limiting magnitude of 6.5. To average out statistical fluctuations, the IMO collects hundreds of reports for each hour, enabling 24 hours a day coverage to be obtained. The magnitude distribution of shower members can then be seen by totalling their number per magnitude class, and variations in the mass distribution during the Earth's journey through the stream will be reflected in day-to-day variations in the magnitude distribution. A graph of Perseid activity in 1988 is shown in Fig. 8.2.

PHOTOGRAPHIC METEOR WORK

Given the necessary patience, it is not difficult to photograph meteors: an ordinary miniature camera with a standard lens, a tripod, a cable release and Tri-X film will do. The faster the lens, and the more sensitive the film, the better. It is not necessary to be an accomplished astrophotographer in order to obtain useful pictures; beautiful meteor pictures can be almost completely useless for positional measurement, or astrometry, if they are not properly documented (see below).

Positional measurement from photographs can easily be done by amateurs, but magnitude measurement, or photometry, requires more specialized equipment. Therefore, the main aim of meteor photography is to derive accurate positional data.

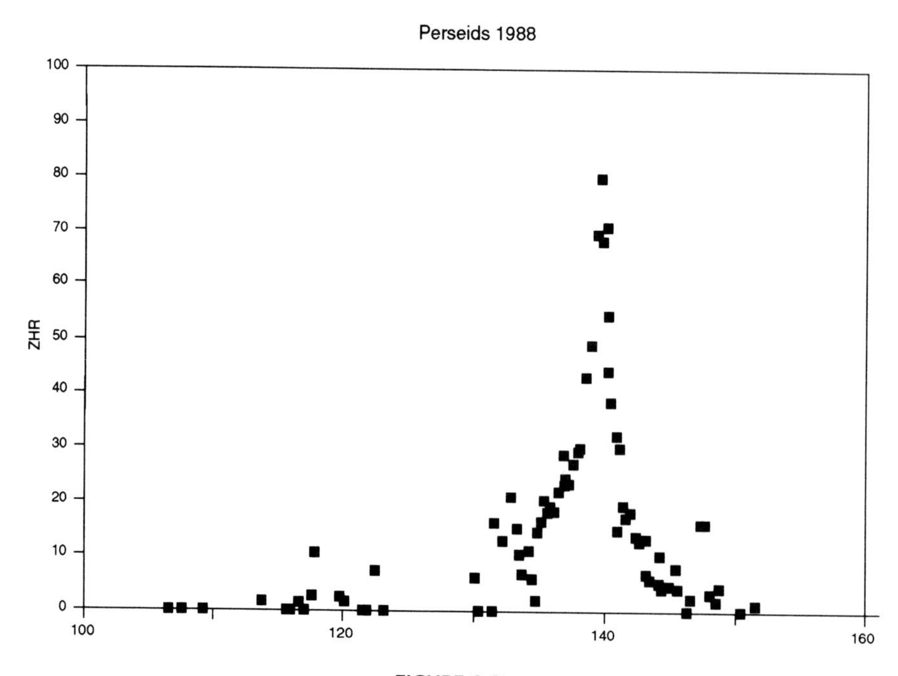

FIGURE 8.2
The Perseid activity profile for 1988, from IMO observations.

The following points should be borne in mind when planning a photographic watch:

(1) Use film with a speed of at least 400 ASA and operate as many cameras as possible. The object is not to achieve aesthetic pictures, but to cover the maximum amount of sky.

(2) Use one or more cameras with a 'B' shutter setting and a cable release which can be locked open. Lenses of 50 mm focal length, working at f/1.8 or f/1.4, are most often used. Wide-angle lenses cover more sky, but their smaller f-ratio will not record such faint meteors. Fast lenses of the kind recommended will easily photograph meteors of magnitude 0 (the brightness of Vega). Magnitude +1 meteors may appear as very faint lines on the negative, but are difficult to find on the film after development, while old cameras with very slow lenses consume a great deal of film with little to show for it.

Almost all meteor photographs are unguided, which means that the stars appear as trails on the film.

The chances of success increase greatly with the number of cameras using fast lenses. With some practice up to eight cameras can be used at a time, and for those who like building equipment the construction of an automated camera battery, containing several cameras mounted on a single compact base, is a challenge (Fig. 8.3)!

FIGURE 8.3
A meteor camera battery. A good example of how up to eight cameras can be mounted. This equipment was built by a Dutch amateur. (Reproduced by permission of The International Meteor Organization).

(3) When using a single camera, a tripod is needed to allow it to be directed at the selected area of sky. Keep the tripod as low as possible to minimize vibration.

(4) The choice of aperture, film, and exposure time must be selected with the aim of photographing as many meteors as possible. The exposure time must be reasonably short so that the star trails do not become too long — a maximum exposure time of 15 minutes is suggested, unless an all-sky lens is being used. Also, the sensitivity of most films decrease after the first few minutes of exposure. The instructions enclosed with the film may be helpful. The maximum exposure is also dictated by light pollution: at perfectly dark sites the most sensitive films (3200 ASA) have been found perfect for meteor photography with exposures of 3 minutes or even more.

(5) Working at night brings the problem of dew, which under certain conditions can ruin the exposures. However, dew cannot form on lens surfaces when they are even slightly warmer than the air. Most amateurs place some electric resistance wire around each lens, or a simple battery-operated lens heater can be obtained. If the lenses become dewed, it is better to end the exposure than to clean them with a handerkerchief, as this may badly damage the optical surfaces.

(6) In addition to astronomical objects, aeroplanes, satellites, and even balloons produce enough light to leave meteor-type trails on the negative. A rotating shutter (Figs. 8.4, 8.5) will help to distinguish them. This consists of a motor with a known and stable number of revolutions per second, carrying a disk with at least two open sectors, placed immediately before the lens. For example, with two open sectors and a motor running at 25 revolutions per second, the exposure will

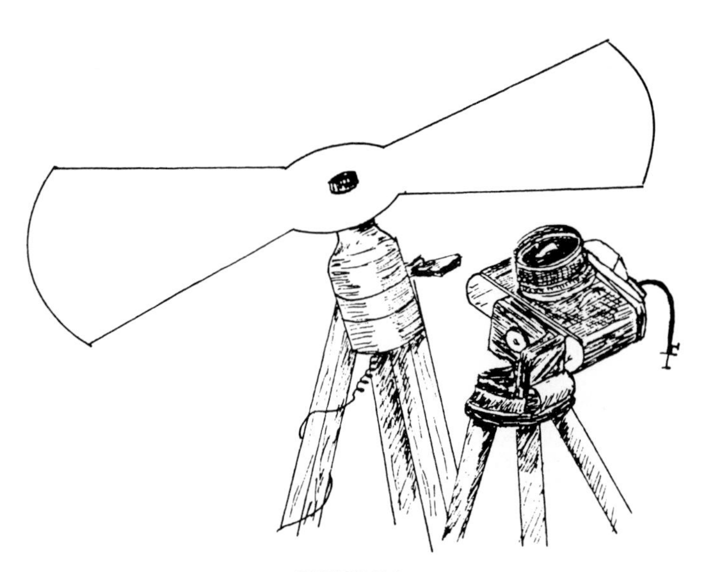

FIGURE 8.4
Principle of the rotating shutter for meteor photography.

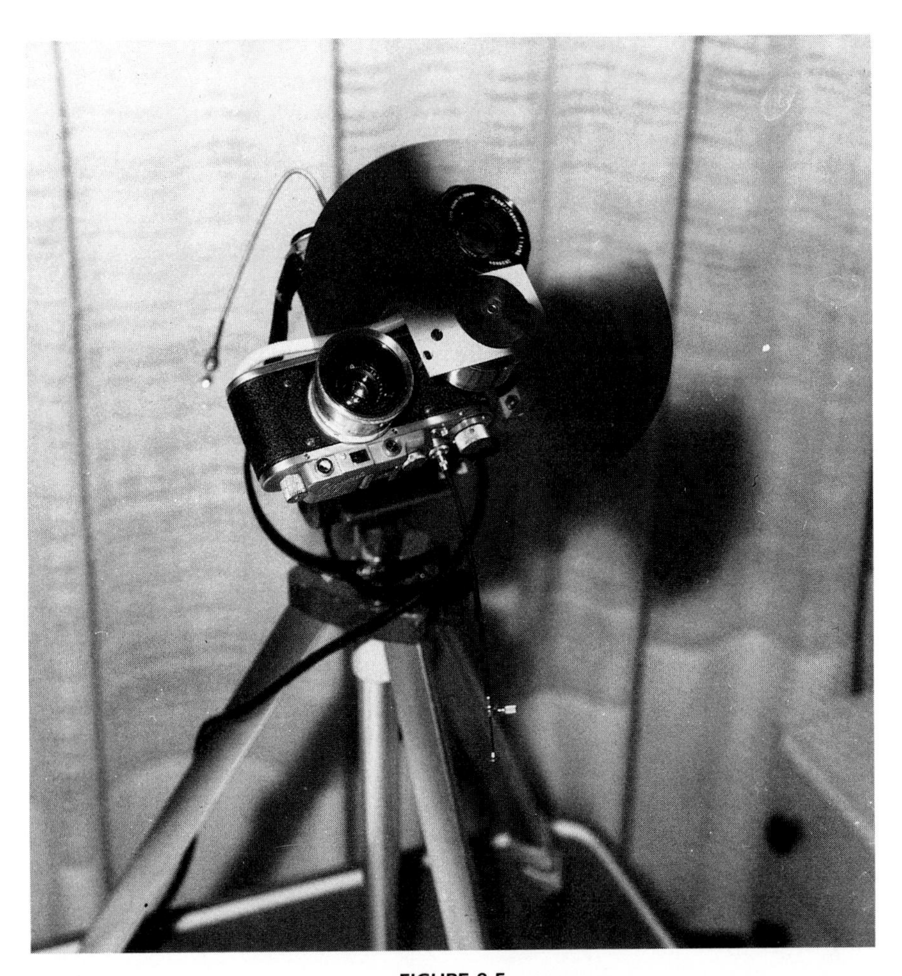

FIGURE 8.5
A rotating shutter serving two cameras, built by an Italian amateur. (Reproduced by permission of The International Meteor Organization.)

be interrupted 50 times per second. A meteor trail will, therefore, show several breaks when photographed — for example, if 15 interruptions could be counted, the meteor took 15/50 or 0.3 second to pass across this arc of sky. If the length of the trail can be calculated, this information enables the velocity of the meteor to be calculated, which is most valuable information.

Aircraft and satellites move too slowly to show an interrupted trail, although some flashing satellites, which produce a distinct light pattern, can sometimes resemble meteor trails: some satellites produce brilliant flashes that resemble a point-source meteor.

For these and other reasons, the meteor photographer has plenty to do during the course of an exposure, noting any activity occurring in the area being photographed. Even so, it is surprising how often unnoticed satellite images, or very short meteor trails, appear in the photograph! The points to record during an exposure include the time (to the nearest second) when the exposure begins and ends, and the position (quoting the constellation or reference star) of every satellite, aeroplane, and meteor brighter than magnitude 2, describing the direction of their paths across the sky.

Accuracy of timing is vital to ensure high-precision positional information for computing the trajectory. It is not good enough to notice a meteor, shine a torch on an old mechanical watch, and try to estimate to the nearest ten seconds when the meteor occurred! To get timings to a precision of a second, a remote-controlled digital clock is recommended.

Each negative should be carefully examined after development: black and white films are easier to examine than colour slides. It is not easy to trace the fainter meteors, which are usually short, weak trails covered up by the star trails, and therefore a quick examination may reveal no meteors even though some have been recorded. The visual notes made during the exposure will help, but additional unnoticed meteors are also likely to have been photographed.

Photographs showing measurable trails must be enlarged before measurement (Fig. 8.6). It is helpful to make a contact print as well. An enlargement of between 10 and 20 times is suitable. All relevant data should be noted on the back of the print.

FIGURE 8.6
A magnitude -2 Perseid photographed by the author on 17 August 1988, using Tri-X film.

METEOR ASTROMETRY

Assuming that the print has not been accidentally made with the negative reversed, that it is sharp, and that sufficient star trails, are visible, it should be possible to identify between five and seven reference stars. The closer these stars are to the trail the better, with the following provisos:

1. Do not use two or more stars whose trails are nearly collinear.
2. Do not use bright stars which have left swollen trails.
3. Avoid double stars or overlapping trails.

Once the reference stars have been identified, the beginning and end point of each trail can be measured, using the left edge of the picture as the y-axis and the bottom edge as the x-axis. Use a transparent ruler of sufficiently good quality to allow tenths of a millimetre to be estimated: for this work good illumination and a magnifying glass are essential. Mark the estimated centre of the negative with a small cross.

The measurements can be recorded on a pro-forma containing the following information:

1. Observer's name and address.
2. Date and time limits of the exposure.
3. Geographical coordinates (accurate to within about 30 metres) of the observing site.
4. Details of camera, lens, film, and development.
5. Information about the meteor: time of appearance, magnitude, general description, probable radiant.
6. An estimate of the field centre on the negative.
7. The x and y coordinates of the reference stars. For example:

Star name	Begin		End		
	x	y	x	y	
δ Cas	59.7	108.7	64.8	110.9	(in mm)

7. Similar coordinates for the meteor trail.

A single photograph of a meteor is not very useful, but when the same meteor is photographed from another location at least 30 km away it is possible to determine the true path of the meteoroid through the atmosphere, its length, the heights of the beginning and end points and their geocentric position, and the radiant. To describe the full method would require as much space as this chapter, but readers with some mathematical knowledge will easily understand the principle involved. Each observing site represents a point from which the directions of the beginning and end points were seen, and these directions can be considered as directional vectors defining a plane through the observing site and the meteor trail. Another plane passing through the other observing site can also be defined. The intersection

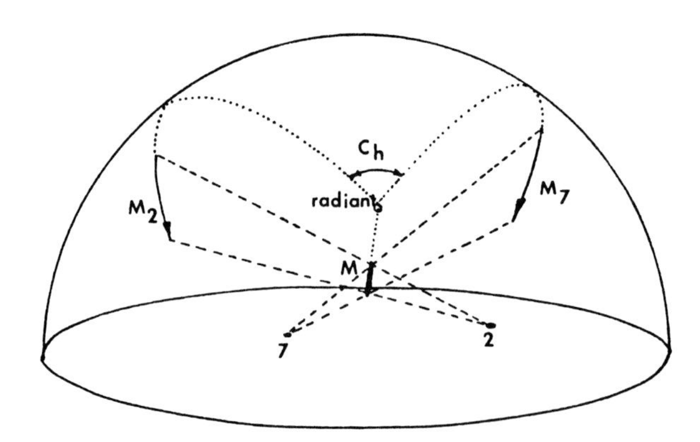

FIGURE 8.7
Two-station meteor observation. The same meteor is seen projected on to a different sky background.

between these planes contains the meteor trail, which is defined by the directional vectors as derived from both observing sites. The directional vector on the intersection line defines the intersecting line radiant of the meteor. (See Fig. 8.7.)

When a meteor is photographed at two stations, at least one of which used a rotating shutter, the velocity of the meteor can be derived. Subsequently the heliocentric orbit can be computed, and the orbital elements will be a new contribution to our knowledge of these small solar-system particles.

Many amateur meteor photographs have remained in the possession of astronomical societies or individuals, which means that the astrometric information has never been made available. To avoid such a waste of effort and potentially useful information, the IMO collects as many photographs as possible and stores the data in a Photographic Meteor Database. The following applications of such data are possible:

1. Statistical research into radiant positions, diameters, and drift.
2. Simulations and computations of individual meteor paths, which may lead to further photographs of a particular meteor being located. Meteor stream associations can be investigated.
3. Searches for double-station meteor photographs.
4. The computation of the atmospheric trajectories of meteors.
5. Heliocentric orbital calculations.
6. Searches among the orbital data for evidence of meteor streams.

All meteor photographers are strongly advised to send all their photographs showing meteor trails, with complete documentation, to an appropriate national or international organization to ensure that they are put to the best possible use.

RADIO OBSERVATIONS OF METEORS

When a meteoroid enters the atmosphere, the particle will be completely ionized at an altitude of between 120 km and 80 km, and the ionized trail will diffuse away in a slightly longer time than the visual trail takes to disappear. This trail will reflect radio waves between two sites provided the meteor trail touches an ellipsoid of revolution with the radio transmitter and receiver at the foci. Amateur radio observations usually use a remote commercial broadcasting station to produce an effect known as *forward scatter*. Professional researchers, using transmitters and receivers operating at the same site, use *back scatter*.

Radio observation has the advantage that it can be undertaken at any time of day. Some radiants rise only during daylight hours, and are therefore undetectable by the night-time observer. These showers were, in fact, discovered by radio techniques. The disadvantage of this method is that it only permits counts of overall meteor activity to be made, so that the contribution of each shower to the total activity is not known.

Although radio observations can usually be made without difficulty, interference can occasionally occur. Sporadic E — strongly ionized clouds formed by winds at a height of 100 km — is a local phenomenon lasting from a few minutes to several hours. Ionospheric disturbances caused by unusual solar activity can also affect observations. Aeroplanes can also reflect radio waves, although the type of reflection is quite different from ordinary meteor reflections, and can therefore be recognized. No observations are possible during ionospheric disturbances.

To get started in meteor radio work, an ordinary Yagi antenna and an FM tuner with a good audio amplifier should serve very well. The antenna consists of a dipole, one reflector, and two or more directors, mounted on a spar. The length of the dipole is equal to about half the wavelength being received (Fig. 8.8). The

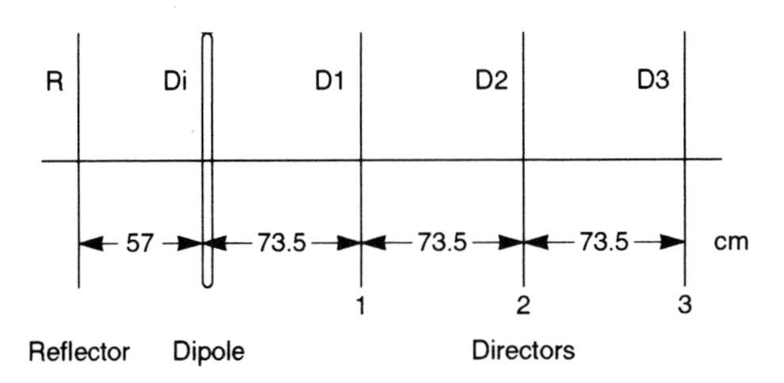

FIGURE 8.8
Diagram of a Yagi antenna for meteor work. The indicated spacings in centimetres are correct for use at 70.3MHz. The lengths of the components are given in the text.

recommended antenna for use at 70.3 MHz would have the following lengths:

Reflector	217 cm
Dipole	197 cm
Director 1	188 cm
Director 2	185.5 cm
Director 3	183 cm

Aluminium is the best material to use, but it is subject to corrosion.

Catalogues of frequencies are available, and a station may be chosen which will fulfil the reflection conditions mentioned above. However, since considerable technical insights are needed when setting up such a station, the reader who is not familiar with the problems involved will need further information. This can be obtained from the IMO, and probably from other organizations as well.

Assuming that the antenna is correctly set up, the background noise of the receiver will suddenly be replaced by fragments of a recognizable signal when a meteor trail reflects transmissions from the distant station. Some reflections are louder than others, and the record can therefore include the strength as well as the duration of the signal. Few reflections will be heard during a period of sporadic activity, but the hourly rate over a 24-hour period will show a distinct diurnal variation, with the lowest rates at 6 p.m. (1800 hours) local time and the greatest activity at 6 a.m. (0600 hours). As soon as a meteor stream begins to contribute to this activity this daily sporadic variation will be disturbed, and higher rates will of course be recorded. The number of events will depend, in addition to the true meteor activity, upon the geometry of the two stations.

The observing record must include relevant technical details about the equipment, the transmitter frequency, the geographical coordinates of the observing site and the elevation and azimuth of the antenna. The time of each reflection to the nearest second of UT must be noted, together with its maximum strength using the following code:

1 = weak, mixed with background noise
2 = weak, but clearer
3 = average
4 = strong
5 = very strong

The strength estimate is rather subjective, and may differ from one observer to another. The duration of the signal can be estimated in tenths of a second, and practice with a chronometer will be very helpful.

Several types of reflection can be distinguished. Very short reflections or 'pings' can be registered with sensitive equipment, while ordinary receivers will only detect 'bursts' or longer reflections. Even if the reflection occurs at a moment when the transmission is silent, the background noise will still be interrupted by this silence and it will be possible to estimate the duration of the signal.

While most observers will listen with headphones and take notes while listening, it is also possible to read all the information directly into a computer or to plot the reflections with a pen recorder. Immediate analysis of the recorded data is also possible. However, while acoustic observers simply concentrate on the background noise, electronic interference becomes significant if more advanced recording techniques are used.

The most obvious way of presenting the results is as a histogram showing the number of echoes recorded per hour. If the observing period was sufficiently long, an hourly-rate histogram will show a maximum during the hour interval when the transmitter–receiver geometry was optimal for the meteor stream being observed. This in turn is affected by the position of the radiant, allowing further analysis to be made. Over a longer period, histograms representing the daily averages of meteor reflection counts will give a very rough idea of the activity profile of the meteor stream: peaks are caused by showers which produce sufficient meteors to exceed the average combined activity of the minor showers and sporadic meteors.

Radio receivers have a wide range of sensitivity, providing information on different populations in the meteoric complex. An ordinary system will record rates comparable to those achieved by the visual observer, but more sensitive equipment will reach the fainter classes of meteor and record rates 10 or 15 times higher than the visual rate. However, meteor showers that are very well known for their impressive activity in the magnitude range from -4 to $+5$ (such as the Perseids) are deficient in meteors in the fainter ranges ($+5$ to $+12$), and since sensitive radio equipment picks up large numbers of non-shower meteors at these faint levels, the extra activity of the Perseids and similar showers is unnoticeable by comparison. Instead, this equipment is valuable for detecting streams too faint to be seen by naked-eye observers, but which could be checked by telescopic observation.

OBSERVING METEORS TELESCOPICALLY

The faintest meteor magnitudes visible with the naked eye are between 5 and 6, and whoever claims to see fainter meteors than this should be viewed with suspicion. The limit of normal cameras is 0 or $+1$, and about $+4$ for Super Schmidt cameras, while CCD video techniques enable meteors as faint as magnitude 7 to be recorded. Radar observations can detect echoes that must correspond to visual meteors of about magnitude 12.

Since the brightness of a meteor is linked to its size, it follows that the particle size range being observed depends upon the sensitivity of the equipment. However, meteors fainter than magnitude 15 are not expected to occur, as the corresponding particle size would not produce enough energy to generate any light at all; these meteoroids descend slowly through the atmosphere and fall to the surface as dust.

With binoculars or a telescope it is possible to study visually the meteor population in the brighter part of the range that is so abundant in radio and radar observations. A pair of 7×50 binoculars are suitable to start with, but as with all other observing techniques careful preparation is necessary. Comfort is an important aspect, since the observer will be looking constantly through the binoculars in a fixed direction for between 10 and 30 minutes. A comfortable seat and a suitable mounting for the binoculars are necessary, but the decision about where to look for meteors depends upon what the observer intends doing.

It is possible simply to determine the general telescopic activity, deriving an observed hourly rate. However, like radio determinations of rate, the results are only comparable with other estimates made using similar equipment. Nevertheless, changes of activity can be recorded. Although the real field of view is restricted, this is compensated to some extent by the greater number of faint meteors — on average the factor is about 3 per magnitude.

The plotting of meteor trails is more useful. The small field of view is now an advantage, since the observer can concentrate on a small region of the sky: if the field is copied from a star map with a suitable limiting magnitude, the trails can be plotted with much smaller errors than are usual with naked-eye work, and radiants can be determined with some precision. Assuming that a known shower is being observed, the telescopic observer should select two fields, each one about 25° away from the radiant, and forming an angle of 90° with the radiant.* *Uranometria 2000.0* [2] is a suitable atlas to use as a basis for telescopic charts. Both fields should be studied before commencing the watch so that they become familiar, and magnitude estimates of some reference stars must be made.

Each field should be watched alternately for about 30 minutes throughout the observing session. When a meteor is seen, remain looking at the field and memorize the position of the train relative to the reference stars before withdrawing the eye and plotting the meteor on the chart. Record the time, the appearance of the meteor, and the magnitude. The following data should be noted with respect to the watch as a whole:

1. Name and address of observer.
2. Place and geographical coordinates of the observing site.
3. Instrument, eyepiece, field diameter, and any other relevant information.
4. Date and time of beginning and ending of the watch.
5. Sky conditions, limiting magnitude, moonlight.

The improved plotting accuracy enhances the precision with which a radiant can be determined, and when a statistically adequate number of telescopic meteors have been observed the daily motion of the radiant may be observed. It may even be possible to resolve the radiant structure, with subcentres and branches of showers that diverge from the main stream. Mapping radiant positions, sizes, and motion

*I once heard the suggestion that attempts should be made to select fields containing a flare star, which would then be monitored, as a bonus, during the watch! — Ed.

requires many reliable observations, and these data are still lacking for the majority of meteor showers. Anyone volunteering for this field of observation has a wealth of work awaiting them.

Telescopically, the sporadic hourly rate is difficult to standardize to a uniform level, since so many factors are involved, each one of which is subject to some degree of uncertainty, that a telescopic Zenithal Hourly Rate is not easy to determine. However, relative activity levels may be derived to complement radio and naked-eye data. In this way the overall picture of a meteor stream may be completed down to the smallest particles it contains.

VIDEO-CCD METEOR OBSERVATIONS

A few years ago, video manufacturers introduced a very sensitive camera for use at night. At that time it was priced beyond the range of amateur astronomers, but several enthusiasts have now built such cameras from second-hand components, although they are still expensive. The main components of a system suitable for meteor work are:

1. A fast objective — for example, a 50 mm f/0.85 lens.
2. A microchannel plate image intensifier.
3. A black and white CCD video camera and recorder.

The components must be properly assembled in a tube and mounted equatorially. Beware of image intensifiers, which can have an internal voltage as high as 7000 V!

Experiments with this type of equipment were very successful, the limiting magnitude for stars being 8.5 to 9, while meteors as faint as magnitude 7.5 were recorded. The test camera had a field of 17° and recorded rates of 30 meteors per hour, while 56 hours of observation during the 1988 Perseid shower, from 5 to 17 August, yielded 2600 meteors on videotape. However, investigating all these images would be very time-consuming, and the current hope is to develop digitizing techniques that will measure recorded meteor positions automatically. This method could also be applied to photographic prints. In this field, amateurs are standing at the technological front-line.

If they can be developed to fulfil their potential, video techniques will rapidly provide all the positional information required about radiants, and computers could derive statistical data about activity levels, from which a 'video ZHR' could be obtained. Human uncertainty would be excluded — however, technical shortcomings or failures would always be possible. Visual work will not be made useless, since very few people will be able to operate such equipment, and visual and video results should be calibrated with each other to facilitate comparison with visual archives of past shower activity. Nevertheless, CCD-video meteor observations are likely to grow into the most productive source of information on meteor streams in the 1990s. [See Chapter 17 — Ed.]

RECORDING FIREBALLS

The most impressive events visible in the night sky are bolides or fireballs. A fireball is a meteor of at least magnitude −4 in brightness (about the brightness of Venus). Bolides are fireballs of magnitude −8 or brighter.* It is unfortunate that witnesses of these spectacular events are usually unable to provide any relevant information, since these meteors are of great scientific interest. A magnitude −4 meteor is only as large as a small stone, but the more brilliant bolides are metre-size bodies, and may end in a meteoric impact if a fragment survives the journey through the atmosphere.

Some fireballs are even bright enough to be seen in full daylight, and on 10 August 1972 a body with an estimated diameter of several metres was observed leaving a trail in the atmosphere and then escaping back into space. The IMO runs a Fireball Data Center (FIDAC) where reports on fireballs from all over the world are collected and stored in a database.

If the reader observes a fireball, or interviews someone who has, the following information should be collected:

1. Date and time of the event, in UT.
2. Information concerning colour, train, fragmentation, and sound.
3. Location and coordinates of the observing site.
4. Position of the trail in equatorial or horizontal coordinates.
5. Any sound impressions, using comparisons such as snow-slip, supersonic bang, rustle, etc., and noting the time lapse between the visual observation and the sound being detected.

CONCLUSION

Since meteors are caused by the smaller bodies of the solar system, they did not receive serious scientific interest until the 1950s, when meteoroids were seen as a potential hazard for space travel. As soon as the danger was proved negligible, professional studies in the Western world were almost totally abandoned. However, although these studies were brief, they did leave many insights into the field of meteor astronomy, showing very clearly that the meteor complex is subject to (astronomically speaking) extremely rapid evolution and unpredictable events. The only way of studying this adequately is through a long-term worldwide effort by amateurs in close cooperation with professional astronomers. The IMO was formed by amateurs and professionals from 30 countries, and everyone is welcome to join this global association.

About the Author

Born in Belgium in 1958, his first attempts at astronomical observations date back to 1972. He started to specialize in meteors in 1975, and since 1978 he has

* Some observers reserve the term 'bolide' for fireballs producing an impression of noise, which may be due to electromagnetic effects caused by an unusually bright object — Ed.

been in contact with most meteor observers around the world. He has amassed well over a thousand hours of effective observing, as well as a good collection of photographs, and to date has recorded 30 000 meteors.

With an intense international correspondence, and being a very regular participant at international astronomical meetings, he was one of the driving forces behind the establishment of the International Meteor Organization, of which he is Secretary-General. A systematic literature search between 1983 and 1987 has led to the Bibliographic Catalogue of Meteors, now also computerized. His interests also include variable-star observing and astrophysics.

Contact address – International Meteor Organization, Pijnboomstraat 25, B-2800, Mechelen, Belgium.

References

1. Roggemans, P. (ed.), *Handbook for Visual Meteor Observations*. Sky Publishing Corporation, 1989.
2. Tirion, W. *et al.*, *Uranometria 2000.0*, in 2 vols. Sky Publishing Corporation, 1987–88.

Bibliography

Bronshten, VA., *Physics of Meteoric Phenomena*. Reidel, 1983.
Halliday, I. and McIntosh, B.A., *Solid Particles in the Solar System*. Reidel, 1980.
Hawkins, G.S., *Meteors, Comets and Meteorites*. McGraw-Hill, 1964.
Hemenway, C.L., Millman, P.M. and Cook, A.F., *Evolutionary and Physical Properties of Meteroids*. NASA, 1973.
Kresak, L. and Millman, P.M. (eds), *Physics and Dynamics of Meteors*. Reidel, 1968.
Lovell, A.C.B., *Meteor Astronomy*. Clarendon Press, 1954.
McKinley, D.W.R., *Meteor Science and Engineering*. McGraw-Hill, 1961.
Moore, P. and Mason, J., *The Return of Halley's Comet*. Warner Books, 1983.

Notes and Comments

When I received the first draft of this chapter, I invited Neil Bone, of Chichester, Sussex, England, to read it. The points he raised have led to further elaboration by the author, and some of these comments follow. — Ed.

Wide-angle versus normal lenses – Bone: Wide-angle lenses do have a place in meteor photography. In my experience, under comparable conditions during the Perseids of 1982 and 1983, I found that a 29 mm f/2.8 wide-angle lens had a superior (about 2–3 fold) capture rate relative to a 50-mm f/1.8 standard lens. It should be mentioned, however, that wide-angle lenses do introduce greater distortion at the edge of the field, so that such photographs are less useful for the astrometric work described later.

Roggemans: Our experience is about the opposite! Under very transparent skies our f/1.4 50 mm lenses photograph about three times as many meteors as wide-angle optics, and I wonder if the generally poor skies in the United Kingdom explain the difference?

Tripods – Bone: From personal experience, I would suggest putting the camera and tripod as high above the ground as possible, to reduce dew condensation on the lens.

Roggemans: We work with compact, stable tripods or camera batteries. Tripods set to their maximum height are very impractical to work with. [I wonder if the European sites are less dew-prone than their United Kingdom counterparts, so that the problem is less? — Ed.]

Meteor spectroscopy – Bone: Spectroscopic photography is still regarded as a principal area in which the amateur can make useful contributions. The method of preference seems to be to use an objective prism of about 30° angle, which restricts one to the capture of the spectra of fireballs that are brighter than about magnitude −6.

Roggemans: Spectroscopic photography would indeed be useful, but so far it has proved very unsuccessful. Despite enormous efforts, few amateurs have so far captured any spectra.

Meteor trains – Bone: Regarding visual work, the recording of persistent train phenomena can be useful. Coupled with magnitude data, such records provide further information on particle composition. The eye has the advantage, in such observations, over video methods — the latter tend to suffer from image persistence, so that every recorded meteor has an apparent train!

Roggemans: I agree that persistent train observations are useful. These are normally noted under 'remarks' on the observing form.

Tape recorders – Bone: I should caution that the successful use of tape recorders is dependent upon good batteries and efficient running. Several observers (including some very experienced ones!) have lost data to flat batteries or tape failure. Especially in cold or damp conditions, even modern tape recorders can malfunction!

Roggemans: I began using a tape recorder in 1980, and since 1985 I have used a small dictaphone which works on two small batteries. I replace these once a year, although they were never totally run down even after 30 nights of service and the recording of about 6000 meteors. In over 1000 hours and well over 20 000 meteors I have only had one failure, which happened when I had pressed the 'pause' button by mistake and lost 30 minutes' worth of data.

Zenithal hourly rate – The following table, based on a formula by J.P.M. Prentice, gives the factor *F* by which the observed rate should be multiplied for different radiant altitudes (A). This table assumes excellent conditions, and in practice F will probably be considerably greater than this, particularly at very low radiant altitudes. — Ed.

A°	0		3		9		15		21		27		35		43		52		66		90
F		10.0		5.0		3.3		2.5		2.0		1.7		1.4		1.25		1.1		1.0	

In critical cases ascend.

CHAPTER NINE · *Comets and Comet Hunting*

DAVID H. LEVY
TUCSON, ARIZONA, USA

HUNTING: A TAIL OF TWO COMETS

'YOU REALLY NEED TO SEE THIS PARTICULAR ECLIPSE', STEVE EDBERG told me one bright morning in 1987. We were in Acapulco, Mexico, getting a first look at the supernova which had just appeared in the Large Magellanic Cloud. Since the total solar eclipse to which he referred was a whole year away, I was not thinking of it too seriously. More importantly, after three days of supernova discussion and eclipse planning, my thoughts were turning back to my first love. 'The trouble with solar eclipses', I offered, 'is that they take place at New Moon, right at the height of comet hunting. Now, why couldn't an eclipse take place at a more convenient time, say two or three days after first quarter? Then all our comet hunting for the lunation would be done, and we could enjoy it.'

That stopped the conversation. In any event, I decided not to go to the eclipse, even though the chances that I would find a comet on the few mornings around New Moon were very low indeed. At the time of our conversation I had only two comet discoveries, 1984t and 1987a (the twentieth comet to be found in 1984, the first for 1987), and six months later I would find 1987y.

When the lunation of March 1988 began — comet hunters define lunations as lasting from one Full Moon to the next — I began by trying to get as much comet hunting as possible done in the hour or two before moonrise. The Full Moon acts as an umpire in the comet-hunting game; for up to two weeks, the sky it has illuminated may have hidden a brightening comet. Now that the Moon was past full, comet hunters the world over would take advantage of the brief hour or two of dark sky in what one person calls a scramble for comets.

My evening comet hunt involves two patterns. One is a 30–40° azimuth sweep near the horizon near the end of twilight, followed by swinging the telescope back to the start point and starting over. This way I take advantage of the Earth's motion to keep the telescope roughly at the same distance above the horizon, and also maintain about 1/3 field of overlap. Unfortunately this method does not allow much checking time for an object before it sets; thus I will sometimes search in strips of altitude, moving down, then up, in a zig-zag pattern. Either approach offers a chance of spotting a comet: the key is not so much how you hunt but the intensity of the searching.

As the Moon approached New in that March 1988 lunation, I began hunting in the eastern sky before dawn. Here the azimuthal sweep was done in reverse, starting at a relatively high altitude three hours before dawn in a pattern that would ideally reach the horizon at dawn. I conducted this sweep with a 150-mm reflector, whose 2° field enabled me to cover a lot of sky quickly in hopes of finding a comet brighter than the 9th magnitude. Then I began with a 400-mm telescope, working in altitude sweeps.

I had long forgotten my discussion with Steve about the March 20 eclipse when I came across what appeared to be an 11.5 magnitude galaxy. In fact, through the 400-mm telescope, most of the 'faint fuzzy' objects I see are easily identifiable. The spiral galaxies appear symmetrical and are elongated; the globular clusters are resolvable; many of the planetary nebulae have very sharp edges. A comet usually does not look like any of these things. It may be elongated, but usually only in the direction away from the Sun. It is usually diffuse, with ill-defined edges. Someone just starting will have to check the identity of every object, but with growing experience a searcher comes to recognize the identity of many faint fuzzies simply by what they tell you about their appearance.

There are exceptions, and my suspect of March 19 was one of them. It looked almost definitely like an elongated galaxy, but the slightest doubt in my mind made me sketch its position. Since dawn was already present, I did not have a chance to check for possible motion.

The question is often asked: At what point does a comet hunter announce a discovery? The answer, of course, is not until the hunter is absolutely certain. A report of M31 as a comet will hardly enhance a searcher's reputation. Not finding the object on a star atlas is not good enough: unless the atlas is the *Palomar Observatory Sky Survey*, it is always possible that it may have omitted an object. (The photographs in the excellent *Atlas Stellarum* are published in contrasty form, so that most deep-sky objects do not appear at all.) My rule is to wait until the object shows definite motion. Even then, it is possible that I could have picked up an already-known comet. A check of recent literature (see below) showed that no known comet was in that position. I decided to wait a night.

In the meantime, the Moon's shadow swung across the Earth, and thousands saw the total eclipse of the Sun that my commitment to comet hunting had caused me to miss. The following morning I turned the 400-mm telescope to the position of the suspect. There was nothing there. I moved the telescope slowly around. Half a degree to the north was a fuzzy object, this time with a very strongly condensed central core. This new sighting looked even more like a galaxy than the previous one: the two sightings did not resemble each other in the slightest. Could I have been that wrong in the position? Thoroughly confused, I decided to wait yet another day.

With mounting anxiety, on the third morning I set the telescope on the first night's field as soon as it cleared the horizon. Still there was nothing there. The real surprise came when I moved to the second field: the galaxy's central core was still there, but the rest of the 'galaxy' was not! Finally my slowly working brain realized what had been going on: there was indeed a comet, and the second night it was passing directly in front of a star. Carefully I moved the telescope north one-half degree. There the comet was sitting. I now had enough time to detect some motion in the following hour, and then I reported it to Dr Brian Marsden of the IAU's Central Bureau for Astronomical Telegrams.

A correctly reported discovery should have as much information about the comet as possible. My report went something like this:

David Levy, Tucson, Arizona, reports his discovery of a comet. Positions are available as follows:

	RA (1950.0)	Dec	
Mar 19 12.00 UT	21 h 30 m	+16°.2	M1 = 11.5*
Mar 20 11.45	21 32.6	+16° 48′	
Mar 21 11.30	21 35.2	+17° 30′	

Comet diffuse, coma diameter 1′, tail 1.5′ in p.a. about 240. (D. Levy, Tucson, Arizona, 400-mm reflector.)

Notice how the report emphasizes the positions, providing enough so that people trying to see the comet in these early days will have an idea where to look. It is always better to report more than one position, although a minimum requirement is a position plus the direction and rate of motion. The magnitude estimate of 11.5 is provided for the first night. The comet was subsequently announced as Comet Levy (1988e).

The main thing necessary after a discovery is the obtaining of 'precise positions' to that an orbit can be calculated. These positions are obtained either photographically or with charged-coupled devices (CCDs). At the time of discovery, Gene and Carolyn Shoemaker and Henry Holt were observing using the venerable

*Comet magnitudes are given either as M1 (total magnitude) or M2 (the magnitude of the central condensation). — Ed.

400-mm Schmidt camera at Palomar Mountain. Built in the 1930s, largely by the famous telescope maker Russell Porter, this telescope has been responsible for all of the Shoemaker comet discoveries as a result of their Palomar Asteroid and Comet Survey.

By August 1989, the Shoemaker survey had produced 17 new comets, proving that a Schmidt camera is an ideal instrument for such work. Using hypered 4415 sheet film, the camera will record a 6° field down to fainter than the 18th magnitude in 8 minutes. However, like visual searching, the secret of success in a photographic patrol lies in taking as many exposures as humanly possible, followed by thorough scanning of each field immediately after the films have been processed.

The Shoemaker procedure involves taking four exposures, usually of 6 minutes each, and then repeating the sequence. Each pair of films is then examined under a specially designed instrument that allows the pair to be studied stereoscopically. This way, a moving comet or asteroid might appear to 'jump out' from the background.

In addition to their survey fields, the Shoemakers often include regions containing already discovered comets and asteroids for the purpose of obtaining precise positions for measurements. The act of measuring these positions is known as astrometry. It turned out that the Shoemakers obtained the first 'astrometric positions' of Comet 1988e using the Schmidt camera.

There is no reason why amateurs equipped with Schmidt cameras cannot make observations in this way. Unfortunately, the scale of 200-mm and smaller Schmidt cameras may not be sufficient to produce very accurate positions, although instruments of this size have been used for the vital positions required for a comet just after its discovery.

In May of 1988 the Shoemakers again put Comet 1988e on their observing list. Their observing preparation involved noting the position of each survey field on a chart. Since by this time Comet 1988e was far to the north of the series of overlapping circles that outlined their observing plan, they marked its position in a circle at the top of their observing chart, with a note that the comet was north of that circle.

Five nights later, the comet field was finally photographed, and the following evening Carolyn examined the films. A comet image was off to the side, more than a magnitude brighter than they had expected. Carolyn quickly realized that something was seriously wrong when she could not identify the stars in the field that had been photographed. It turned out that at the end of an exhausting week of observing they had pointed the telescope, not to the comet's actual position, but to the position of the plotted circle, forgetting that the comet was actually almost 20° further north!

That explained the incorrect field, but what was the comet they had recorded? It turned out to be a new discovery. The comet was announced as Comet Shoemaker–Holt (1988g). Although this comet was found quite by accident, it was Carolyn's care in scanning that revealed its presence. Astrometric positions were obtained that very night.

When a preliminary orbit was calculated, the really unlikely conclusion of this story was reached. At the Harvard–Smithsonian Centre for Astrophysics, Conrad Bardwell noted that the orbits of Comets 1988e and g appeared to be close to identical, except that g was following e by 76 days!

A pair of comets in the same orbit? Although rare, it was not unheard of: a series known as the 'Kreutz sungrazers' share the same orbit. They are remnants of earlier comets that broke up as they made hairpin turns around the Sun. The group was first identified by Kreutz late in the nineteenth century. Also, occasional periodic comets have been seen to split, the most notable example being the periodic Comet Biela, which divided into two pieces in 1846, returned as two comets in 1852, and was never seen again. Instead, a major meteor storm known as the Andromedids appeared in 1872, the meteors being related to this comet.

The interesting thing about the comet pair Levy and Shoemaker–Holt is that their orbit never carries them closer to the Sun than the Earth is. Thus, the stress of solar encounter should not have been that great. Is it possible that the two comets split far from perihelion, perhaps after encountering an object out there? Brian Marsden suspects that the two comets may have split at their last perihelion two orbits in the past, when they rounded the Sun some 12 000 years ago. It is also possible that after the split the comets stayed together, perhaps orbiting each other for a portion of one orbit, before starting to move apart.

The story of these two comets represents an interesting series of coincidences that illustrate how comets really are found, the story of an exception that proves the rule. Had I gone eclipse-chasing, it is possible that 1988e, faint and at the eastern horizon before dawn, might not have been found. In that case it is possible that 1988g might not have been found either! Still, both comets were discovered in systematic searches that provide examples of how one might search for comets visually or photographically.

VISUAL COMET HUNTING NOTES

No matter what the method, persistence in searching is the best guarantee of success. When planning a comet hunt, think of the following factors.

Field of view – The more sky you cover, the greater your chance of discovering a comet. Unfortunately, if you cover too much sky at once, you might sweep right over a faint comet. Thus, you need a balance between power and area coverage. My 150-mm reflector covers more than 2°, but it would never have picked up 1988e. With its 1° field, the 400-mm reflector did have the light-gathering power to spot the comet, but comet hunting proceeds about five times more slowly.

You should try for a minimum of a 1° field. Although higher powers and smaller fields might let you see fainter objects, I do not think the exchange is worth all the extra comet-hunting time. A wider field might snatch a bright comet, but will provide you with too superficial sky coverage for efficient sweeping. I recommend a power of about $\times 50$ as a reasonable magnification for spotting comets. Moreover, higher power does not necessarily make a diffuse comet stand out better among the background stars. Faint comets are known for their low surface brightness; that is, a 10th-magnitude comet spread out over a fifth of a degree will be very hard to spot in a high-power eyepiece.

Mounting – Although most observers use altazimuth mounts to coincide with their horizons, there is no real reason why altazimuth mounts are better than equatorial ones. For his four discoveries between 1978 and 1984, Canadian amateur Rolf Meier used a 400-mm reflector mounted on an equatorial mount; he swept along arcs of right ascension and declination.

Overlap – Although Meier claims that overlapping is not necessary when using a good-quality equatorially mounted telescope, most observers overlap fields. Because the performance of many telescopes deteriorates near the edges of a field, and because searchers tend to concentrate on a field's centre, overlapping by about a third of a field is usually a good idea.

Frequency of search – A regular, nightly, no-excuses patrol programme is far more likely to produce a comet than occasional long nights at the eyepiece: 2 hours on each possible clear and moonless night is a better plan than a single 10-hour marathon. Since competition is heavy, consistent searching is the key to success. This can involve some personal sacrifice: I cut short a very enjoyable date one night when the sky suddenly cleared, went home, and discovered my first comet.

Reporting discoveries – The telegram or telex address of the IAU's Central Bureau for Astronomical Telegrams is TWX 710-320-6842 ASTROGRAM CAM. There also are three computer addresses, if you have access to one of three major networks: EASYLINK 62794505, MARSDEN or GREEN@CFA.BITNET, and MARSDEN or GREEN@CFAPS2.SPAN.

The IAU *Circulars* are available as index-card-sized announcements that keep observers advised of recent comet discoveries and recoveries, as well as other important astronomical observations. If you are serious about comets, you should subscribe to these circulars, for they also contain recent comet magnitude estimates made by visual observers. These circulars are also available via modem, in a programme that also allows you to report discoveries and other observations directly to the Bureau. For information about the cost, write to the Bureau at the Smithsonian Astrophysical Observatory, Cambridge, MA 02138, USA. [See Chapter 18 for further information about communicating discoveries and receiving news. — Ed]

Is comet-hunting a science? – That is a matter for some debate. Obviously, if no one looked for comets, few would be discovered and our knowledge of them would suffer. However, to spend your life searching for a comet, and then find one three hours before someone else, is a pursuit with questionable scientific benefit. On the other hand, your faint comet may never have turned up if you did not find it.

There is absolutely no question that comet hunting is a marvellous pastime, and a good sport with lots of competition. It is also a chance to study the sky on rather personal terms: you never know what the next field will bring.

OBSERVING COMETS

Of the many types of objects that we can observe, comets are among the most challenging. They share the uncertainties of moving objects like planets and asteroids: before we observe them we have to find where they are at a given time. They also share the changes in magnitude that make variable stars so appealing. Also, they change in physical appearance, sometimes from one night to the next.

Locating known comets – Finding which comets are visible can be difficult, for not all publications list all comets. *Astronomy* and *Sky & Telescope* magazines publish ephemerides for comets likely to be visible through binoculars and small telescopes. However, many interesting comets never get brighter than 10th magnitude and might miss being referred to in these sources. The *International Comet Quarterly Comet Handbook* is an annual publication that lists the positions of all comets expected to be visible in a particular year, with predictions of their magnitudes. This book encourages observers to report their observations to the *International Comet Quarterly*, which publishes them along with interesting news and research about comets. Edited by Daniel Green, the ICQ can be reached at 60 Garden Street, Cambridge, MA 02138, USA.

Annual predictions for periodic comets are also published in the *Handbook* of the BAA. Their selection includes all comets that are likely to be visible with 400-mm or smaller telescopes. Remember that a magnitude prediction of 15 might be based on photographic observations from the comet's last return. Since visual values are as much as two magnitudes brighter, the 15th-magnitude comet might turn out as 13th through your telescope.

Another reason to be sceptical of a comet's predicted magnitude is that the value does not necessarily show how easily visible a comet might be: the more diffuse it is, the harder it will be to spot.

Based on these predictions, reports of other observers, the quality of your observing site, and the size of your telescope, decide which comets to observe. If the predictions are to equinox 2000, you need to use an atlas designed around the same equinox, or precess the positions. Partly for that reason, the Sky Atlas 2000.0 by Wil Tirion is very useful for comet observers.

With plotted positions, try to find the comet, making sure that the object you see is indeed the comet and not some nearby galaxy. During the course of your observation the object should move slightly; if it does not, it may be a galaxy or other fuzzy object, and not the comet.

Recently discovered comets – What if the known comet you are trying to find has just been discovered? It is a fact that the less that is known about a comet, the more useful are the visual observations. It is also a fact that freshly discovered comets might be hard to find, since little is known about their orbits.

Any announcement of a new comet is also a request for observations. You should estimate the magnitude and provide as good positions as you can, based on the star images in an atlas such as *Stellarum*. When reporting positions, include the equinox (1950 or 2000) of your atlas. Approximate positions are useful in the first week or two after an announcement. By that time precise astrometric positions have usually been obtained.

Why not try astrometric measurements yourself? The work begins with a good photograph on a field wide enough to show both the comet and a collection of stars whose positions are listed in the *Smithsonian Astrophysical Observatory Catalog*. The reductions that must be done later are described in Chapter 16 by Brian Manning, who has done much astrometric work with comets and asteroids. Though time-consuming, astrometry is a highly productive activity for amateurs interested in observing newly found comets.

When at least three astrometric positions are submitted, the minimum set of orbital elements are calculated, to be refined later with more positions. These include the comet's perihelion date (T), the eccentricity (e, usually assumed at first to be 1, or parabolic), the closest distance to the Sun (q), the inclination to the ecliptic (i), the longitude of the ascending node (Ω), and the argument of perihelion (ω). From these elements the comet's future course can be calculated.

The longitude of the orbit's ascending node, the point where the orbit of the comet intersects the ecliptic, is measured from the vernal equinox (First Point of Aries) counterclockwise to the node. The argument of perihelion is measured in the plane of the comet's orbit, from the ascending node to the point of perihelion.

Physical appearance – The complex physical appearance of a comet is challenging to record. Not only is every comet different from every other comet, but also a comet changes its appearance in response to many factors. As it approaches the Sun, it should brighten as solar energy makes its gases more active. The Sun also will affect a comet's dust output: in addition to a curved dust tail, some comets even have jets of material that come from their nuclei. If a comet is making its first appearance from the hypothetical Oort cloud of comets located 40 000–50 000 AU (almost a full light year) from the Sun, it may show unusually rapid brightening farther from the Sun, with less brightening as it approaches periheion; this is why Comet Kohoutek in 1974 did not fulfil its 'comet of the century' predictions. [Comet Austin, another 'young' comet expected to put up a fine show at Easter 1990, was also several magnitudes fainter than the original predictions. — Ed.] As a comet approaches the Earth, the geometry by which we view it changes rapidly, thereby causing rapid changes in apparent brightness and appearance. Periodic Comet Halley did that in April 1986, showing a bright coma and fanned tail in mid-month, and a very long tail only two weeks later. Since comets have different ratios of gas and dust, the output of these materials helps determine their appearance, as well as how bright they will be. Finally, as a comet rotates it might change in brightness as well as produce dust jets as the Sun shines on the fissures of a comet's nucleus that produce these jets. With all these variables, it is no wonder that comet observing is such a challenge!

Magnitude estimates – For variable-star observers, the changing brightness of a star is an intensely interesting thing. This change is what attracts people to variable stars: they are not just there, they perform. Over many years, a number of methods have been devised to produce an accurate visual estimate of a star's brightness. If so many people put in so much energy to estimate a star — a point source of light — that varies, how much more difficult would it be to estimate the magnitude of a comet?

It is actually far more difficult. We cannot compare a comet to another fuzzy object, for there are too many differences in size and physical appearance for such an estimate to be reliable. Comparing comets to out-of-focus star images has proved a much more reliable approach, but a comet is not a point source and the estimate must be made with extra care.

In each of the following methods, we must match a coma's average brightness — not the brightness of the coma's brightest part — with those of defocused stars. The *Sidgwick* method is also known as the 'In–Out', because the comet is in focus while the stars are out of focus; the *Bobrovnikoff* method is 'Out–Out'; and the *Morris* method is 'Equal Out'.

The Sidgwick (In–Out) method has been popular since its first description by the British astronomical writer J.B. Sidgwick. Begin by choosing nearby stars of known magnitude that appear to be in the range of the comet's brightness. It is quite possible that, as you proceed, you will find that all your chosen stars are too bright or too faint and that you will have to choose other stars. Variable star charts or the AAVSO *Star Atlas* offer convenient comparison stars. A more complete source is the *Guide Star Photometric Catalog* with its list of stars from 8.5 to 15.5. Another source of photoelectric magnitudes was prepared by Arlo Landolt and can be found in the November 1973 issue of the *Astronomical Journal.*

First, memorize the 'average' brightness of the coma in focus. The next step is critical — defocus one of the comparison stars until it appears to be the same diameter as that of the comet's coma. Is it brighter or fainter than the comet? If it is brighter, use a fainter star for your second choice, and defocus it in the same manner.

The magnitude of the comet's coma — the tail is generally not included — can be determined by comparing it with those of the two stars *between* which its brightness lies. If the stars nearest to the comet's brightness are all either brighter or fainter than the comet, extrapolation will produce an unreliable magnitude estimate.

Two other methods are commonly used. An approach devised by Nicholas Bobrovnikoff, one of the twentieth century's best-known comet observers, prescribes this procedure. Defocus your telescope until the comet and surrounding stars have a similar apparent size. (Notice how the star images enlarge much faster than does that of the comet.) Then interpolate the brightness of the comet between a pair of stars brighter and fainter than the comet.

In a method developed by Charles Morris, a comet observer now living in California, you match the diameter of a defocused star with a less-defocused comet. Defocus the coma just enough to smear out the image to make it more uniform: the image will be slightly larger than the in-focus image. Memorize the image you have obtained and then match its size (not brightness) with defocused comparison stars. To do this, you will have to defocus the stars much more than the comet.

Whichever method is used, record the magnitude you derive to the nearest tenth. In most cases, observations done by several observers under equal conditions reveal a range over about half a magnitude, which is certainly good enough to provide a general indication of how bright a comet is on a given night. If you are uncertain of your result, because of clouds or other interference, put a colon (:) after the estimate.

One reason why visual magnitude estimates are useful is that a wide variety of observers around the world can observe a comet virtually at a moment's notice. This sort of coverage is rarely available with photometers or CCD images taken with professional observatory telescopes. Although visual magnitudes lack the accuracy that other approaches can achieve, they do provide a good general picture of a comet's behaviour.

Very often a comet appears brighter or fainter than predictions indicate. Often these predictions are made before the comet's pattern has been fully established. If at discovery the comet was in an outburst, or slightly fainter than average, the magnitude predictions will be off, although usually not by more than a magnitude.

Probably the faintest magnitude estimate of a comet was completed by Stephen O'Meara in January 1985. While atop Hawaii's Mauna Kea, he made the first sighting of Comet Halley at magnitude 19.6. In late 1987 and early 1988, I made a series of observations that apparently were the last of this apparition, recording a final estimate of 16.8.

A COMET'S APPEARANCE

Coma diameter – Determining a comet's magnitude is only one part of an observation; measuring the coma diameter is important too. Consider, for example, a case of two observers who report magnitudes of 11.4 and 12.5 for the same comet. With no other information, it is hard to believe which estimate is the correct value.

However, one observer might be using a darker site, or have better eyes, than the other, and report a much larger coma diameter. That information would quickly explain the magnitude discrepancy.

An easy way to measure the coma diameter is to draw the coma and surrounding stars in your telescope's field of view. Then, using an atlas such as *Stellarum,* measure the diameter in relation to the surrounding stars. A large comet can simply be allowed to move out of the eyepiece field as you measure the time elapsed between the preceding and the following edges of the coma passing the edge of the field of view. For example, if the time interval is 20 seconds, the coma diameter is 5′ arc.

Degree of condensation – As an active comet can change its appearance over several nights or weeks, the aspect that is likely to change most dramatically is the coma's condensation. If a jet of dust is erupting, or if the comet is undergoing an outburst of some kind, the initial appearance might be a sharply condensed inner coma.

A comet's degree of condensation (DC) is measured on a 10-point scale, where 0 represents a diffuse 'blob' without the slightest evidence of condensation near the centre. If the comet has a slightly condensed inner coma, you might judge the DC to be 2 or 3. DC = 6 implies a fairly strong central condensation. If the coma is so strongly condensed that it appears star-like, record DC = 9.

Tail – With some experience you will notice some sort of tail on many of the comets you watch, even the faint ones. Comet tails consist of two types of emission: gas and dust. A gas tail typically stretches out straight behind the coma, pointing away from the Sun. The dust tail, consisting of particles that move more slowly, tends to curve somewhat. Since comets have different ratios of gas and dust, the visibility of these tails varies widely. Also, gas tails are far more likely to change their appearançe quickly than are their dusty counterparts.

Two measurements describe a comet's tail: the length and the position angle. The length is measured in the same way you have just recorded the coma diameter. The position angle, best figured with the help of a detailed star atlas, is recorded in degrees, with 0° at north and 90° to the east. If the tail is a broad fan, record the position angle of each edge.

Occasionally a comet will produce a sunward-pointing feature. It is misleading to call this an 'anti-tail' because the feature is not really a tail. A comet is surrounded by a stream of particles in the plane of its orbit, and if the Earth happens to cross that plane we might see particles as a sunward feature or spike.

Jets – One of the more subtle features of Comet Halley was its production of dust jets on a regular basis. In 1835, 1910 and 1986, these jets appeared as dust erupted from the nucleus and were recorded by visual drawings (1835), photographs (1910) and CCDs (1986). Work by Zdenek Sekanina and Stephen Larson, who

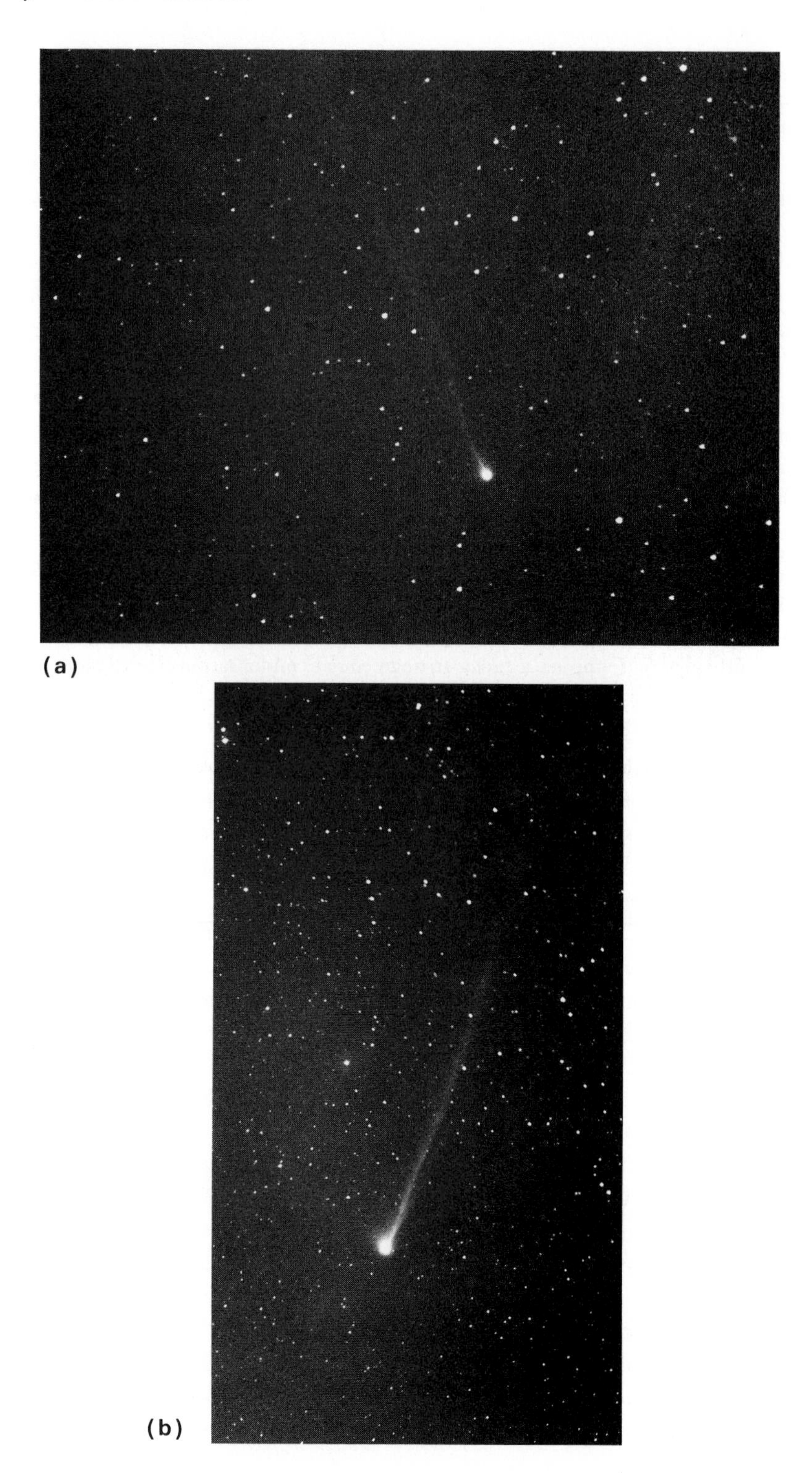
(a)
(b)

(c)

FIGURE 9.1
Three views of Comet Okazaki–Levy–Rudenko (1989r), photographed by Jonny Horne (Raeford, North Carolina) using a 200-mm Celestron Schmidt camera, exposure 5 minutes on hypered TP 2415. (a) 22 October 1989. (b) 10 November 1989. (c) 11 November 1989.

computer-enhanced many 1910 photographs, showed that these jets come from discrete sources like fissures in the comet's nucleus, and that they tend to erupt as the comet rotates and sunlight first strikes them.

If jets appear at all, they are usually subtle and difficult to see through a telescope. They might appear as short, curved features that fan out in a curved fashion, usually (though not always) less than the diameter of the coma. Although it is unlikely that you will see more than one or two jets, Comet Halley once produced at least five that could be observed visually in one night.

As the particles from jet eruptions move away from the comet, they curve outward into broad fans or hoods that might stretch to one or more coma diameters from the comet. If you see these jets, record their lengths and position angles.

Filters – To enhance a comet's appearance, observers often use filters that emphasize their bright 'C2' emissions. Sometimes known as 'Swan band' filters, these pieces of glass will enhance a comet relative to its sky background if it is sufficiently gaseous. If the filter does not affect the comet's appearance, or makes it appear to fade, then the comet is more dusty than gaseous.

Since they add an element of uncertainty to the process, do not use filters when making magnitude estimates.

PHOTOGRAPHING COMETS

Because comets are so complex in appearance, photographing them is a very useful exercise that will provide important quantitative evidence of a comet's physical appearance: rapidly-changing features like jets, gas tails and sunward spikes can be measured accurately.

A large, bright comet can be photographed through any camera and telephoto lens. Mounting such a camera on a motor-driven telescope will help to produce well-guided exposures. Active guiding, during which you use slow-motion controls to keep the comet in the centre of the field of view, will enhance the accuracy further.

Thanks to their wide fields, Schmidt cameras provide an ideal format for comet photography. After hypering a fine-grained film like Kodak TP 2415, you might try photographing a bright comet with 5- or 10-minute exposures each dark night, and note the changes that take place. If the sky is completely clear, a condition known as 'photometric', photograph some standard star field, like an open cluster or a field in Vehrenberg's *Atlas of Selected Areas*. This way, measurements of the comet's brightness can be compared with the standard field.

For any guided photography, remember to align the mounting accurately on the pole. Correcting for bad polar sighting by moving the declination slow-motion control will result in poorly guided star images at the edge of the film. Also, unless the comet is moving faster than half a degree a day, it is better to guide on the comet rather than on a nearby star; otherwise the comet image might trail.

CONCLUSION

Whether you observe them visually or photographically, over many years you will count a long series of comets: a few observers have seen more than a hundred. Observing lots of comets is important both to those who want to learn about how comets behave and to those who hunt for them. The more comets you see, the better you will be able to identify a new one.

Comets offer the elusiveness and mystery of distant deep-sky objects, yet are close enough that we can enjoy watching them change as they move past us. Periodic comets return every few years, and on a few occasions amateur astronomers have been the first to spot them: Charles Morris's recovery of Periodic Comet Faye in 1984 is an example. Whether they are bright and dramatic, or faint and diffuse, comets offer a field of observation filled with excitement and surprise.

About the Author

Although he has been interested in astronomy since 1960, when he gave a three-minute lecture on comets to his sixth-grade classmates, David Levy has never actually taken a course in astronomy. His formal undergraduate education from Acadia University had a major in English literature, and for his Masters degree from Queen's University he did a thesis on the astronomical writings of the nineteenth-century English poet Gerard Manley Hopkins.

Levy began comet-hunting in 1965, and independently discovered periodic Comet Hartley–IRAS in November 1983 just a few days after it was found. His first actual discovery came a year later with Comet Levy–Rudenko 1984t. In 1987 came two more comets, Levy 1987a and Levy 1987y, and in March 1988 Levy 1988e arrived. Levy's fifth comet, Okazaki–Levy–Rudenko 1989r, was found in August 1989, his sixth, Levy 1990c, came home the following May, and his seventh periodic comet Levy 1991g was found in June 1991. In addition Levy is part of the photographic observing team that includes Eugene and Carolyn Shoemaker; by mid 1992 this team has discovered ten comets; thus Levy's total is 17.

Levy's astronomical writings include articles on variable stars, astronomy and children, comets, telescopes, and astronomical history. He writes StarTrails, a monthly column in *Sky & Telescope* about the observing experience. For amateurs he has written *Observing Variable Stars* (Cambridge: Cambridge University Press), 1989, *The Sky: A User's Guide* (Cambridge), 1991, and a biography *Clyde Tombaugh: Discoverer of Planet Plat*, Univ of Arizona Press, 1991. His awards include the Chant Medal of the Royal Astronomical Society of Canada, the Astronomical League's Leslie L. Peltier Award, the G. Bruce Blair Award from the Western Amateur Astronomers, and asteroid 3673 Levy, named by the IAU in 1988.

Contact address – 120 William Carey St, Tucson, AZ 85747, USA.

Notes and Comments

Potential comet discoverers are committing themselves to many hours of technically fruitless searching (fruitless, that is, measured in tangible results, although the delights of deep-sky objects, meteors, the strengthening dawn and the first cheeps of birdsong as the telescope is packed away may be sufficient compensation). Therefore, decisions about the aperture, magnification, and style of sweeping likely to bring about the most rapid discovery are of more than academic interest.

It is interesting that in the late fifties and early sixties, when Mrkos in Czechoslovakia, Ikeya and Seki in Japan, and Alcock in England were gathering the amateur garlands, large-aperture binoculars were widely used. Even so, none was larger than 150 mm, and the comets tended to be bright (about magnitude 9) at discovery. The 1980s have seen a trend towards much larger apertures, higher magnifications, and fainter discoveries. On the other hand, since almost all the amateur-found comets do attain much greater brightness after discovery, the use of larger telescopes is not so much allowing more comets to be discovered as enabling them to be picked up at greater distances from the Sun.

It is impossible, however, to separate instrumental factors from climatic ones. Observers in the maritime English and north-western European climate, where cloud cover is on average something like 70%, must adopt a different approach. It is simply not realistic to plan total sky coverage on a regular basis using the relatively narrow fields and slow sweeping speeds inseparable from large-aperture work. One must be ready to go out whenever the sky clears and do as much as possible before it clouds over again. On the face of it, the best hope under these conditions does seem to be to pick up a comet that has suddenly brightened, or been hidden by moonlight; for the observer with, say, a 2° field is four times as likely to sweep up a comet as someone else with a 1° field, assuming that both instruments are capable of showing it.

I once discussed sweeping procedures with two successful discoverers, Jack Bennett (Pretoria) and Eric Alcock (Peterborough, England). Bennett made his sweeps parallel with the horizon, whereas Alcock preferred to work vertically, so that every sweep sampled the good higher transparency as well as the horizon haze and illumination.

[Since writing his chapter, David Levy discovered his sixth comet. I requested a brief account, which follows. — Ed.]

> On the morning of 20 May 1990 I discovered my sixth comet using Miranda, my 400-mm reflector. The discovery was somewhat surprising since I had just a few hours earlier had dinner with Walter Scott Houston of the Deep Sky Wonders column of *Sky & Telescope*. As we parted I told him I was planning to do a short comet hunt in the morning until moonrise, and he replied that he hoped I would find one, for the sixth comet would tie my record with that of Mellish, the successful sweeper from the midwestern United States earlier this century.
>
> As I began hunting, the sky cleared to a beautiful night, and even after moonrise I continued; one should not waste a night such as this. I though of Scotty's comment and how it would probably be a long wait for number 6. I passed α Andromedae (Alpheratz) and suddenly spotted a rather bright diffuse object. I was thoroughly familiar with this part of

the sky and knew instantly that something was wrong: a 'red alert' went off in my mind. I looked again and detected a faint tail. Although I detected no motion I reported it to the IAU Central Bureau, and then drove to town to check the Palomar Sky Survey, which showed no bright galaxy with a tail in that position!

The following day I drove as planned to the Texas Star Party, near Fort Davis. There I got a real surprise: the comet had moved only 1/8°, which is much slower than the $\frac{1}{2}$–1° daily motion which is typical of comets. Hours later I finally reached Central Bureau Associate Director Daniel Green from a pay telephone on Main Street in Fort Davis. As cars passed by and noisy trucks roared through town, he could hardly hear me report the new position, and I could barely hear him tell me that the comet would be named Levy 1990c.

Bibliography

Atlases

Becvar, A., *Atlas Coeli 1950.0.* Cambridge, Mass.: Sky Publishing Corp., 1962.

Scovil, C. (ed.), *The AAVSO Variable Star Atlas.* Cambridge, Mass.: Sky Publishing Corp., 1980.

Tirion, W., *Sky Atlas 2000.0.* Cambridge, Mass.: Sky Publishing Corp., 1982.

Vehrenberg, Hans, *Atlas Stellarum.* Cambridge, Mass.: Sky Publishing Corp.

Catalogues

Guide Star Photometric Catalogue: I. Astrophysical Journal Suppl. Series 68, 1, Sept. 1988.

Landolt, A.U., UBV photoelectric sequences in the celestial equatorial selected areas 92–115. *Astronomical Journal,* 78(9), 959–1021 (1973).

CHAPTER TEN · *Observing Variable Stars*

JOHN ISLES
FORMER DIRECTOR, BAA VARIABLE STAR SECTION

THE STUDY OF STARS THAT CHANGE IN BRIGHTNESS IS ONE AREA above all where amateur astronomers can make a useful contribution to science. Since the beginning of the Space Age, variable stars have become one of the main fronts of research. Far from diminishing the amateur's role, this has enhanced it. Well over 30 000 variable stars are known, poorly in most cases, and thousands more stars are suspected of change but have never been properly studied. There are thus far more variable stars than there are astronomers, and for many of the 70-odd types of variable so far recognized we can learn as much or more from a long run of data such as amateurs can provide than a professional can glean in a few allocated nights on a large telescope.

Amateurs can provide continuous coverage of the magnitude changes in a select list of important variables, provided their work is pooled through one of the organizations that collate the results reported by a widely spread network of observers. Enterprising individuals can also attempt to discover new variables, and to identify the nature of newly discovered or poorly studied objects. Such work, as well as the refinement of the details for the known variables, the determination of their current light elements and their monitoring for unusual behaviour, all helps the world's astronomers in planning their observations and in directing scarce resources where they will yield greatest benefit. Amateurs can thus help, directly or indirectly, towards the solution of problems associated with almost every type of object.

TYPES OF VARIABLE

Table 10.1 lists the types of variable star currently recognized (1992). Full definitions are given in the *General Catalogue of Variable Stars* (GCVS, [1]), compiled in Moscow. Two new types (BE, R) and some other modifications to the classification have been introduced in the subsequent name-lists of new variables, which are published in the IAU Information Bulletins on Variable Stars issued by Konkoly Observatory, Budapest. A convenient and up-to-date summary of all the types is available in reference [2]. For an account of what is known of the physical nature of each class of object, see reference [3] and the further books and review articles cited below.

Numbers – The numbers of objects in the GCVS, which lists variables named up to 1982, are given in Table 10.1. These do not indicate the true proportions that actually exist, for several reasons. Stars that are intrinsically bright and have a large range of variation, such as the Mira stars, are the most likely to be discovered. Those that are only occasionally active are harder to detect; bright long-period eclipsing binaries are still being found, and almost all nova outbursts occur in previously unstudied objects. The most numerous type in our Galaxy is probably the red dwarf UV Cet class. The GCVS does not aim to include variables in globular clusters or external galaxies. Some stars show more than one type of variation, so the figures for the individual types add up to more than the total for all named variables.

Scope for useful work – All the types offer potential to observers, especially by means of photoelectric photometry. Visual and photographic observations can be attempted on stars with a range as small as 0.4 magnitude, but are most productive on those that vary by a magnitude or more. In the following notes, what is said about visual observation applies equally to photographic observation. Only the most important areas for study are mentioned. Some recent light curves from amateur observations of a few representative stars are given in Figs 10.1 and 10.2. Useful work does not necessarily mean observation, however. There is a great need for statistical analysis of the data already existing in the files of variable-star organizations.

Eruptive variables – The R CrB stars (type RCB) are the most important stars in this class. They should be checked once or twice a week for the onset of their deep minima, which are due to the formation of a dark cloud of sooty dust in front of the star. Daily observations should be made throughout the minima. At maximum, many of these stars show oscillations with periods that may be changing as they evolve through the short-lived RCB stage [4].

The young Orion variables (type IN) and T Tau stars (INT) show irregular light changes that are difficult to interpret if only visual observations are available.

TABLE 10.1

VARIABLE-STAR TYPES

Type	Description	No.
Eruptive variables		
BE	B-type emission-line star with small-scale variations	–
FU	FU Ori: gradual rise by several magnitudes to maximum lasting years	3
GCAS	γ Cas: shell star with irregular fades	108
I	Irregular. Suffixes (often in combination): A = early spectral type, B = intermediate or late spectral type, N = nebular (Orion variable), S = rapid variations, T = T Tauri (intense emission of Fe(I) at 4046 and 4132 Å)	1 447
RCB	R CrB: cyclic pulsations and irregular deep fades	37
RS	RS CVn: binary with starspots and active chromosphere	14
SDOR	S Dor: high-luminosity irregular variable	15
UV	UV Cet: flare star	747
UVN	Flaring Orion variable	398
WR	Wolf–Rayet variable: unstable mass outflow	9
Pulsating variables		
ACYG	α Cyg: non-radially pulsating supergiant	24
BCEP	β Cep: hot pulsating star	86
BCEPS	Short-period β Cep star	3
CEP	Cepheid: radially pulsating giant	180
CW	W Vir: Population II (old) Cepheid. Suffixes: A = period > 8 d, B = period < 8 d	173
DCEP	δ Cep: Population I (young) or classical Cepheid	418
DCEPS	Short-period young Cepheid with symmetrical light curve	42
DSCT	δ Sct: pulsating Population I main-sequence star	86
DSCTC	Small-amplitude δ Sct star, commonly found in open clusters	120
L	'Long': slow irregular variable. Suffices: B = late spectral type, C = supergiant of late spectral type	2 393
M	Mira: long-period variable giant	5 829
PVTEL	PV Tel: pulsating helium supergiant Bp star	3
RR	RR Lyr: 'cluster-type' variable; radially pulsating Population II star. Suffixes: AB = steep ascending branch in light curve, C = symmetrical light curve	6 112
RV	RV Tau: radially pulsating supergiant with alternating deep and shallow minima. Suffixes: A = mean magnitude constant, B = mean magnitude varies	122

TABLE 10.1

VARIABLE-STAR TYPES

Type	Description	No.
Eruptive variables		
SR	Semiregular: noticeable periodicity, but with irregularities. Suffixes: A = late-type giant with persistent periodicity, B = late-type giant with weak periodicity, C = late-type supergiant, D = giant or supergiant of intermediate spectral type	3385
SXPHE	SX Phe: Population II subdwarf resembling type DSCT	15
ZZ	ZZ Cet: non-radially pulsating white dwarf. Suffixes A, B, O indicate DA, DB, DO spectra	22
Rotating variables		
ACV	α CVn: rotating magnetic main-sequence star	159
ACVO	Rotating magnetic main-sequence star with rapid non-radial oscillations	5
BY	BY Dra: rotating red dwarf with starspots and active chromosphere	34
ELL	Ellipsoidal: close binary with distorted components showing changing surface area as system rotates	42
FKCOM	FK Com: rapidly rotating giant with starspots	4
PSR	Pulsar: rapidly rotating neutron star with narrow beam of optical radiation	1
R	Non-eclipsing binary showing reflection of light of hot component on surface of cool one; brightness varies as system rotates	–
SXARI	SX Ari: hot rotating helium variable resembling type ACV	15
Cataclysmic variables		
AM	AM Her or 'polar': interacting binary with magnetic white dwarf; accretion on magnetic poles gives rise to emission of polarized light	1
N	Nova: thermonuclear runaway on white dwarf component of close binary. Suffixes: A = fast, AB = intermediate, B = slow, C = very slow	200
NL	Novalike: outbursts like novae or resembling old novae	29
NR	Recurrent nova	8

TABLE 10.1 continued

VARIABLE-STAR TYPES

Type	Description	No.
Cataclysmic variables		
SN	Supernova: catastrophic explosion of star. Suffixes: I = type I (thermonuclear detonation of white dwarf component in interacting binary), II = type II (collapse of massive star to form neutron star)	7
UG	U Gem: dwarf nova with cyclic outbursts. Suffixes: SS = SS Cyg, SU = SU UMa (short outbursts and supermaxima), Z = Z Cam (standstills)	332
ZAND	Z And: symbiotic star; close binary comprising cool star and hot one exciting extended envelope	47
Eclipsing binaries		
E	Eclipsing binary. Suffixes: A = Algol; B = Beta Lyr, W = W UMa	5 022
X-ray sources		
X	X-ray binary. Suffixes (often in combination): B = burster, D = dwarf companion, F = rapid fluctuations, G = giant companion, I = irregular, J = relativistic jets, N = nova-like, P = pulsar, R = reflection effect, M = strong magnetic field	44
Other symbols		
BLLAC	BL Lac: variable galaxy nucleus (flat radio spectrum, no strong emission lines)	4
CST	Constant	158
GAL	Variable galaxy nucleus	6
QSO	Quasar	2
S:	'Short': rapid variable of unknown type	182
*	Unique object	42
	Unknown type	428
	All named variables (1982)	28 435
	(1988)	30 099
	(1990)	30 264

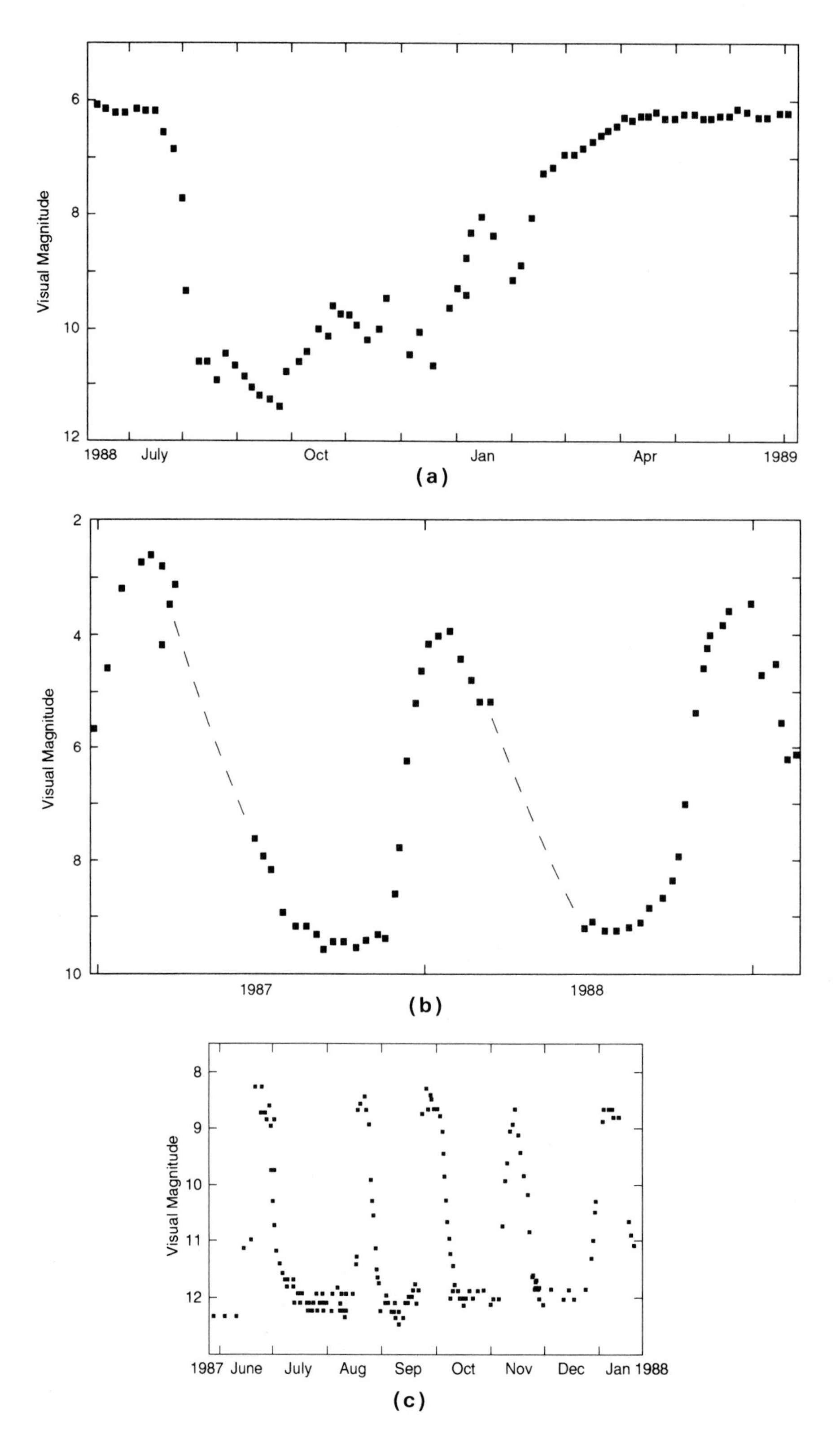
Visual Magnitude
6
8
10
12
1988
July
Oct
Jan
Apr
1989
(a)
2
4
6
8
10
1987
1988
(b)
8
9
10
11
12
1987
June
July
Aug
Sep
Oct
Nov
Dec
Jan 1988
(c)

Moreover, estimates of those enmeshed in nebulosity, such as the ones in the Orion Nebula itself, are notoriously inaccurate. Multicolour photoelectric photometry can help to establish whether the changes are intrinsic to a given star, or due to obscuration by dust clouds. Five INT stars have shown long, bright outbursts to become FU Ori stars or 'fuors' (type FU). Given the short time the sky has been closely monitored, this must be a recurrent phenomenon among INT stars. It would be useful to detect further examples, for which photography of the dark clouds in Auriga, Taurus, and elsewhere, would be most effective [5].

Unstable stars such as γ Cas (type GCAS) and S Dor (type SDOR) show slow changes at intervals of decades or centuries, usually associated with the ejection of a gas shell. Except when they are active, they are worth checking only about once a month for any marked fading or brightening. Visual observations of UV Cet stars (type UV) are probably no longer of value, but amateurs can do useful photoelectric photometry on these objects [6].

Pulsating variables – With cycles typically a year long, the Mira stars (type M) can make up an excellent programme for observers in cloudy climates. The study of their changing periods, which are not understood, requires long runs of the timings of maxima and minima that can readily be derived from visual estimates, which should be made every five days or so [7].

The periods of most semiregular variables (type SR) are uncertain; these stars often show multiple periods, but few cases are well studied [8]. The RV Tau stars (type RV) may be considered semiregular variables with two simultaneous periods, one twice the length of the other, and often a third, slower change in the mean magnitude; this behaviour is unexplained [9]. Many of the so-called irregular red variables (type L) are probably semiregular stars or variables of other types, whose periods have not yet been determined. Estimates of SR and L stars may be made about every ten days, and every five for type RV.

The maxima of Cepheid variables (types CEP, CW, DCEP) and of RR Lyrae stars (type RR) can be timed, providing a useful check on their periods and hence their rate of evolution. The brightest stars of these types are mostly well studied, so such work is most productive in the case of the fainter variables. Observations should be made at intervals ranging from a few minutes to one day, depending on the length of the period.

Cataclysmic variables and X-ray binaries – These are interacting binary stars, most of which show outbursts at intervals ranging from a few days in some dwarf

---

FIGURE 10.1

Light curves of three variable stars, from visual observations. (a) A multiple minimum of R CrB (type RCB) from 5-day means of estimates by BAA observers. (b) o Cet (Mira, type M) from 10-day means of estimates by Junior Astronomical Society observers. Gaps occur where the Sun was too close to permit observation. (c) SS Cyg (type UGSS) from estimates by John Isles.

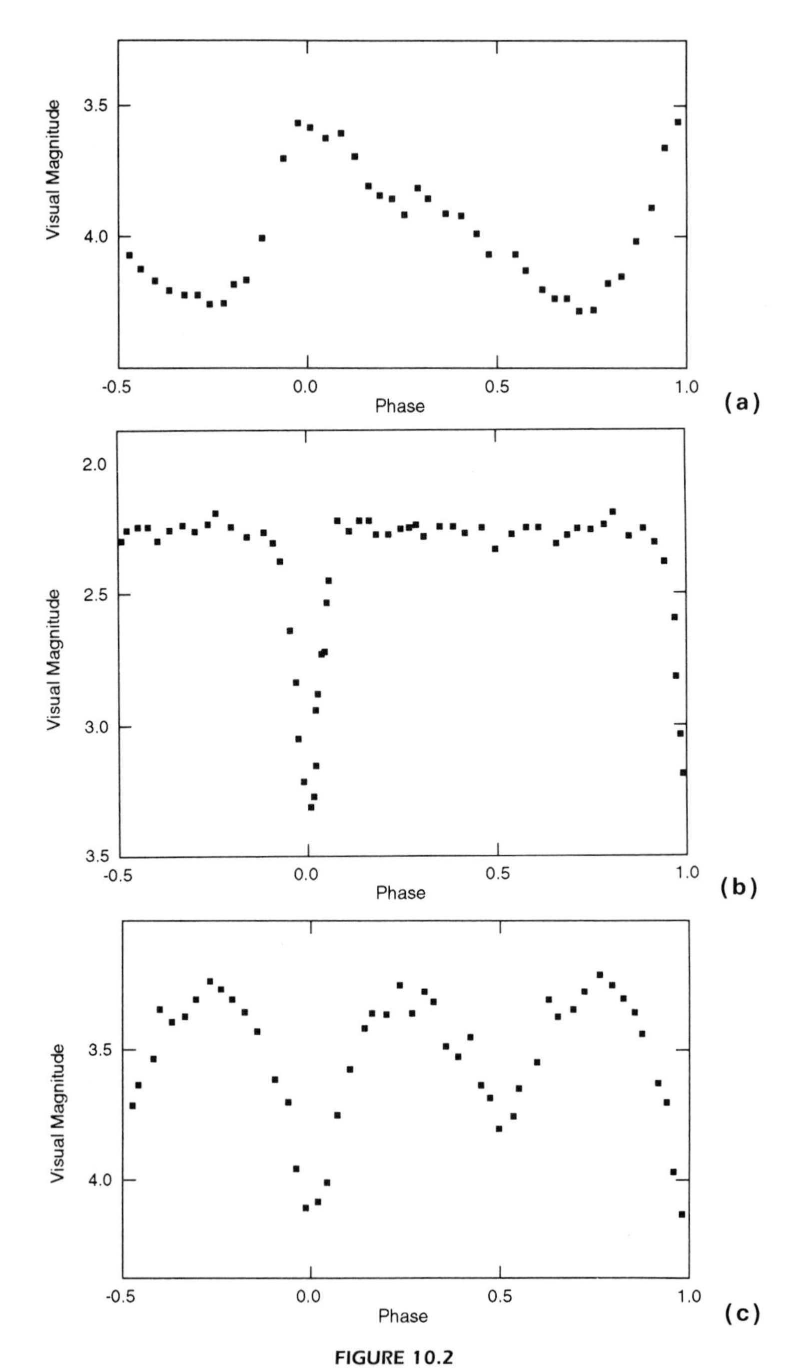

FIGURE 10.2

Light curves of (a) γ Aql, (b) β Per, (c) β Lyr. The observations have been arranged according to phase, using the light elements given in the GCVS. Each point is the mean of 10 or 11 visual estimates by John Isles in 1987–88. The curve of γ Aql has a bump at about phase 0.3, when the decline is temporarily reversed.

novae (type UG) up to perhaps thousands of years in the classical novae (type N, discussed in Chapter 12). The outbursts have different causes. Those of novae are due to the thermonuclear detonation of hydrogen that has accumulated on the surface of a white-dwarf component. In dwarf novae, the outbursts are the pulsed release, in the form of light and heat, of gravitational energy in an unstable 'accretion disc' surrounding the white dwarf. Neither type is completely understood, and many of the other variables in this broad class are a total mystery.

Continuous light curves of these objects are generally available only from the pooled work of amateurs. The long-period dwarf novae such as WZ Sge, and the recurrent novae, are especially important to watch for their rare outbursts, as are the SU UMa stars (type UGSU) for their rare supermaxima.

Many of the 'nova-like' variables (type NL) resemble novae at minimum brightness, but some may be Z Cam stars (type UGZ) stuck in an indefinite standstill state between maximum and minimum. Thus TT Ari, formerly regarded as a bright nova-like variable, had three faint minima in the early 1980s, when its spectrum revealed that it was a UGZ star. Several ex-novae resemble other types of cataclysmic variable, and astronomers believe that all these objects are potential future novae [10].

The Z And or symbiotic stars (type ZAND) are also interacting binaries, but with a red-giant component. They may show eclipses, semiregular variations, minor outbursts and very slow nova eruptions (type NC). All are worth monitoring, including the ones in which no noticeable activity has yet been recorded [11]. Supernovae are discussed in Chapter 13.

The frequency of observation required by these objects depends on the rapidity of their variations. Most should be observed visually at least once a night, though novae in their final decline, and ZAND stars, need normally be observed only about once a week. Estimates should be made with great care at all times. Their value lies not only in the detection of outbursts: the analysis of minor fluctuations and of rates of rise and fall has also yielded important results. Photoelectric photometry of dwarf novae, nova-like variables and ex-novae can be of great value as the means of detecting variations through their orbital cycles, and hence of modelling their structure. They are mostly faint objects at minimum, but several are brighter than magnitude 12 at that phase.

Eclipsing binaries – Timing the minima of eclipsing binaries enables their period changes to be studied. These are due to mass exchange between the components, to its ejection from the system, to the disturbing effect of a third component, or to the stars being distorted and not attracting one another as if they were point masses. Even a change of a fraction of a second in the length of the period can become obvious after a few years as a deviation of minutes or hours in the times of minima. Photoelectric timings are a good project for the beginner in this field, as they require only differential photometry (Chapter 15). Visual timings are of lower precision, and should preferably be made on systems that have not yet been studied photoelectrically. Magnitude estimates should be made every 15 minutes or so through the course of a predicted minimum, or less often in the case of longer eclipses [12].

Unstudied variables – The GCVS lists 117 slow (type L:) and 182 rapid (type S:) variables of unknown type, as well as 428 variables about which nothing at all is known. The *New Catalogue of Suspected Variable Stars* (NSVS, [13]) lists nearly 15 000 further stars that had been suspected of variation up to 1979. The scope for study is thus enormous, though probably many of these objects have been reported as variable on insufficient grounds, and are actually constant. Even a few photoelectric measurements, showing that a particular suspect is certainly variable, would be worth publishing so that the star can be investigated by others.

Most of the noticeably variable bright stars may have been discovered by now, but new ones are turning up all the time, often as a by-product of photographic patrols for novae. The southern sky is less thoroughly explored. Of the 24 variable stars that are known to have a range of at least 0.4 magnitude and are at least as bright as magnitude 4 at maximum and 5 at minimum, 17 are in the northern hemisphere; so southern variables even of naked-eye brightness may still await discovery. As recently as 1990, γ Persei was discovered to be an eclipsing binary with magnitude range 2.9–3.2 and a period of 14.6 years.

METHODS OF OBSERVATION

The analysis of starlight involves measuring the distribution of energy through the optical spectrum. The resolution that can be achieved in terms of wavelength or frequency depends on several factors, of which the amount of light available for analysis is the most important. With the small instruments available to most amateurs, detailed spectrophotometry is possible only for the brightest stars. In narrow-band photometry, the light is measured in a small number of spectral regions (usually fewer than ten) centred on specified wavelengths; in each measurement, most of the star's light is rejected, so large apertures are usually required. In broad-band photometry, as in the UBV system, a substantial proportion of the available light is included in each reading; useful photoelectric photometry in B and V can be done with telescopes as small as 100 mm aperture. Using the eye or a photographic emulsion as the detector, almost all the light is used; much work can be done without any telescope, but information about the spectral distribution is virtually lost.

Visual estimates are uncertain by typically ± 0.2 magnitude, or more in the case of red variables. Estimates from photographs have a similar precision, but are less subject to systematic errors. Microdensitometer measurements from photographs, or measurements with a visual photometer, can in favourable cases be accurate to ± 0.05 magnitude, for considerably greater effort. These methods have largely been superseded by photoelectric photometry, in which errors can be reduced to ± 0.01 magnitude or less.

The contribution an observer can make obviously depends in part on how much time he or she is prepared to spend on the hobby. While it is certainly true that more information per hour spent observing can be acquired by the more instrumental methods, those who find more enjoyment in a simpler approach may finish up accomplishing more in the long run. At all events, an observer should make every effort to perfect the technique adopted, as this is the key to successful and rewarding observation. Visual, photographic and spectroscopic methods are discussed below, and photoelectric photometry in Chapter 15.

VISUAL OBSERVATION

As in other forms of photometry, visual estimates of a variable are made by comparison with field stars that are presumed to be constant in brightness, and whose magnitudes have often already been measured. Usually one comparison star is chosen a little brighter than the variable, and one a little fainter: the magnitude of the variable is then found by interpolation. Sometimes a variable will appear equal in brightness with a comparison, giving an immediate estimate of its magnitude; but it is always desirable to check this by reference to other comparison stars.

In the method used by the AAVSO, the estimated magnitude of the variable is written down directly. In the Variable Star Section of the BAA, the light estimate is recorded in terms of the comparison stars, which usually form a lettered sequence (A, B, C, . . .). Most other organizations have adopted one or other method, each of which has its advantages; an observer should obviously use the methods laid down in the instructions supplied by the group to which he belongs.

To show how estimates are actually made, let us suppose that the variable appears fainter than a star P of magnitude 9.5, but brighter than a star Q of magnitude 10.0. Using the *AAVSO method*, an observer should automatically divide the interval between the comparisons into five brightness steps (their difference in tenths of a magnitude), and place the variable on this scale. If the variable appears three parts fainter than P (rather than two or four) and consequently two parts brighter than Q, the observer records the magnitude as 9.8. A BAA observer, using the *fractional method*, might write 'P(3)V(2)Q'. This means exactly the same thing, and gives the same final result; but the observer would also have the option of writing other fractions such as P(1)V(1)Q (giving 9.75 for the variable, which would be rounded to the fainter tenth, 9.8), or P(2)V(1)Q (giving 9.83, rounded to 9.8), if one of these seemed more satisfactory. At a later date, the comparison star magnitudes might be revised: photoelectric V sequences are progressively being introduced. If this happens, the revised magnitude of the variable can be calculated from such a record of the light estimate.

In *Pogson's step method*, the eye is trained to recognize magnitude intervals of one, two, three, four, and five tenths of a magnitude. A record such as 'P − 3,

Q + 1' means the variable appeared 0.3 magnitude fainter than P and (independently) 0.1 magnitude brighter than Q. This gives magnitudes of 9.8 and 9.9 for the variable, which would be averaged to give 9.85 and then rounded to the fainter tenth, 9.9.

Argelander's step method is similar, except that the step is not fixed at 0.1 magnitude, but is the smallest perceptible brightness interval. The value in magnitudes of one step is personal to the observer. An estimate 'P(4)V, V(3)Q' means that the variable appeared four steps fainter than P and three brighter than Q. The estimate is reduced as if it were the fractional observation P(4)V(3)Q, giving 9.8 for the magnitude of the variable.

Occasionally only one comparison star is available, and a step method is then all that can be used. Step methods are also particularly useful where the magnitudes of comparison stars are unknown, as the implied step intervals between the comparisons can be used to set up an arbitrary scale of brightnesses on which the estimates can be reduced. This can be quite sufficient if the purpose of the observation is merely to find the time when the variable was at maximum or minimum brightness.

The fractional method requires no previous training in recognition of steps, so it is more suitable for the beginner, but it involves intercomparing three stars. In step methods, only two stars are compared at a time, so these are preferred by many experienced observers. Step intervals greater than 0.5 magnitude cannot reliably be estimated, so where there is a large gap in the comparison star sequence it is best to use the fractional method.

If the variable is invisible, a *negative observation*, recording the faintest comparison star that can be seen with certainty, can be very useful.

Sources of Error

There are many potential sources of error in visual work, listed at length in [2]. Probably the most important are as follows.

Careless mistakes – Observations should not be rushed. The observer should be in no doubt about the identity of variable and comparisons. At least two comparisons should always be used. Records should be checked at the end of an observing session.

Bias – Every effort must be made to eliminate preconceptions about a variable's magnitude. Previous results should never be consulted before observing a star. Observers should expand their programmes to the point where they seldom remember at what magnitude a variable appeared the last time it was seen.

Colour errors – There is no fully satisfactory way to estimate the magnitude of a red variable using white comparison stars, and observers can differ from one another by a magnitude or more. The best that can be done is to use a consistent approach. The variable should not be brighter than about four magnitudes above the threshold. Recommended methods are to use quick glances by direct vision; to use averted vision, which is less sensitive to colours; or to throw the stars out of focus so that their colours are less perceptible. The cells in the retina that are responsible for colour vision (cones) and the detection of faint light (rods) have a different response to light of different wavelength — consequently, different methods can give different results. One method should be adopted by the observer and then used on every occasion.

Position angle error – Of two stars of equal brightness, one above the other, the lower one appears brighter. Errors can be reduced by keeping the line joining the observer's eyes parallel with the line joining the stars being compared.

Bright sky background – In twilight or moonlight, estimates of red stars may be too bright, as white comparison stars lose contrast against the light sky. Fortunately, most red variables change only slowly: they can be observed before moonrise or after moonset, and not observed at Full Moon. Twilight skies should be used only when observations of a star are otherwise impossible. Artificial lights can produce errors in either direction; for example, red variables appear fainter in skies polluted by low-pressure sodium lighting. Light-pollution rejection filters should not be used for variable-star observation, as they may not transmit equally the light of the stars to be compared. Observers who suffer from light pollution should therefore consider finding a darker site, or confining their work to variable stars that are not strongly coloured.

Atmospheric extinction – When a variable is low, comparison stars should be chosen at nearly the same altitude. For some bright naked-eye variables this may be impossible, as in the case of α Ori at the beginning or end of an apparition. Corrections may be applied, using tables given in various books (such as reference [2]), but these cannot be relied on, especially if haze is present.

OBSERVATIONAL RECORDS

These must obviously include all the details required by any group to which the observer belongs, but it is useful to keep an even fuller record in case the reliability of any observation comes into question.

Table 10.2 gives an extract from the writer's observing book, showing observations made for the BAA. Entries at the telescope are made in pencil, and any subsequent additions or corrections are in ink (bold type in Table 10.2). The record starts with the calendar date, weekday and location, the times of beginning and end of the session, and sky conditions (class 1, on a three-point scale with 1 the best). Subsequent lines record the time of each observation, variable star name, light estimate and any remarks. The symbol '[' means 'fainter than': AH Her was invisible but comparison star B was glimpsed ('gl'); '×100', '×40' record changes

TABLE 10.2

EXTRACT FROM LOG OF VISUAL OBSERVATIONS OF VARIABLE STARS

7502	1988 Dec 6 Tuesday Akrounda, Cyprus						
	0340-0515 Fine Cl 1						
.16	0348	AH	Her	[B(g1)	low	× 100	**[12.1**
	0352	AY	Lyr	[K			**[12.9**
	0354	CH	Cyg	E(1)V(1)F	tree	× 40	**8.3**
.17	0359	WZ	Sge	[W [X(gl)		× 100	**[13.3**
	0402	VY	Aqr	[O [R			**[13.4**
	0403	SS	Cyg	N(1)V(3)P			**12.1**
	0406	RU	Peg	G(l)V(1)H			**12.4**
	0409	EF	Peg	D-5 =2			**12.9**
	0411	η	Aql	$\delta(2)V(1)\beta$		NE	**3.6**

of magnification. A note in the front of the book records that the main instrument was a 127-mm Schmidt–Cassegrain. η Aql was observed with the naked eye ('NE').

At the end of the session, the calendar date is converted to Julian date (see below), of which the last four figures are given at top left, and times, originally in GMAT (GMAT = UT − 12 hours), are converted to decimals of a day in the left margin. The light estimates are also converted to magnitudes, given in the right margin. The details may subsequently be transferred to report forms or entered on a microcomputer for storage and output in whatever formats are required.

PHOTOGRAPHY

Photography offers several advantages over visual observation. One exposure can provide a permanent, impersonal record of the brightnesses of hundreds of stars that can be examined at leisure. With a given aperture, stars can be photographed up to two magnitudes fainter than the visual limit. The main disadvantage is probably the continuing cost of photographic materials.

The photography of star fields is discussed in Chapter 14, and photographic nova patrol work is dealt with in Chapter 12. In this section we consider the problem of deriving magnitude estimates from photographs.

If there is a sequence of stars of known magnitude in the field, estimates of a variable can be made by comparison, much as discussed earlier in relation to visual work, provided the colour response of the photographic emulsion is reasonably compatible with the magnitude system used to measure the sequence stars. Ideally, only black-and-white emulsions should be used for photometry, though colour film is often preferred for patrol work as it is less prone to produce star-like flaws. Panchromatic emulsions are more sensitive than the human eye to longer wavelengths, so red stars appear too bright on them. A yellow filter can be used to eliminate this oversensitivity, but exposure times must then be increased to reach a given magnitude. Tri-X with a Wratten 8 filter is fairly satisfactory, but systematic errors of a few tenths of a magnitude may still arise for very red variables or those with strong emission lines.

If there is no ready-made comparison star sequence in the field, there are various methods for setting one up.

Comparison with another field – Comparison is made with another field photographed with the same exposure and in the same conditions (especially of altitude above the horizon). Comparisons between photographs are difficult to make by eye, however. Even if they are measured by a microdensitometer, zero-point differences of a few tenths of a magnitude between photographs are commonly encountered.

Double exposure – If a star field is photographed twice on the same negative, with one exposure three times as long as the other, then (over a range of seven or eight magnitudes) corresponding images of the same star will differ by typically one magnitude. The exact difference will depend on the amount of reciprocity failure, and must be determined by a similar double exposure of a standard sequence. Once it is known, the brighter and fainter images of various stars in the field can be placed in order of brightness, their intervals in steps estimated, and the magnitudes of the other stars can be found.

Secondary images – These can alternatively be produced by means of a grating, or a narrow-angle wedge of glass placed in front of part of the objective. The method of analysis is exactly the same as in the double-exposure method.

For precise work, a microdensitometer may be used. This is a photoelectric device in which a beam of light is traced across each star image in order to measure its optical density. A calibration curve may then be drawn relating densitometer reading to magnitude. Some microdensitometer designs are described in references [14] and [15].

SPECTROSCOPY

A spectroscope is an instrument that spreads out the light of a star into a line or band, in which the various wavelengths are displayed in order. In principle, it provides much more information than broad-band photometry or estimation of the integrated light from a star. Spectroscopy has hitherto been a neglected field among amateurs, because small telescopes collect enough light for analysis from only the brightest stars; but with the progressive increase in the size of affordable telescopes it has become a field ripe for exploitation.

The spectra of many variables have never been recorded, yet a single spectrum can often be sufficient to allocate a variable star to its probable type. From visual or photographic observations it is often difficult to distinguish, for example, an eclipsing binary of W UMa type that has nearly equal minima from an RR Lyrae star with a period half as long and with a symmetrical light curve; but both types of variable have well-known relations between their period and spectral class, so one spectrum could be enough to settle the matter.

Rapidly fading objects are also often difficult to classify without a spectrum. To take another example, PQ And, known as Nova And 1988, was discovered at magnitude 10 in March of that year by D. McAdam in the course of a photographic search for novae. A spectrum was recorded at La Palma but subsequently lost owing to a computer failure. Casual inspection of it had suggested the object was a classical nova, but further spectra obtained at Steward Observatory in July, when it had faded below magnitude 18, resembled that of the long-period dwarf nova WZ Sge, so the true status of this object remained uncertain. A handful of spectroscopes on large telescopes in amateur hands about the world could add considerably to our knowledge of objects that undergo such short-lived outbursts.

Interesting views of the spectra of the brighter variables can be obtained visually with a hand-held pocket spectroscope or grating placed over the eyepiece of a telescope, but for scientifically useful work the spectra must be recorded photographically or electronically. In slitless spectroscopes, the star's light is dispersed by a prism or grating in the optical train, usually in front of the telescope objective or camera lens. In a slit spectroscope, the spectrum of a laboratory comparison source is recorded at the same time as that of the star, to provide a reference for measuring wavelengths and hence radial velocities.

A slitless spectrum can be examined qualitatively, or it can be calibrated in intensity (but not in wavelength) by comparison with a standard star, using a microdensitometer [16]. Such observations of novae can give useful measurements of the equivalent widths of spectral lines, especially of the intense emission lines that may be saturated on the spectrographs used on large professional telescopes. Reference [17] gives a series of spectra of V1500 Cyg (Nova 1975) that were obtained with a 35-mm camera, and show the changing intensity of the continuum and emission lines as the star faded from magnitude 2 to 9.

DATA ANALYSIS

Practical work in the field of variable stars includes data analysis as well as observation. An observer can derive much interest from analysing his own work, but it is also possible to contribute to science by helping in the analysis of the observations in the files of variable-star organizations. Only a brief account of some important techniques can be given here; more details are available elsewhere [2]. Reference [18] includes some useful papers, and several further references are cited below.

Much can be learnt from graphical methods, but elaborate statistical analyses can also be carried out using a microcomputer. To interpret their result correctly, the user should first acquire some grounding in statistics. Many textbooks are suitable: reference [19] is a readable one which covers most of the basics.

The analysis of variable-star observations is by no means a matter of subjecting them to a standard set of statistical recipes. Depending on what we know about the star and others of its type, there are always different questions to be asked of the data, so the first step should be to study the literature and see what others have done. The publications of many of the organizations listed later contain informative papers on the analysis of variable-star data.

Julian date – Before analysis, it is convenient to convert the time of observation, which may be recorded in years, months, days, hours and minutes, to days numbered consecutively. The Julian date (JD), universally used in variable-star work, is the number of days elapsed since January 1, 4713 BC. It may be calculated as follows:

1. Let the year, month number and day be Y, M, and D.
2. Convert the time in hours and minutes GMAT to a decimal of a day (hours/24 + minutes/1440) and add this to D. (If UT is used, subtract 0.5.)
3. If $M > 2$, subtract 3 from M; otherwise add 9 to M and subtract 1 from Y.
4. Then $\text{JD} = 1721104 + \text{INT}(365.25Y) + \text{INT}(30.61M + 0.5) + D$.

[INT( ) is the integer function: the expression inside the brackets is evaluated, any fractional part is discarded, and the whole number part is retained. For negative values, take the numerically larger or more negative whole number.]

Example: 1991 July 17, 06.00 GMAT gives $Y = 1991$, $M = 7$, $D = 17.25$. $M > 2$, so M becomes $7 - 3 = 4$. $\text{JD} = 1721104 + \text{INT}(727212.75) + \text{INT}(122.94) + 17.25 = 2448455.25$. This formula gives correct results for all dates since the introduction of the Gregorian calendar.

Light-time – A further adjustment is necessary for some short-period variables, such as eclipsing binaries or RR Lyrae stars, to allow for the fact that the light-time from the star to the observer can vary by up to 500 seconds, according to the Earth's position in its orbit. To eliminate the consequent annual variation in the star's apparent period, it is customary to correct the times to those that would have been recorded by a hypothetical 'heliocentric' observer at the centre of he Sun.

The following algorithm is adapted from reference [20]. Let G be the observed Julian date, and let α and δ be the star's right ascension (in degrees) and declination for epoch 2000.0. Calculate in turn:

$$T = (G - 2451545)/36525$$

$$L = 280^\circ.4659 + 35999^\circ.371946\,T$$

$$M = L - 282^\circ.9405 - 0.322204^\circ\,T$$

$$
\begin{aligned}
\lambda &= L + 1^\circ.9148 \sin M + 0^\circ.02 \sin 2M \\
v &= \lambda - L + M \\
R &= 0.99972/(1 + 0.01671 \cos v) \\
l &= \cos\delta \cos\alpha \\
m &= 0.91748 \cos\delta \sin\alpha + 0.39778 \sin\delta \\
\tau &= 0.005775R\,(l \cos\lambda + m \sin\lambda) \\
H &= G - \tau
\end{aligned}
$$

H is then the heliocentric time corresponding to the observed time G.

Tests for variation – When dealing with observations of a variable star of small amplitude, or one only suspected of variation, the first point to consider may be whether the observations really show any evidence for change in the star's magnitude. Variation may be revealed by the fact that the standard deviation of estimates of its magnitude is significantly greater than the standard error of the individual estimates, as determined by the F-test described in statistics textbooks (e.g. [19]). The standard error must obviously be known independently, for example from a similar series of observations of a constant star. In visual work it is usually very difficult to estimate the standard error of an observation, but in photoelectric work it can be estimated from measurements of the check star. Another method, using Spearman's rank-correlation coefficient, tests to see whether bright or faint observations tend to come in groups or at random [21].

Light curve – This is a graph of magnitude against time, as in Fig. 10.1, or against phase as in Fig. 10.2. If T is the Julian date of an observation and P is the star's period, the phase is the fractional part of $(T - T_0)/P$, where T_0 may be the Julian date of a known maximum or minimum in the past, or an arbitrary point in time. When the work of several observers is combined, it is often found that some produce consistently brighter or fainter estimates than the mean, and it may be helpful to apply small corrections to their work. A curve representing the probable course of variation may be drawn through the data by eye, or by fitting an appropriate function on a computer.

Maxima and minima – As a step towards reducing a large volume of data to a manageable size, it is often convenient to find the times when a star was at maximum or minimum brightness. Two graphical methods are shown in Figs 10.3 and 10.4. Pogson's method of bisected chords is generally applicable, but is most appropriate for stars whose light curves may be asymmetric, such as Mira stars. The tracing-paper method is used for variables whose curves can be assumed to be symmetrical, for example eclipsing binaries. An alternative method to Pogson's, used by the AAVSO, is to prepare on tracing paper a mean light curve of the variable from previous observations, with the estimated points of maximum and minimum

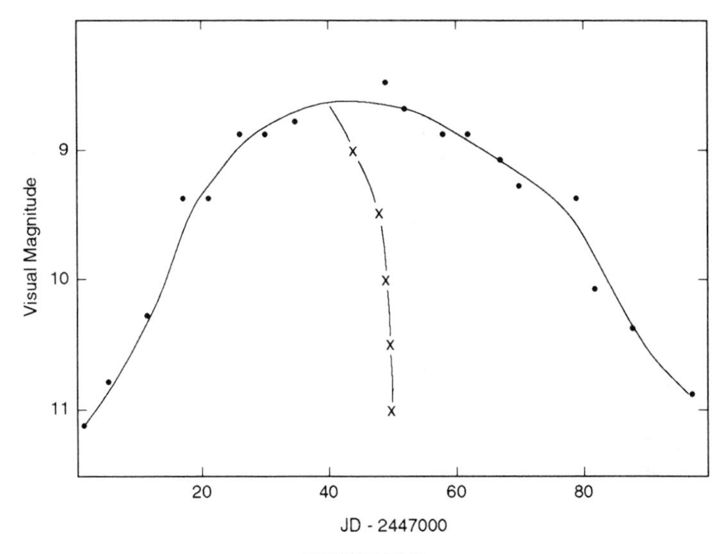

FIGURE 10.3
Pogson's method of bisected chords. Dates of rising and falling past various magnitudes are averaged, and a cross is plotted at the mean date. The maximum is taken as the point where a curve drawn through the crosses cuts the light curve. (Observations of W CrB by John Isles.)

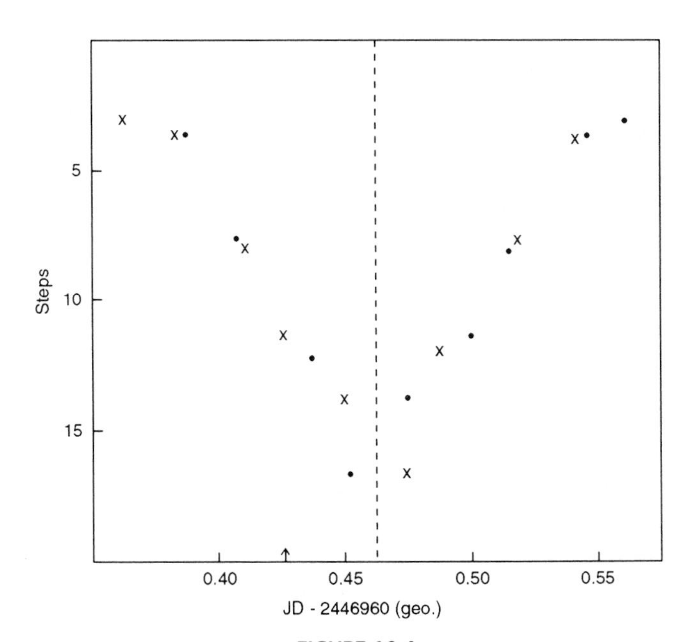

FIGURE 10.4
Tracing-paper method. The observations are plotted (dots) and a tracing of the plot is made (crosses). The tracing is reversed and slid along until the best fit is found. The arrow marks the position of midnight (0.50d) on the tracing, which falls over 0.426d on the original plot. The axis of symmetry, here shown by a dashed line, is at the mean of these two times, 0.463d. This is taken as the time of minimum. (Observations of TY Del by John Isles.)

marked, and to place this over the observed curve to find the position of best fit: the date of maximum or minimum is then read off.

Computer methods are also used. A mathematical curve can be fitted, such as a low-degree polynomial, and the time of maximum or minimum is taken as that when the function has its maximum or minimum [22]. A mean light curve can be fitted by computer as described in reference [23]. For eclipsing binaries, a standard method is that of Kwee and van Woerden [24], but in visual work more stable results are given by the polygonal-line method [25]. Both methods find an axis of symmetry about which the data can be reflected with maximum agreement between observations on the falling branch and reflected observations on the rising branch.

Period estimation – There are many features in a star's light curve that may be important, but above all a recognizable period in its variations may (though need not necessarily) correspond to that of some physical process in the star. Provided a star's period is known well enough to say how many cycles have elapsed between each observed maximum or minimum, its length can be estimated by dividing the time from the first to the last by the total number of cycles. In less-regular variables such as the dwarf novae and Mira stars, this gives almost the best estimate of the period that can be derived from the data, and more complicated methods may actually be misleading. If the variations are very regular, as for example in most eclipsing binaries, RR Lyrae stars, and Cepheids, a better estimate is given by linear regression.

If the period is unknown, very poorly expressed, or possibly multiple, computer methods can be used to search for periodicities in the data. In this case, the original observations are used, rather than times of maxima and minima. The most widely-used methods are fourier analysis [26, 27], the string-length method devised by Dworetsky [28], and the binning method devised by Jurkevich [29, 30]. Some advantages and disadvantages of each are discussed in reference [2].

Period change – To check whether a star's period has changed, a useful tool is the $O-C$ diagram. This plots the difference between observed and calculated dates or times of maxima and minima, against date or cycle number (Fig. 10.5). If the true period is longer than the period assumed in the calculation, the $O-C$ diagram will slope up; if shorter, it will slope down, and if the period has changed during the interval covered by the observations, the diagram will show a curve rather than a straight line.

To test whether an $O-C$ curve really deviates from a straight line (or some other curve), the standard method is to compare two fits to the data. For example, suppose an eclipsing binary is suspected of showing a steady increase in its period, so that the $O-C$ curve may be part of a parabola. To check whether this is really a better fit than a straight line, we find (a) the best straight-line fit by least squares, and (b) the best parabolic fit. The parabolic fit is accepted if the coefficient of the

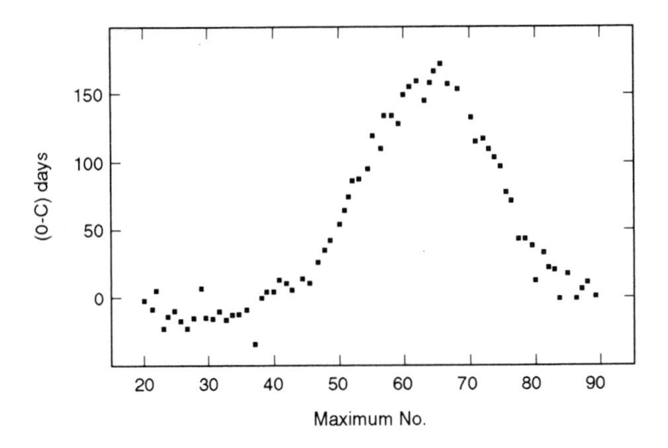

FIGURE 10.5
O—C diagram for the Mira star T Cep, from BAA observations. O—C is the difference between observed and calculated dates of maximum. The star's period first lengthened, causing the O—C curve to slope up (as maxima were occurring progressively later than calculated), then shortened, causing the curve to slope down.

square term in (b) is more than twice its calculated standard error (strictly, the test should use Student's *t*); or we may compare the variance of the individual timings of maxima or minima about fit (b) to see whether it is significantly less than with (a), using the *F*-test [31].

This procedure is valid only for the more regular variables: eclipsing binaries, RR Lyrae stars, Cepheids, and so on. In others, such as Mira stars, with individual cycle lengths that appear to vary at random about a mean value, the time gained or lost when a cycle is shorter or longer than the mean may not be compensated in the next cycle, so that the O – C curve can wander about, suggesting period changes when in fact the mean period has not varied. Some Mira stars do certainly have variable mean periods, but the best way to test for this is still a matter of controversy.

The traditional method of Sterne [32] is criticized in reference [33], which introduces a new one called the span test. This is most sensitive to a single, sudden period change in the middle of the observed interval, but can miss a case where the period changes more than once in such a way that the O – C value never becomes very large. Another method is introduced in reference [34], the lag-2 correlation test. The performance of the various tests is compared in references [35], [36] and [37]. There is no uniquely best test, as it is always possible to think of a way a star's period might vary that would elude any given test.

VARIABLE-STAR ORGANIZATIONS

The world's most active group is the AAVSO, 25 Birch Street, Cambridge, MA 02138, USA. Over half the observations now reported annually to the AAVSO are made by observers outside the United States, including the work forwarded by other national organizations. Ultimately, the AAVSO aims to maintain a world archive of variable-star observations, although there is much work to be done on harmonizing comparison-star sequences before this can be fully achieved. The AAVSO's own main programme covers well over a thousand objects, and is particularly strong in Mira stars and cataclysmic variables. There are additional programmes on eclipsing binaries, RR Lyrae stars and Cepheids, and searching for novae and supernovae.

The oldest group, formed in 1891, is the Variable Star Section of the BAA, Burlington House, Piccadilly, London W1V 9AG, UK. Most members are in the British Isles and northern Commonwealth countries. It runs programmes on binocular and telescopic variables, recurrent objects (known or suspected long-period dwarf novae and recurrent novae), eclipsing binaries and searching for novae and supernovae.

In the southern hemisphere, the leading group is the Variable Star Section of the Royal Astronomical Society of New Zealand (RASNZ), PO Box 3093, Greerton, Tauranga, New Zealand.

There are many other national and regional organizations; some of the more active are:

France: Association Française d'Observateurs d'Etoiles Variables (AFOEV), 16 rue de Plobsheim, 67100 Strasbourg, France.

France, Italy and Spain: European Group of Star Observers (GEOS), 3 Promenade Venezia, F-78000 Versailles, France.

Germany: Berliner Arbeitsgemeinschaft für Veränderliche Sterne (BAV), Munsterdamm 90, D-1000 Berlin 4, Germany.

Hungary: Pleione Változócsillag-észlelö Hálózat, 1114 Budapest, Bartók Béla út 11-13, Hungary.

Japan: Japan Astronomical Study Association, National Science Museum, Ueno Park, Taito-Ku, Tokyo, Japan.

Netherlands: Werkgroep Veranderlijke Sterren, Postbus 800, 9700 AV Groningen, The Netherlands.

Scandinavia: Scandinavian Variable Star Observers, SF-36280 Pikonlinna, Finland.

Switzerland: Bedeckungsveränderlichen Beobachter der Schweizerischen Astronomischen Gesellschaft (BBSAG), Rebrain 39, 8624 Grüt, Switzerland. (Specializes in eclipsing binaries.)

Mention should also be made of the International Amateur-Professional Photoelectric Photometry Organization (IAPPP), Dyer Observatory, Vanderbilt University, Nashville, TN 37235, USA.

GENERAL LITERATURE

The official catalogue of variable stars is the GCVS [1], to which the NSVS [13] is a companion volume. Lists of the brighter variables and suspects are given in *Sky Catalogue 2000.0*, vol. 2 [20]. For observers of eclipsing binaries and

RR Lyrae stars, the Cracow yearbook [38] gives current light elements and predictions.

For all but the naked-eye variables, identification charts are needed showing the comparison stars to be used. Many of these are available through the organizations listed earlier, and extensive selections are given in references [2] and [39]. The *AAVSO Variable Star Atlas* [40] plots all bright variables and also all fainter objects in the programmes of the AAVSO and RASNZ, and gives many magnitudes of comparison stars. For identifying variables brighter than magnitude 10, the best atlas is *Uranometria 2000.0* [41]. The Papadopoulos atlas [42] comprises photographs to magnitude 13 or 14 that closely match the sky's visual appearance.

The most useful books about variable stars have already been mentioned, and are included in the list of references. Most of the societies listed earlier have their own series of publications. *Sky & Telescope* [43] often includes articles about variable stars. *The Astronomer* [44] includes monthly reports of observations, and issues alerts to active observers by means of circulars, the telephone, and electronic mail.

About the author

John Isles is one of the world's most active variable-star observers. He read astronomy at Edinburgh University and statistics at London University, before working as a statistician for the UK government. In 1987 he moved to the clear skies of Cyprus to concentrate on studying and writing about the stars, and he now concentrates on photoelectric photometry.

He is a former Director of the Variable Star Sections of the BAA and the Junior Astronomical Society. He has contributed articles and papers to many journals and magazines, and is author of Volume 8 (Variable Stars) in the Webb Society's *Deep-Sky Observer's Handbook* series.

Contact address – 1821 Gulfview Drive, Kalamazoo, MI 49001-5224, USA.

References

1. Kholopov, P.N. (ed.), *General Catalogue of Variable Stars (GCVS)*, 4th edn., in 4 vols. Nauka, Moscow, 1985–90.
2. Glyn Jones, K. (ed.), *Webb Society Deep-Sky Observer's Handbook*, vol. 8: Variable stars. Enslow, Hillside, NJ, 1990.
3. Hoffmeister, C. *et al., Variable Stars*. Springer-Verlag, Berlin, 1985.
4. Fernie, J.D., *AAVSO Journal*, **15**, 206 (1986).
5. Herbig, G.H., *AAVSO Journal*, **16**, 1 (1987).
6. Page, A.A., *Astronomy Now*, **2**(10), 25 (1988).

7. Willson, L.A., *AAVSO Journal,* **15**, 228 (1986).
8. Whitney, C.A., *AAVSO Journal,* **13**, 31 (1984).
9. Wing, R.F., *AAVSO Journal,* **15**, 212 (1986).
10. Warner, B., *AAVSO Journal,* **15**, 163 (1986).
11. Kenyon, S.J., *The Symbiotic Stars.* Cambridge University Press, Cambridge, 1986.
12. *Variable Star Section Eclipsing Binary Programme Handbook.* BAA, London 1988.
13. Kukarkin, B.V. *et al., New Catalogue of Suspected Variable Stars (NSVS).* Nauka, Moscow, 1982.
14. Cerchio, F., *The Astronomer,* **16**, 99 (1979).
15. Guinnebert, J., *Journal AFOEV,* **12**, 57 (1978).
16. Kaila, K., *BAA Journal,* **87**, 525 (1977).
17. Pennell, W.E., *BAA Journal,* **86**, 246, 337 (1976).
18. Percy, J. (ed.), *The Study of Variable Stars Using Small Telescopes.* Cambridge University Press, Cambridge, 1986.
19. Moroney, M.J., *Facts from Figures.* Penguin, Harmondsworth, 1978.
20. Hirshfeld, A. and Sinnott, R.W., *Sky Catalogue 2000.0,* vol. 2: Double stars, variable stars and nonstellar objects. Cambridge University Press, Cambridge, 1985.
21. Isles, J.E., *BAA Variable Star Section Circular,* **67**, 12 (1988).
22. Roney, J.M. and McCallum, R.G., *Journal of the Royal Astronomical Society of Canada,* 77, 231 (1983).
23. Belserene, E.P., *AAVSO Journal,* **15**, 243 (1986).
24. Kwee, K.K. and van Woerden, H., *Bulletin of the Netherlands Astronomical Institute,* **12**, 327 (1956).
25. Isles, J.E., *BAA Variable Star Section Circular,* **58**, 3 (1984).
26. Belserene, E.P., *Sky & Telescope,* **76**, 288 (1988).
27. Deeming, T.J., *Astrophysics & Space Science,* **36**, 137 (1975).
28. Dworetsky, M.M., *RAS Monthly Notices,* **203**, 917 (1983).
29. DuPuy, D.L. and Hoffman, G.A., *IAPPP Communication,* **20**, 1 (1985).
30. Jurkevich, I., *Astrophysics & Space Science,* **13**, 154 (1971).
31. Pringle, J.E., *RAS Monthly Notices,* **170**, 633 (1975).
32. Sterne, T.E. and Campbell, L., *Harvard Annals,* **105**, 459 (1936).
33. Isles, J.E. and Saw, D.R.B., *BAA Journal,* **97**, 106 (1987).
34. Isles, J.E. and Saw, D.R.B., *BAA Journal,* **99**, 121 (1989).
35. Lloyd, C., *The Observatory,* **109**, 146 (1989).
36. Isles, J.E., in J.R. Percy *et al.* (eds.), *Variable Star Research, An International Perspective.* Cambridge University Press, Cambridge, 1991.
37. Lloyd, C., in J.R. Percy *et al.* (eds.), *Variable Star Research, An International Perspective.* Cambridge University Press, Cambridge, 1991.
38. Rocznik Astronomiczny Obserwatorium Krakowskiego, *International Supplement.* (Annual, published by the Astronomical Observatory, Jagiellonian University, ul. Orla 171, 30-244 Kraków, Poland.)

39. Burnham, R., *Burnham's Celestial Handbook,* in 3 vols. Dover Publications, New York, 1978.
40. Scovil, C.E., *The AAVSO Variable Star Atlas.* Sky Publishing Corporation, Cambridge, MA, 1980.
41. Tirion, W. *et al., Uranometria 2000.0,* in 2 vols. Willmann-Bell, Inc., Richmond, Va, 1987–88.
42. Papadopoulos, C. and Scovil, C.E., *True Visual Magnitude Photographic Star Atlas,* in 3 vols. Pergamon, Oxford, 1980.
43. *Sky & Telescope,* PO Box 9111, Belmont, MA, 02178-9111, USA.
44. *The Astronomer,* 177 Thunder Lane, Norwich NR7 0JF, UK.

Notes and comments

The author's comment that new noticeably variable bright stars are turning up all the time is well illustrated by the discovery of a bright Algol-type eclipsing binary in Perseus, SAO 23229, at 2 h 21.2 m, +54° 31′ (2000), which varies from magnitude 6.9 to 7.4 in 2.11 days. It was the twelfth new variable of 24 found to date (1992) by Indiana amateur Dan Kaiser, who compares photographs taken using a 35-mm camera as part of what began as a nova-hunting project.

His method is to take two photographs of standard fields on every possible occasion, and to compare the photographs taken on different nights using two ordinary slide projectors aimed at the same screen, with a rotating cardboard shutter to uncover each projector lens alternately. Any changing object will appear to 'blink'. The reason for taking two photographs on each occasion is to eliminate the possibility of a flaw affecting the appearance of one of the star images.

There are still many slowly-varying variables of small amplitude, within range of binoculars or short-exposure photography, to be discovered, but for Kaiser to find a relatively bright eclipsing binary was quite unexpected.

In the United Kingdom, similar discoveries have been made by Mike Collins (Sandy, Bedfordshire) in the course of photographic searches using TP 2415 film, for the UK Nova Patrol. Like Kaiser, he uses a 135-mm focal length lens with a 35-mm camera. One of these finds is of an Algol-type variable in Hercules, magnitude range 10.7–11.4, period 8.47 days. Collins has now generated a substantial list of certain and probable variables, some of which show a substantial range, including a star in Cassiopeia which apparently brightened from magnitude 11.1 to 9.0 in the course of 10 weeks at the end of 1988. The star was subsequently identified as a suspected variable, but with a much fainter range: clearly there is a vast amount of work waiting to be done with modest equipment, either systematically checking suspected variable stars or scanning star-fields wholesale. Collins reports his finds regularly in *The Astronomer,* which also publishes provisional sequences.

I invited Tom Saville, who edits variable-star observations for *The Astronomer,* if he had any additional comments to make on variable stars and the amateur. A summary follows. — Ed.

Amateurs can contribute to variable-star astronomy in two ways:

1. *Long continuous runs as a team (the Mira observers).* This suits observing each star once a week to give full coverage of a set programme.
2. *Detecting outbursts of cataclysmic types and notifying professionals.* To do this effectively, an observer needs to observe each star every clear night (including the difficult Full Moon period) and to be part of a suitable organization. This type of contribution may get the observer 'mentioned in despatches' (the IAU Circulars), which some people find rewarding. The need for priority makes it somewhat akin to hunting and discovering novae. It also requires experience and definite accuracy — wrong alerts are useless!

Much work also remains to be done in the southern sky, with its larger number of variables — the galactic centre is in the southern sky. Hence the opportunities for amateurs in that hemisphere.

I would reinforce the need to identify the variable clearly on every occasion when it is observed. Many fainter variables have close companions that can only too easily be confused with the variable itself. Wrong identification may result in false alarms for outbursts or in an entire sequence of one observer's results being thrown into doubt. This problem is behind several dubious observations I receive each month.

It is also a good idea for observers to make estimates of the sequence stars on each chart for their programme stars on a regular basis, preferably using the Pogson step method. This results in (1) a check of the sequence for accuracy (many sequences are thought to be unreliable), and (2) a check on the observer's own accuracy in the use of the step method.

The message that quality is what counts is often lost, especially as many variable-star organizations list the number of observations made by each observer and not the quality of them. A smaller number of good-quality consistent observations of a continuous programme is what counts in the end.

With regard to telescopes, I prefer to observe stars no more than 3 magnitudes above the limit of the instrument — perhaps 2 magnitudes is even better. This means that the observer covering stars of a wide range of magnitude needs a variety of instruments. For example:

8 × 30 binoculars — mag. 5–8.

20 × 80 binoculars–mag. 8–10.5.

Telescope — mag. 10.5–14.

A really good finder is essential.

Several dozen semiregular and irregular variables can be studied throughout their entire range with small binoculars (8 × 30 or 10 × 50), while larger binoculars can bring hundreds of interesting stars into range, including a number of the important Z And type, and a dozen Mira stars over their entire range.

Large telescopes (300 mm and upwards) are particularly valuable for studying faint types such as dwarf novae. Larger Dobsonians (450 mm or so) can make a substantial contribution by covering stars in the magnitude 13–15 range, including the maxima of some more-recently discovered stars and the minima of the brightest dwarf novae and novae.

The observer will of course need a chart for each variable. As for atlases, I would recommend *Nortons 2000.0* (Longman) for small binoculars, *Sky Atlas 2000.0* (Cambridge University Press) for large binoculars, and *Uranometria 2000.0* (Willmann-Bell, Inc) for telescopes, although you can't beat having all three!

Summing up, I would stress the following necessary characteristics of a variable-star observer:

1. A studious interest in accuracy above all else.
2. An attempt to cover all programme stars continuously throughout the year, including the summer months and morning work.
3. Use of consistent methods, with continuous checking.
4. Working as a member of a team.
5. Sending reports in a clear and legible format on a regular basis and in good time.

CHAPTER ELEVEN · *Observing Double Stars*

GLENN CHAPLE
TOWNSEND, MASSACHUSETTS, USA

NO OTHER CLASS OF SKY OBJECT CAN MATCH DOUBLE STARS FOR sheer numbers and diversity of appearance. More than 73 000 are listed in the *Washington Double Star Catalogue* [1]; indeed, astronomers estimate that perhaps one-half of all stars in our night sky are actually double, triple or multiple. Visually, double stars run the gamut from the colourful topaz and sapphire jewels that comprise Albireo, to stunning snow-white twins like γ Arietis, to delicately beautiful, unequal pairs like Rigel. Many are wide enough to be separated by binoculars, while others challenge the resolving power of the largest telescopes. Some are bright like Castor, while legions more lie at the magnitude limit of large telescopes.

Of great interest to the professional astronomer is the fast-moving visual binary, whose component stars undergo noticeable orbital motion in a few decades. Patient study of the motion of such a system will yield its orbital elements, which, in turn, will provide the masses of its component stars — the only means astronomers have of determining stellar mass.

A century ago, when the mainstay of the astronomer was the long-focus refractor, double stars enjoyed unparalleled popularity. Today, the astronomical scene is dominated by the short-focus reflector, whose wide field is more suited for extended objects like clusters, nebulae and galaxies. Modern-day discoveries about the nature of these celestial bodies, coupled with their popularity as targets for the astrophotographer, have pushed double stars out of the limelight.

As a result, a small band of professional astronomers faces the monumental task of shepherding an overwhelming flock of known double stars. Some visual binaries go neglected during critical phases of their orbits. There is a definite need for help from the amateur ranks. It is the author's hope that this chapter will inspire the reader to embark on a programme of double-star study.

HISTORY

A detailed history of double star astronomy would surely encompass this entire chapter, therefore a brief summary is provided. For detailed information, refer to the historical sketch in Robert G. Aitken's classic *The Binary Stars* [2], or the biographical notes in the *Webb Society Deep-Sky Observer's Handbook* [3].

Considering the sheer numbers of double stars, it is hard to believe that over four decades elapsed between the invention of the telescope and the discovery of the first double star in 1650. The pair, Mizar, was accidentally found by Jean Baptiste Riccioli. Subsequent finds were likewise accidental and undoubtedly considered to be mere celestial curiosities.

By 1779, when Christian Mayer at Mannheim released the first double-star catalogue — a modest listing of some 80 pairs — double stars had become something to be reckoned with. Were they chance alignments of stars actually light-years apart (an opportunity to determine the parallax, then distance, of the near star) or true binary systems (a chance to test the workings of gravity in deep space)?

Working on the first assumption, William Herschel began a systematic search for double stars that uncovered over 800 pairs (and led to the serendipitous discovery of Uranus in 1781). These were the first doubles to be measured with a micrometer — a fortunate occurrence, because subsequent measures by Herschel proved that some were undergoing orbital motion and were gravitationally bound.

In the early decades of the nineteenth century more double star discoveries were made by Herschel's son, John (whose finds, designated by the letter 'h', were mostly from the southern skies), and by F.G.W. Struve (Σ), who resurveyed the northern sky with a 230-mm refractor. The two alone accounted for over 8000 new pairs.

A period during the mid-1800s when attention turned to accurate measures of known double stars was followed by a rebirth of discovery. In 1873, S.W. Burnham published a list of 81 new pairs found with a 150-mm refractor. During the next 40 years, Burnham worked with various refractors to uncover over 1340 new doubles, and published a complete catalogue [4]. Catalogues were also produced by Hough, Espin, Aitken and Jonkheere [5–8].

Double-star astronomy continues to the present, although on a much reduced level. Working with a 500-mm refractor at Nice, Paul Couteau has discovered over 2000 double stars since 1951 [9]. Couteau also supplies many double-star measures, as do Charles Worley [10] and Wulff D. Heintz [11]. Their work is augmented by observations from amateur astronomers, usually under the guidance of groups like the Webb Society. Cooperation between the professional astronomer and his amateur counterpart may well underscore the success of future double-star astronomy.

THE NATURE OF DOUBLE STARS

Visual double stars fall into two categories: optical pairs which are aligned by chance, and the far more numerous binary systems whose member stars are gravitationally bound. To the casual observer, the distinction is a moot point; the aesthetic appeal of two stars in close proximity is all that counts. Optical pairs, however, are of no use to the astronomer who wishes to calculate an orbit and thus derive the masses of the stars involved.

A description of a double star includes four pieces of information; the magnitudes of the member stars, separation, position angle, and year the separation and position angle were measured. All help the observer select which pairs to observe and make for positive identification at the eyepiece.

Magnitudes – The magnitudes describe the brightness of the member stars. The brighter component is called the primary, or 'A' star; its dimmer companion is the secondary 'B' star. Some books refer to an extremely faint secondary as a *comes*. If the stars are identical in magnitude, the discoverer designates the primary. The magnitudes in a double-star catalogue help you avoid pairs too faint or unequal for your telescope.

Separation – The separation is the apparent distance between the component stars, expressed in seconds of arc. One arc-second (1″ arc) equals 1/3600 of a degree — equal to the apparent diameter of a garden pea viewed from a distance of over one kilometre! A telescope's aperture determines the closest double star it can resolve. The Dawes Limit defines this as $R = 115.8/A''$ arc, where R is the limit of resolution in arc seconds and A is the telescope's aperture in millimetres. A 150-mm telescope, for example, will separate double stars as close as 0.8″ arc. The Dawes Limit assumes stars of equal brightness and about 6th magnitude. Brighter stars have larger images that tend to merge before the resolution limit is reached. Faint or unequal pairs have component stars that are difficult for the eye to detect and also fail to adhere to the Dawes Limit.

Position angle – The position angle defines the location of the fainter star, relative to the primary. It is measured eastward from an imaginary line traced from the primary towards the North Celestial Pole to a second line passing from the primary through the secondary. Along with the separation, the position angle is critical to the astronomer who wishes to compute the orbit of a binary star. It also lets the casual observer confirm the sighting of a faint companion.

Epoch – Because the components of a double star system are in motion, you must know the year that the published separation and position angle were measured. Do not use an 1860 handbook to obtain the separation and position angle of a double star whose orbital period is 200 years. By now, the companion may be on the opposite side of the primary! Most wide pairs (separations of 20″ arc, or more) have orbital periods of many centuries, so that early measures are still reasonably valid. Close binaries, however, demand current data. The double-star list in this chapter uses measures that are as recent as possible and includes predicted locations for fast-moving binaries.

Orbits – Studies of the ever-changing separation and position angle of a binary star allow astronomers to plot its orbit. The greater the portion of the orbit observed, the more accurate its orbital plot. At present, the binary stars with well-known orbits are those that have completed the better part of one full cycle since measures of separation and position angle were first made — namely, those with orbital periods of less than one or two centuries.

A complete description of the orbit of a binary star requires seven elements.

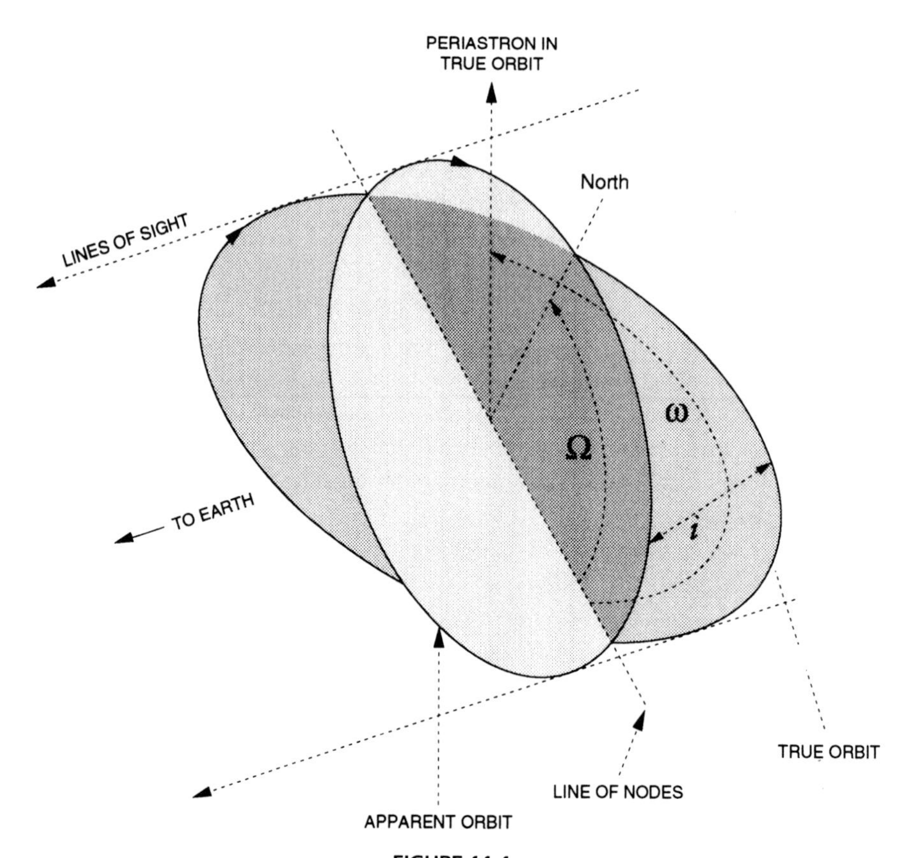

FIGURE 11.1
True and apparent orbits of a binary star. The heavy ellipse represents the true orbit, while the light ellipse is the apparent orbit as seen from the Earth. The constants i, ω, and Ω are shown (see text).

Four describe the true orbit, and include the semimajor axis (a), eccentricity (e), date of the companion's passage through periastron or its closest approach to the primary (T), and period of revolution (P) in years. Since most binary stars have orbits that are tilted to our line of sight, we must add the inclination of the true orbit to the plane of the sky (i), the position angle of the ascending node, which is the direction of the orbit plane's line of intersection with the plane of the sky (Ω), and the argument of periastron, or the angle, in the orbit plane, between the ascending node and the periastron point, measured in the direction of the companion's motion (ω).

Once the orbital elements of a binary star are known, the masses of its component stars can be determined through the use of the equation $m_1 + m_2 = a^3/P^2\pi^3$, where π is the parallax of the system. The determination of the individual masses of the component stars — their mass ratios — requires separate study.

TELESCOPES FOR DOUBLE-STAR OBSERVING

For casual observing of double stars, virtually any quality optical device will work. Even a common pair of binoculars will capture several dozen wide pairs. This number increases dramatically as we move up to small-aperture telescopes. A 75-mm refractor can resolve several thousand pairs.

The requirements for serious work, such as double-star photography or micrometer measures of separation and position angle, are similar to those for lunar and planetary work. A minimum aperture of 150-mm and an accurate motor drive are essential. Long-focus instruments (those with high focal ratios) work better than rich-field telescopes.

Owners of large Dobsonian-mounted reflectors are often disappointed with the spurious star images such instruments produce, especially when high magnifications are used. The solution is an off-axis aperture mask that effectively reduces the aperture while increasing the focal ratio. In this way, a 300-mm f/4 reflector can be converted into a 100-mm f/12 system. The loss of brightness is offset by increased image sharpness.

An optically perfect telescope used on a night of ideal seeing will be rendered useless if your eyepiece is of poor quality. An eyepiece for double-star observing should produce a crisp, ghost-free image at high magnification. The field does not have to be wide or uniformly sharp to the edge, since the typical double star will take up only the very centre of the field. Experiment to see which eyepiece design best suits your telescope while providing the greatest eye comfort. Orthoscopic, Plossl, and the 'solid' Tolles and monocentric designs seem to work best.

OBSERVING CONDITIONS

As with any astronomical endeavour, the choice of observing site is an important consideration. Select an area that is open and removed from sources of air turbulence. Double stars do not require dark skies. It is seeing, not transparency, that is critical for double-star work. In fact, the air over a city, though often awash with light pollution, is often more stable than country air.

Seeing conditions vary dramatically from one location to the next, and from moment to moment. The best rule of thumb is simply to observe whenever you can. A rough estimate of seeing conditions can be made by observing bright stars with the unaided eye. The more they twinkle, the poorer the seeing.

A more accurate method of judging the seeing is by aiming the telescope at a bright star and looking into the eyepiece holder, minus the eyepiece. You will see a large blob of light (the star's out-of-focus image spread across the objective). The oval-shaped patches drifting across the light are seeing cells. The larger they are and the slower they move, the better the seeing.

The ultimate test of the seeing is the appearance of a star's image viewed with high power. Under poor seeing, all you will see is a boiling blob of light. As seeing improves, the image becomes crisper, and is surrounded by a ring of light. The appearance of two more such rings arranged concentrically around the spurious Airy disk indicates excellent seeing conditions.

THE EYE

Without question, your most important piece of observing equipment is your eye. Be sure you are well-rested at the onset of an observing session and have allowed your eyes ample time to dark-adapt. While outside, avoid exposure to bright lights; cover your flashlight with red cellophane to reduce glare. High-power observing can quickly tire the eye, so take an occasional break.

If you suffer from astigmatism, star images will be distorted and close pairs hard to resolve. Use eyepieces that provide plenty of eye relief, allowing you to wear eye-glasses while observing. If you prefer not to wear glasses, ask your optometrist about contact lenses, or experiment to find eyepieces that are least affected by your astigmatism.

Finally, watch what you eat and drink! Some astronomers include carrots in their diet, because the vitamin A in carrots is claimed to promote the production of visual purple — a substance that improves night vision. Avoid smoking and alcohol prior to going out. The former retards dark adaptation, while over-consumption of the latter may make every star you observe appear double!

DOUBLE STARS AS TEST OBJECTS

Perhaps you are thinking of using close double stars to test your telescope's resolving power. It certainly was the fashion until quite recently to publish lists of 'test objects' for different apertures, based on the fact, as was mentioned earlier, that a telescope's limit of resolution is mathematically predicted by dividing its aperture, in millimetres, into 115.8″ arc. However there are numerous objections to using double stars for this purpose:

Magnitude of the object: If the stars are too bright, they dazzle, their excess light revealing imperfections that have nothing to do with the telescope's optical quality (dust on the optical surfaces, flaws in the eye, and of course air currents), all of which spread light beyond the Airy disk. But if they are too faint, the eye's ability to resolve is reduced. Optimum resolution occurs at about three or four magnitudes above the telescope's limit.

Relative brightness: A difference of only a magnitude between the components of a close pair makes it significantly harder to resolve. Increase the difference still more, and the textbook limit certainly will not apply.

Seeing conditions: Without periods of exceptional steadiness, the telescope will have no chance of resolving to its theoretical limit.

What this means is that a telescope failing to resolve to its theoretical limit is not necessarily a bad one! But it is also a fact that successful resolution of a close double star is not the 'ultimate' test. A far better test of optical quality is a telescope's ability to reveal fine planetary detail. This is because a really good telescope provides an image of high contrast as well as good resolution, and since planetary markings are of much lower contrast than a double star projected against a black sky, their clear visibility requires even better optical quality than simply dividing two adjacent stars. Having said that, it is also true that some experience is necessary to know just what sort of detail on, say, Jupiter, should be expected with telescopes of different aperture.

(One very simple and excellent test is to view a fairly bright star with a high-power eyepiece moved alternately a few millimetres inside and outside best focus. If the expanded disks look similar, the telescope is a good one.)

Long-focus instruments are more likely to achieve high-quality star images than are short-focus telescopes. It has already been mentioned that for double-star work aperture masks can be used on the latter.

Close double stars require the highest possible magnification — typically, × 1.5–2 per millimetre of aperture. This mandates an evening when the seeing is excellent. Even then, you may have to wait patiently for the atmosphere to settle, turning that elongated blob into two distinct points of light.

DIVIDING UNEQUAL PAIRS

When one member of a double-star system is substantially fainter than the primary, it is often lost in the primary's glare. It may even be hidden by one of the diffraction rings of the primary.

High magification is recommended for unequal pairs. This widens the gap between the two stars and increases contrast, making the fainter star more visible.

The best way to observe a wide, unequal double is with a sidewards glance — a technique known as averted vision. When you are looking slightly away from the double star, its image falls on a more sensitive part of the retina. A faint companion invisible by direct glance may suddenly pop into view with an averted glance.

If a close, unequal pair continually eludes you, try observing when the sky is not completely dark. Evening twilight, haze, even city lights or moonlight tend to reduce the glare of the primary and aid detection of the secondary, provided it is not too faint. The bright but unequal pair δ Cygni is a well-known candidate for this technique.

A hexagonal mask affixed to the top of the telescope tube also helps to resolve close, unequal pairs. The mask creates a spiked image of the primary, and, if properly rotated, allows the secondary to appear in the darkened sky between two of the spikes. [12]

COLOURS

Because of the vagaries of human vision, reports of double-star colours vary widely among individuals. The human eye is too easily tricked by illusions of contrast to be a reliable detector of double-star colour.

A remarkably simple, yet effective means of determining star colour by means of photography was developed and successfully tested by Joseph Kaznica and three other amateur astronomers at the Mt Cuba Astronomical Observatory in Wilmington, Delaware, USA. The technique, described in detail in the *Webb Society Deep-Sky Observer's Handbook* [13], ultimately places each star into one of nine colour categories from 'strong bluish purple' through 'strong red' by matching the tint against a standard colour reference chart. The results closely match the stars' spectral classes.

DOUBLE-STAR DISCOVERY

During routine observing, you may encounter a double star that cannot be identified in any of the major catalogues. Probably, it will be a rather wide, faint pair — one that was ignored by earlier generations of double-star seekers. Such double stars usually turn out to be optical, and are of little interest to the professional astronomer.

There are binary stars that await discovery, but most are extremely close and observable only in the largest telescopes. For this reason, double-star discovery is primarily the realm of the professional astronomer, although amateurs with suitable photoelectric equipment still have a chance of detecting very close binaries by observing the way a star's light fades when it is occulted by the Moon.

ESTIMATING SEPARATION AND POSITION ANGLE

You may need to determine separation and position angle in order to confirm the sighting of a faint companion or to identify positively a pair in a field inhabited by other double stars. While it is true that a micrometer is needed for accurate measures of separation and position angle, good estimates of both can be made by the experienced observer with no special equipment whatever.

Usually, separation can be estimated to an accuracy within a few arc seconds simply by comparing the double with nearby pairs of known separation. An experienced observer often has such a 'feel' for separations that even these comparisons are not needed. He intuitively knows that the given pair is so many arc seconds apart.

Position angle is found by centring the double star in the field and allowing it to drift out of view. The point at which the pair exits the field is 'west' (270°). Going clockwise around the field in 90° increments gives south (180°), east (90°), and north (0°). Remember that the use of a right-angle viewer in refractors and catadioptrics given a mirror image, reversing east and west. Once the cardinal points have been established in the field, you can estimate the position angle to within an accuracy of 10° or 20°.

MICROMETER MEASUREMENTS

Micrometer measurement of the separation and position angle of double stars is an area where the amateur can make important astronomical contributions. Unfortunately, very few amateur astronomers pursue this activity, believing that a large observatory instrument, equipped with an expensive micrometer, is necessary. However, like most things, there are different ways of achieving results.

Double-star micrometers can be divided into three main classes:

1. Those where the separation between the two stars is directly measured, in the focal plane.
2. Those where the real stars are compared with an artificial double star of known separation.
3. Those where the star images are optically modified in some way.

Direct measurement – The filar micrometer, consisting essentially of two threads (spider web) that can be closed and opened by a micrometer screw, is the 'classical' measuring instrument, and was used by the great observers of the past. However, a good micrometer is hard to come by, and can only be used on a telescope with a precise motor drive, since the star images must be fixed in the focal plane while the measurement is being made. The threads need to be made visible against the sky background, either by 'bright field' illumination or by illuminating the threads themselves. Either alternative can affect the visibility of faint stars.

Comparison method – The image of an artificial 'double star' of adjustable separation is projected into the field of view. If desired, the appearance can be altered to make it resemble the real star as closely as possible. The artificial image could be located next to the real star, or fed into the observer's other eye in a binocular arrangement. Such a device is readily home-made by a handy amateur, does not need a precisely-driven telescope, and does not require field illumination. It can be used on close and wide pairs. Some professional observers have preferred this system to the filar micrometer when measuring very close (or merely elongated) pairs.

Image modification – By dividing the lens of a refracting telescope into equal halves, and sliding one against the other, the two components of a double star can be merged, and the separation calculated. The same result can be achieved more conveniently by using adjustable prisms placed near the eyepiece. The construction of a 'double image' micrometer demands some workshop facilites, but among current professional observers it enjoys the same status as the filar micrometer did in former years. A precise motor drive is unnecessary.

A number of amateurs have derived satisfaction from measuring double stars with a 'diffraction micrometer'. Placing a series of parallel opaque strips known as a 'grating' in front of the telescope aperture produces a sequence of faint diffraction images to either side of each star, at right angles to the direction of the strips. Rotating the grating causes the lines of images to rotate, and by measuring the angle of rotation needed to bring the images of the two components into a certain configuration, the separation between the stars can be calculated. Normally, the strips are separated by an amount equal to their own width.

The coarser the grating, the closer the diffraction images, their separation being a function of the width of the strips and independent of the telescope aperture or focal length. For example, if the strips are 10-mm apart the diffraction images will be separated by 11.5″ arc, but if the strips are 50-mm apart the separation will be only 2.3″ arc.

Drawbacks of the diffraction micrometer include the amount of light lost from the star images through obstruction and diffraction, and the fact that it is not well suited to measuring very close pairs. Despite this, it has done good work in the hands of amateurs. More information about micrometers can be found in references [14] and [15].

Many observatories, particularly if they were established several decades ago, may well have a filar micrometer collecting dust in an out-of-the-way drawer, and if you can develop a good working relationship with a staff member you may be able to borrow it (or even make use of one of the telescopes!). Should money be no problem, you may opt to purchase one. Companies selling micrometers are listed at the end of this chapter.

CALCULATING AN ORBIT

Accurately computing the orbit of a binary star typically requires data on separation and position angle made over a period of many decades. Most amateur astronomers are therefore content to turn their measures over to the professional astronomer to be added to the pool of earlier measures and eventually used for orbital calculations. Couteau's *Observing Visual Double Stars* [16] and Aitken's classic [17) provide detailed information on computing orbits.

It is possible to take the orbital elements of a binary star and calculate the position angle and separation for any date of interest. This is especially useful if the data in your handbook are out of date, and you wish to know beforehand whether a particular binary star is resolvable with your telescope. Several computer programs have been created to help eliminate the messy mathematics that such calculations demand. A particularly handy program [18], written in Integer Basic and adaptable to most home computers, allows you to plug in the seven orbital elements of a binary star (orbital elements for over 500 binaries appear in Volume 2 of the *Sky Catalogue 2000.0* [19]) and then derive the separation and position angle for whatever year you desire.

SKETCHING DOUBLE STARS

It is encouraging to note that in this age of astrophotography overkill, many amateurs are discovering the simple pleasure of making astronomical sketches at the eyepiece. Astronomical sketching is kinder to your bank account, it hones your observing skills, and it allows you to portray detail that eludes the camera.

Sketching requires a medium-soft lead pencil, notebook or file cards, and a clipboard. Circles representing the eyepiece field should be traced out beforehand in the notebook or on the file cards. For illumination, use a standard flashlight, covered with red cellophane to preserve night vision.

Before beginning a sketch, you should compose the field much as an astrophotographer would. In most cases, this means centring the double star in the field. If there is something of note close by (a bright star, another double, or a galaxy or cluster), you may prefer to place the double off-centre.

Use a magnification that comfortably separates the component stars while retaining important field objects. Sketch the double star first, portraying relative brightnesses by the size of the pencil dots.

Once the double is sketched, add nearby field stars. Draw in the brightest ones first, paying attention to their distance and location, relative to the double star. Once these are in place, add the fainter ones. If the field is cluttered with faint stars, you may want to leave them out. Finish your sketch by noting the location of north. Add pertinent notes. These can be placed beneath the sketch, and should include the identity of the double star, the telescope used, magnification, field diameter, date and time, sky conditions, and descriptive notes (colours of the components, ease of location and resolution). As your collection of astronomical sketches grows, these notes will be of increasing value.

DOUBLE-STAR PHOTOGRAPHY

Glance at the astrophotos in any astronomy magazine. How many show double stars? Probably none. Besides being less glamorous than nebulae, clusters and galaxies, double stars are more difficult to capture on film. The problem is that a double star, being a much smaller target, requires a high magnification to increase the effective focal length of the system by the necessary amount. To gain this large image scale, you will need to use eyepiece projection — an astrophotography technique that demands an extremely solid mount and accurate drive.

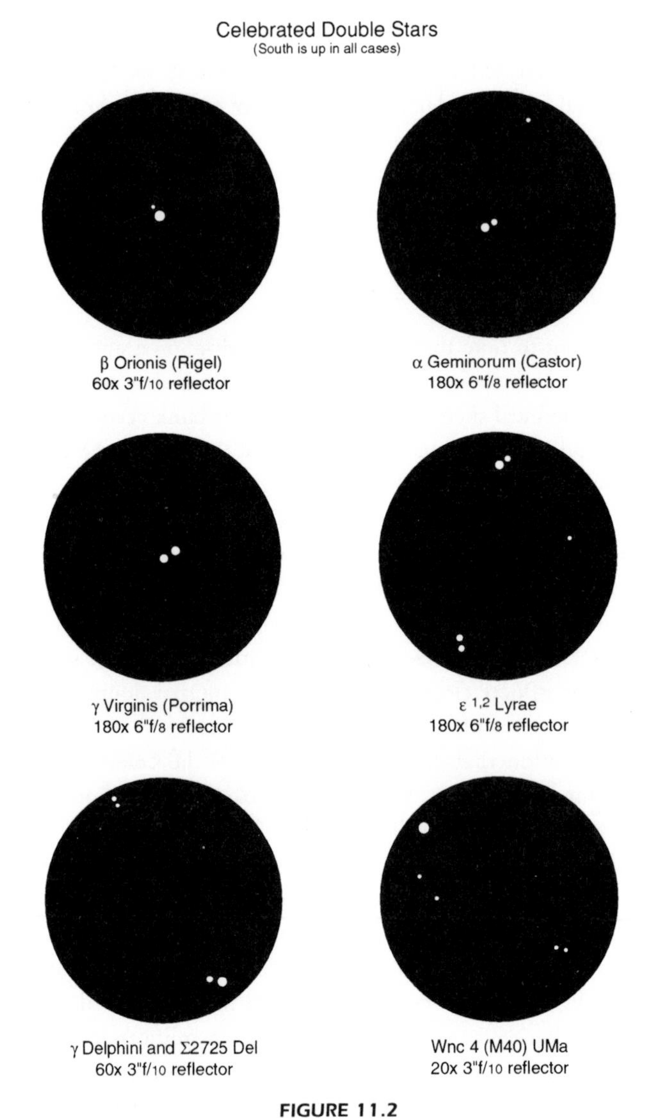

FIGURE 11.2
Drawings of some double stars, made by the author using small telescopes. (South is up.) (a) β Ori (Rigel); 75-mm f/10 reflector, ×60. (b) α Gem (Castor), 150-mm f/8 reflector, ×180. (c) γ Vir (Porrima); 150-mm f/8 reflector, ×180. (d) ϵ^1 and ϵ^2 Lyr: 150-mm f/8 reflector, ×180. (e) γ Del and ϵ 2725 Del; 75-mm f/10 reflector, ×60. (f) M40 in Ursa Major; 75-mm f/10 reflector, ×20.

Because of the numerous kinds of films available on the market today, you may have to experiment on your own. 'Fast' films tend to be grainy, while 'slow' films require longer exposures. Colour films vary in their sensitivity to certain wavelengths — something to consider when capturing rich-hued pairs like Albireo. Do not shy away from black and white film: the results can be quite attractive.

Exposure times will depend on the effective focal length of your system, the film type, magnitudes of the double star, and atmospheric conditions (sky glow can 'fog' film quickly). Turn to Chapter 14 for more information about this branch of amateur astronomy: if you persist to the point of success, you will be the proud owner of a unique set of astrophotos.

OCCULTATIONS OF DOUBLE STARS

Occasionally, the Moon passes in front of a binary star with startling results. Instead of instantly blinking out, as would be the case were it single, the star disappears in two stages as each component is occulted. Predictions for occultations of bright binary stars appear annually in the January issue of *Sky and Telescope* and are available from several sources [20].

PHOTOMETRY OF DOUBLE STARS

In many cases, the magnitudes published in double-star catalogues are visual estimates made by the discovering astronomer. If you own a large telescope (a reflector of long focal length is recommended) and a photometer, you can determine exact magnitudes of double stars. In the process, you may uncover a component that is variable in nature.

The redetermination of the magnitudes of wider, fainter, and therefore neglected pairs would be a useful programme. These are almost all optical doubles, and among the many pairs included in the older catalogues there may well be some variable companions. A study of the literature will also reveal instances where observers disagree about the difference in magnitude between the two components of a double star, and these could be worth investigating.

As mentioned above, a photoelectric photometer would be useful for accurate observations of lunar occulations of double stars. Besides collecting data on known binaries, astronomers using photoelectric equipment have discovered several extremely close systems. Photoelectric photometry is discussed in Chapter 15.

NON-SCIENTIFIC ACTIVITY

Hobbyists tend to become obsessive, and what was supposed to be an escape from the troubles of the day becomes a second job. While the gathering of useful data is a rewarding experience, we need to break away now and then and pursue frivolous activities that make astronomy a fun hobby. Observer Sirius when the great star is near the horizon. Sure, air currents near the ground act like vaporous prisms to distort the image, but look at the beautiful colours that spew forth! On a summer night, leave the charts inside and 'cruise' along the Milky Way with a low power eyepiece. What glorious surprises await the eye as one rich starfield after another parades by! Never mind trying to identify what you encounter. Enjoy the voyage.

Have you ever experienced a 'pendulum star?'. All you need to do is train your telescope on a wide, unequal double star (ε Pegasi is one of the best examples). While viewing the pair gently rock your telescope in a direction perpendicular to a line joining them. As if by magic, the lesser star seems to sway back and forth like a pendulum! When the telescope changes direction, the bright star follows suit, but the fainter star seems to continue on for a moment, always a step behind its master. The explanation, offered by John Herschel, is that the light from the lesser star takes more time to register on the retina. The effect is a surprise and delight to behold — a welcome respite from the rigours of 'serious' observing.

About the Author

Glenn Chaple was born in 1947, and became interested in astronomy in 1963 after being shown telescopic sights by a high-school friend. He took an astronomy degree at the University of Massachusetts, and in 1973 gained an MEd in Science Education. He now teaches science in a secondary school.

In 1977 he started a ten-year stint as double-star columnist for *Deep Sky Magazine*, and currently writes for the children's astronomy magazine *Odyssey*. He has contributed to several astronomy books, and in 1988 published *Exploring with a Telescope.*

He specializes in 'small-telescope' astronomy, having observed all the Messier objects, over 1500 double stars, and over 100 asteroids with a 75-mm reflector. With this and larger instruments he has logged over 20 000 variable-star observations. He is married with two children, enjoys fishing, and is an avid jogger and road racer.

Contact address – 82 South Harbour Road, Townsend, Massachusetts 01469, USA.

Societies

The Webb Society (Double Star Section): R.W. Argyle, c/o RGO (Private mail), Apartado 368, Santa Cruz de la Palma, Tenerife, Canary Islands.

Societe Astronomique de France (Double Star Section): Secretary, Pierre Durand, 34 Avenue Marcel Sembat, 18000 Bourges, France.

International Occulation Timing Association: David W. Dunham, 1177 Collins Avenue SW, Topeka, KS 66604, USA.

IOTA/ES: Hans Bode, Bartold-Knaust Strasse 8, D-3000 Hanover 91, Germany.

Suppliers of micrometers

Ron Darbinian, 1681 12th Street, Los Osos, California 93402, USA.

L.D. Reynolds, Retel, Ferndown Industrial Estate, 51 Cobham Road, Wimborne, Dorset BH21 7QZ, UK.

Notes and comments

Binary star obits – It may be useful to print the fairly simple computer program described in reference [18], which can be used to calculate the current appearance of any binary whose orbital elements are known.

Although both stars in a binary system are in motion around each other, it is customary to assume that the brighter star is stationary and that the fainter component describes an elliptical path around it. To define the shape of this path and the position of the companion on it, the seven orbital elements referred to in this chapter on pages 228–9 must be known. They are published in binary-star catalogues such as reference [19].

If this information is put into New Zealand amateur Michael Greaney's BASIC program, reproduced below, the position angle and separation are derived for any date. The program works in radians rather than degrees for evaluating trigonometric functions. — Ed.

```
10 REM        BINARY STAR ORBIT
12 REM
14 P1=4*ATN(1): R1=180/P1
16 C=2*P1: A1=0.0000005
18 REM
20 INPUT "SEMIMAJOR AXIS";A
22 INPUT "ECCENTRICITY  ";E0
24 IF INT(E0)=0 THEN 30
26 PRINT "NOT VALID; REENTER"
28 GOTO 22
30 INPUT "EPOCH OF PERIASTRON";T
32 INPUT "PERIOD IN YEARS       ";P
34 INPUT "INCLINATION           ";I
36 INPUT "ARG OF PERIASTRON     ";W
38 INPUT "P.A. OF ASC NODE      ";L
40 I=I/R1: W=W/R1: L=L/R1
42 REM
44 INPUT "DATE OF OBS (YR)      ";D
46 T1=D-T: M=C*T1/P
48 E=M+E0*SIN(M)+E0*E0*SIN(2*M)/2
50 N=E-E0*SIN(E): G=M-N
52 F=G/(1-E0*COS(E)): E=E+F
54 IF ABS(G)>A1 THEN 50
56 U=(1+E0)/(1-E0): H=COS(E/2)
58 IF H=0 THEN V=P1: GOTO 62
60 V=2*ATN(SQR(U)*SIN(E/2)/H)
62 K=V+W: Y=SIN(K)*COS(I)
```

```
64 X=COS(K): Q=ATN(Y/X)
66 IF X<0 THEN Q=Q+P1
68 IF X<0 THEN 72
70 IF Y<0 THEN Q=Q+C
72 P2=Q+L: IF P2>C THEN P2=P2-C
74 R=A-A*E0*COS(E)
76 S=R*X/COS(Q)
78 P3=INT(P2*R1*10+0.5)/10
80 S3=INT(S*100+0.5)/100
82 PRINT "P.A.   ";P3;" DEG"
84 PRINT "SEP.   ";S3;" ARC SEC"
86 PRINT
88 INPUT "ANOTHER (Y OR N)";Q$
90 IF Q$<>"N" THEN 44
92 END
```

References

1. Worley, Charles, *Washington Double Star Catalogue*. Washington DC: US Naval Observatory, in press.

2. Aitken, Robert G., *The Binary Stars*. New York: Dover, 1964. A classic reference which includes an in-depth history of double-star astronomy.

3. Glyn Jones, Kenneth, (ed.), *Webb Society Deep-sky Observer's Handbook*, vol. 1: Double Stars. Enslow Publishers, Hillside, NJ, 1986. An informative reference which includes detailed descriptions of micrometers and astrophotography techniques.

4. Burnham, S.W., *A New General Catalogue of Double Stars within 121° of the North Pole*. Publication No. 5 of the Carnegie Institute, Washington, 1906. Contains data on 13 665 double stars for epochs 1880.0 and 1900.0.

5. See ref. [2].

6. See ref. [2].

7. Aitken, Robert G., *A New General Catalogue of Double Stars within 120° of the North Pole*. Publication No. 417 of the Carnegie Institute, Washington, 1932. Includes measures of 17 181 doubles prior to 1927: epochs 1900.0 and 1950.0.

8. Jonkheere, R., *Catalogue Générale de 3,350 étoiles doubles de faible éclat observées de 1906 à 1962*. Observatoire de Marseilles, 1950. Also: Catalogue and measures of double stars discovered visually from 1905–1916 within 105° of the North Pole and under 5″ separation, *Memoirs of the RAS*, **17** (1917).

9. See ref. [16].

10. Worley, C.E. and Heintz, W.D., *Fourth Catalogue of Orbits of Visual Binary Stars*. Publications of the US Naval Observatory, 2nd series, 24, part 7, Washington, 1983.

11. See ref. [10].
12. The use of a hexagonal mask as an aid in resolving unequal pairs is mentioned on pages 17–18 of ref. [3].
13. See ref. [3].
14. See ref. [3].
15. Sidgwick, J.B., *Amateur Astronomer's Handbook*, 4th edn (revised by James Muirden). Enslow Publishers, Hillside, NJ, 1980. Excellent guide which includes detailed information on micrometers.
16. Paul Couteau, *Observing Visual Double Stars*. MIT Press, Cambridge, Mass., 1981. Similar to ref. [2], but more up-to-date.
17. See ref. [2].
18. Greaney, M.P., 'The orbit of a binary star'. *Sky & Telescope*, 71 (July 1987).
19. Hirshfeld, A. and Sinnott, R.W., *Sky Catalog 2000.0*, vol. 2: Double Stars, Variable Stars, and Non-stellar Objects. Sky Publishing Corp., Cambridge, Mass., 1985. Contains up-to-date data on over 8000 double and multiple stars whose combined magnitude is 8.0 or brighter. Also provides orbital data for 518 visual binaries.
20. Occultation predictions appear in *Sky & Telescope*, but are also issued by IOTA (International Occultation Timing Association), David W. Dunham, 1177 Collins Avenue SW, Topeka, CA 66604, USA.

Double-star list

Listed in this section are double and multiple stars of interest to the amateur astronomer of intermediate to advanced skill level. Each entry includes the designation, right ascension and declination (epoch 2000), magnitudes, separation in seconds of arc, position angle, and the year when the separation and position angle were measured. An 'o' following the year indicates that the separation and position angle were calculated from orbital elements. Pertinent descriptive notes are included where applicable.

The following conventions and abbreviations are used:

- *Right ascension:* 5-figure number giving hours, minutes and tenths of minutes. For example 00180 = 00 hours 18.0 minutes.
- *Declination:* 4-figure number prefixed n (north) or s (south), giving degrees and minutes. For example n4400 = +44° 00′.
- *Orbital period:* given as, for example, 3000Y (3000 years).
- Bin = binary; cpm = common proper motion; D = distance (from Sun); dec = decreasing; inc = increasing; sep = separation; sl = slowly.

Andromeda

Grb 34	00180n4400	8.1 + 11.0	35″.9	63°	1990o
	One of the nearest double stars; D = 11.7 LY Bin- about 3000Y; Sep dec, PA inc (34″.6, 66°, 2020)				
36 (Σ 73)	00550n2328	6.0 + 6.4	0″.8	292°	1990o
	Bin- 165Y; sep + PA inc (1″.2, 338°, 2020)				
γ (Σ 205)	02039n4220	2.3 + 5.1	9″.8	63°	1967
	Spectacular gold and blue colours; fainter star is a close binary (OΣ 38, 5.5 + 6.3, 0″.5, 106°, 1990o) with a highly elongated orbit and a period of 61Y, closing to less than 0″.1 (PA = 290°) in 2013				

Aquarius

(Σ ζ 2909)	22288s0001	4.3 + 4.5	1″.9	208°	1990o
	Bin- about 850Y; sep inc, PA dec (2″.7, 170°, 2020)				
107 (H II 24)	23460s1841	5.7 + 6.7	6″.6	136°	1971
	Sep inc, PA dec. (5″.5, 143°, 1823)				

Aquila

π (Σ 2583)	19487n1149	6.1 + 6.9	1″.4	110°	1960
	PA sl dec (1″.5, 121°, 1829)				

Ara

h 4949 — 17269s4551 — 6.0 + 6.7 — 2″.2 — 256° — 1953
Sep and PA sl dec (2″.9, 267°, 1836)

Aries

ε (Σ 333) — 02592n2120 — 5.2 + 5.5 — 1″.4 — 203° — 1966

Auriga

θ (OΣ 545) — 05597n3713 — 2.6 + 7.1 — 3″.6 — 313° — 1976
Sep inc, PA dec (2″.2, 6°, 1871)

Bootes

ζ (Σ 1865) — 14411n1344 — 4.5 + 4.6 — 1″.0 — 303° — 1990o
Bin- 123Y; sep and PA dec (0″.1, 263°, 2020)

ε (Σ 1877) — 14450n2704 — 2.5 + 4.9 — 2″.8 — 339° — 1971
PA inc (2″.6, 321°, 1829); Izar; a small scope challenge; yellow and blue-green

ξ (Σ 1888) — 14514n1906 — 4.7 + 7.0 — 7″.0 — 326° — 1990o
Bin- 152Y; sep and PA dec (5″.2, 296°, 2020); yellow and reddish

44 (Σ 1909) — 15038n4739 — 5.3 + 6.2v — 0″.9 — 33° — 1980o
Bin- 225Y; sep inc until 2008 (2″.4, 56°), then dec for several decades (0″.3, 132°, 2030)

Camelopardalis

2 (Σ 566) — 0400n5328 — 5.6 + 7.3 — 0″.8 — 229° — 1980o
Bin- 425Y; sep dec until 2000 (0″.7, 194°), then increasing (1″.0, 118°, 2050)

Cancer

ζ (Σ 1196) — 08122n1739 — 5.6 + 6.0 — 0″.6 — 182° — 1990o
+ 6.2 — 5″.9 — 76° — 1990o
Grand triple! AB is binary- 60Y; sep inc, PA dec (1″.2, 4°, 2020); AC bin- about 1150Y; sep inc, PA dec (6″.0, 64°, 2020)

57 (Σ 1291) — 08542n3035 — 6.0 + 6.5 — 1″.4 — 316° — 1960
PA sl dec (1″.5, 333°, 1829)

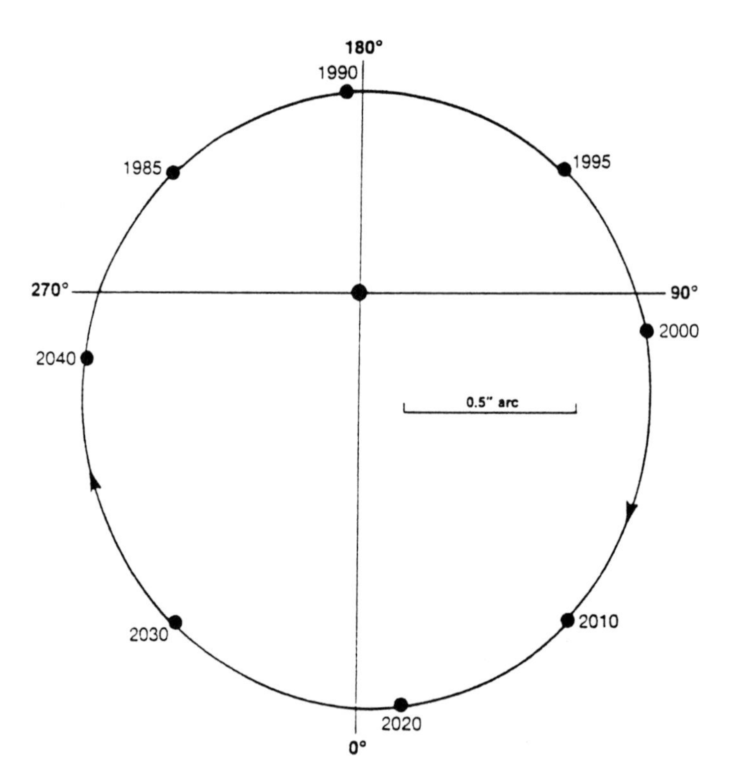

FIGURE 11.3
The orbit of ζ Cnc.

Canes Venatici

25 (Σ 1768)	13375n3618	5.0 + 6.9	1″.8	102°	1990o

Bin- 240Y; PA dec (1″.8, 94°, 2020)

Canis Major

α (AGC 1)	06451s1643	−1.5 + 8.5	4″.5	5°	1990o

Sirius; bin- 50Y, closing to 2″.5 (PA 295°) in 1993, greatest separation (11″.3, PA 63°) occurs in 2023; companion is white dwarf; 5th nearest star (D = 8.7 LY); classic light test for medium to large scopes

Cassiopeia

λ (OΣ 12)	00318n5431	5.5 + 5.8	0″.6	187°	1990o

Bin- 640y; PA sl inc (0″.5, 200°, 2020)

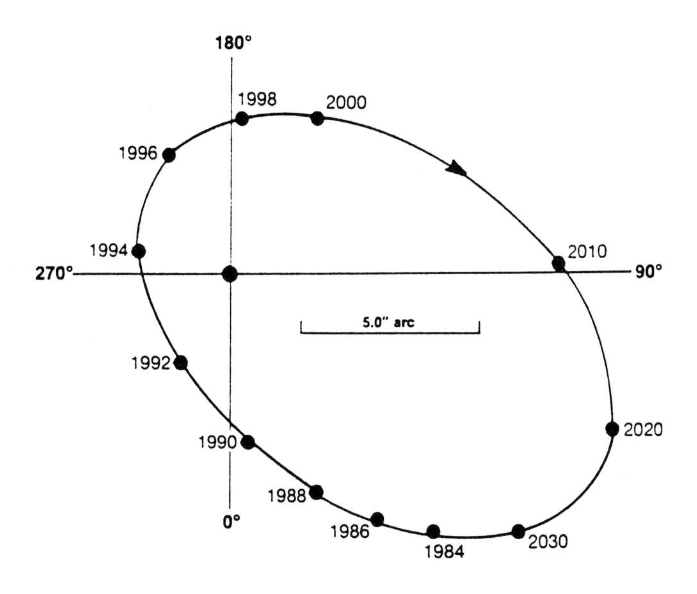

FIGURE 11.4
The orbit of α CMa (Sirius).

η (Σ 60)	00491n5749	3.4 + 7.5	12″.5	312°	1990o

Bin- 480Y; sep and PA inc (13″.5, 326°, 2020); beautiful yellow and red colours

ι (Σ 262)	02291n6724	4.6 + 6.9	2″.5	233°	1990o
		+ 8.4	7″.2	114°	1968

AB bin- 840y; sep inc, PA dec (2″.6, 226°, 2020); fine triple

Centaurus

γ (h 4539)	12415s4858	2.9 + 2.9	1″.4	353°	1990o

Bin- 85Y; sep dec until 2013 (0″.2, 268°), then inc (1″.6, 2°, 2050)

α	14396s6050	0.0 + 1.2	19″.7	215°	1990o
		+ 11.0	(131′)	sp	1915

Bin- 80Y; sep dec until 2016 (4″.0, 302°); then inc (10″.4, 21°, 2030); nearest star to our solar system, and the finest visual binary in the heavens; C is Proxima Centauri

Cepheus

Σ 460	04100n8042	5.5 + 6.3	0″.7	120°	1990o

Bin- 415Y; sep dec, PA inc (0″.5, 173°, 2020)

ξ (Σ 2863)	22038n6438	4.4 + 6.5	7″.7	277°	1974

Bin- about 3800Y; sep inc, PA dec (8″.5, 273°, 2020); pretty pair

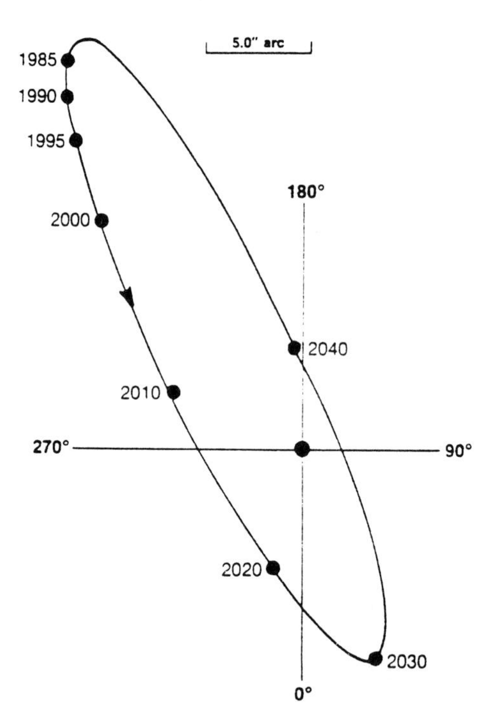

FIGURE 11.5
The orbit of α Cen.

Kr 60	22281n5742	9.8 + 11.5	3″.3	134°	1990o
		+ 10.1	73″.2	62°	1941
	AB bin- 45Y; sep inc until 1992 (3″.4, 126°), then dec (1″.4, 301°, 2015); red dwarf pair, one of the nearest visual binaries (D = 13.1 LY); C is optical				
π (OΣ 489)	23079n7523	4.6 + 6.6	1″.2	346°	1990o
	Bin- 150Y; sep and PA inc (1″.2, 6°, 2010)				
o (Σ 3001)	23186n6807	4.9 + 7.1	2″.9	217°	1980o
	Bin- about 800Y; sep dec, PA inc (2″.3, 239°, 2050)				
	Cetus				
β395	00373s2446	6.4 + 6.4	0″.8	111°	1990o
	Rapid bin- 25Y (0.1, 234°, 1997; 0″.6, 293°, 2001; 0.2, 2°, 2006; 0″.8, 109°, 2014)				
γ (Σ 299)	02433n0314	3.5 + 7.3	2″.8	294°	1955
	Chamaeleon				
δ' (I 294)	10453s8028	6.1 + 6.4	0″.6	76°	1946
	Cpm; PA increasing (0″.6, 61°, 1901)				

ε (h 4486)	11596s7813	5.4 + 6.0	0″.9	188°	1941
	Cpm; sep dec, PA inc (1″.7, 178°, 1835)				
	Circinus				
γ (h 4757)	15234s5919	5.1 + 5.5	0″.9	49°	1949
	Cpm; sep and PA dec (1″.2, 108°, 1836)				
	Columba				
β755	06354s3647	6.0 + 6.8	1″.3	258°	1959
	Sep sl inc (0″.9, 255°, 1887)				
	Coma Berenices				
35 (Σ 1687)	12533n2114	5.1 + 7.2	1″.1	173°	1990o
		+ 9.1	28″.7	126°	1958
	Bin- about 500Y; sep and PA inc (1″.3, 198°, 2020)				
α (Σ 1728)	13100n1732	5.0 + 5.1	0″.2	12°	1990o
	Rapid bin- 26Y (0″.0, 192°, 1989; 0″.5, 12°, 1993; 0″.0, 12°, 2000; 0″.6, 192°, 2010); orbit nearly edge-on				
	Corona Australis				
h 5014	18068s4325	5.7 + 5.7	1″.1	0°	1990o
	Bin- 191Y; sep and PA dec (0″.5, 237°, 2030)				
γ (h 5084)	19064s3704	4.8 + 5.1	1″.3	109°	1990o
	Bin- 120Y; sep inc, PA dec (1″.9, 293°, 2030)				
	Corona Borealis				
η (Σ 1937)	15232n3017	5.6 + 5.9	1″.0	28°	1990o
	Bin- 42Y; sep dec, PA inc (0″.5, 110°, 2005)				
γ (Σ 1967)	15427n2618	4.1 + 5.5	0″.6	118°	1990o
	Bin- 91Y; sep inc until 2001 (0″.8, 114°), then decreasing (0″.1, 20°, 2021)				
σ (Σ 2032)	16147n3352	5.9 + 6.6	6″.9	233°	1990o
	Bin- about 1000Y; sep and PA inc (7″.5, 239°, 2020)				
	Crux				
α	12266s6306	1.4 + 1.9	4″.4	115°	1955
		+ 4.9	90″.1	202°	1913
	Acrux; AB sep and PA sl dec (5″.6, 121°, 1826)				

Cygnus

Star	Position	Magnitudes	Sep	PA	Date
δ (Σ 2579)	19450n4508	2.9 + 6.3	2″.4	227°	1990o
	Bin- about 830Y; sep inc, PA dec (2″.7, 211°, 2020); magnitude difference makes this a difficult object for small scopes				
λ (OΣ 413)	20474n3629	4.9 + 6.1	0″.9	11°	1990o
	Bin- about 390Y; PA dec (0″.9, 6°, 2000)				
61 (Σ 2758)	21069n3845	5.2 + 6.1	29″.7	148°	1990o
	Bin- about 650Y; sep and PA inc (32″.1, 161°, 2050); one of the nearest binary systems (D = 11.1 LY); both golden yellow				
μ (Σ 2822)	21441n2845	4.9 + 6.1	1″.5	307°	1990o
	Bin- about 500Y; sep dec, PA inc (0″.8, 0°, 2015)				

Delphinus

Star	Position	Magnitudes	Sep	PA	Date
β (B151)	20375n1436	4.0 + 4.9	0″.3	181°	1990o
	Rapid bin- 27Y (0″.2, 268°, 1994; 0″.6, 357°, 2004; 0″.2, 87°, 2013; 0″.3, 181°, 2017)				
γ (Σ 2727)	20467n1607	4.5 + 5.5	9″.6	268°	1976
	Slight sep and PA dec (11″.9, 274°, 1830); slight colour contrast; Σ 2725 (7.6 + 8.4, 5″.8, 8°, 1959) in same field; beautiful!				

Draco

Star	Position	Magnitudes	Sep	PA	Date
17 (Σ 2078)	16362n5255	5.4 + 6.4	3″.4	108°	1958
		+ 5.5	90″.3	194°	1956
	AB sep and PA sl dec (3″.7, 116°, 1831); all cpm				
μ (Σ 2130)	17053n5428	5.7 + 5.7	1″.9	25°	1990o
	Bin- 482Y; sep inc, PA dec (2″.9, 310°, 2050) Beautiful twin binary				
Σ 2398	18433n5933	8.9 + 9.7	15″.3	163°	1962
	Bin- about 350Y; sep and PA inc (12″.0, 134°, 1832; one of the nearest binary systems (D = 11.3 LY)				
ε (Σ 2603)	19482n7016	3.8 + 7.4	3″.1	15°	1957
	Sep and PA sl inc (2″.8, 354°, 1832)				

Equuleus

Star	Position	Magnitudes	Sep	PA	Date
ε (Σ 2737)	20591n0418	6.0 + 6.3	1″.0	285°	1990o
		+ 7.1	10″.7	70°	1967
	AB bin- 101 Y; sep and PA dec (0″.0, 127°, 2019)				

Eridanus

p (Dun 5)	01398s5612	5.8 + 5.8	11″.3	193°	1990o
	Bin- about 480Y; sep inc, PA dec (11″.7, 186°, 2020); fine binary for small scopes				
f (Dun 16)	03486s3737	4.8 + 5.3	7″.9	212°	1957
	Sep and PA sl inc (7″.3, 200°, 1836)				
o^2 (Σ 518)	04152s0739	4.4 + 9.5	83″.4	104°	1970
	B is a binary (9.5 + 11.2, 9″.0, 335°, 1990o) that pairs a white dwarf with a red dwarf, P about 250Y, closing after 1990; One of the nearest multiple systems- D = 16 LY				

Fornax

α (h 3555)	03121s2859	4.0 + 7.0	4″.5	298°	1990o
	Bin- about 300Y; sep and PA inc (6″.0, 301°, 2030)				

Gemini

38 (Σ 982)	06546n1311	4.7 + 7.7	7″.1	146°	1990o
	Bin- 3000 + Y; sep inc, PA dec (7″.5, 137°, 2050)				
δ (Σ 1066)	07201n2159	3.6 + 8.2	5″.9	223°	1990o
	Bin- about 1200Y; sep dec, PA inc (4″.9, 237°, 2050); light test for 7.5-cm scopes				
α (Σ 1110)	07346n3153	1.9 + 2.9	3″.1	79°	1990o
		+ 8.8	72″.5	164°	1955
	Castor: AB bin- about 500Y; sep inc, PA dec (5″.6, 55°, 2020); brightest binary pair in northern sky; all three are spectroscopic binaries, making Castor a sextuple system!				
κ (OΣ 179)	07444n2424	3.6 + 8.1	7″.1	240°	1971
	Sep and PA sl inc (6″.2, 233°, 1853)				

Hercules

ζ (Σ 2084)	16413n3136	2.9 + 5.5	1″.6	85°	1990o
	Bin- 34.5Y; SEP dec until 2002 (0″.5, 302°), then inc (1″.6, 83°, 2025)				
α (Σ 2140)	17146n1423	3.5 + 5.4	4″.7	107°	1968
	Rasalgethi; PA sl dec (4″.6, 118°, 1829); primary reddish				
ρ (Σ 2161)	17237n3709	4.6 + 5.6	4″.1	316°	1958
	Sep and PA sl inc (3″.6, 307°, 1830)				

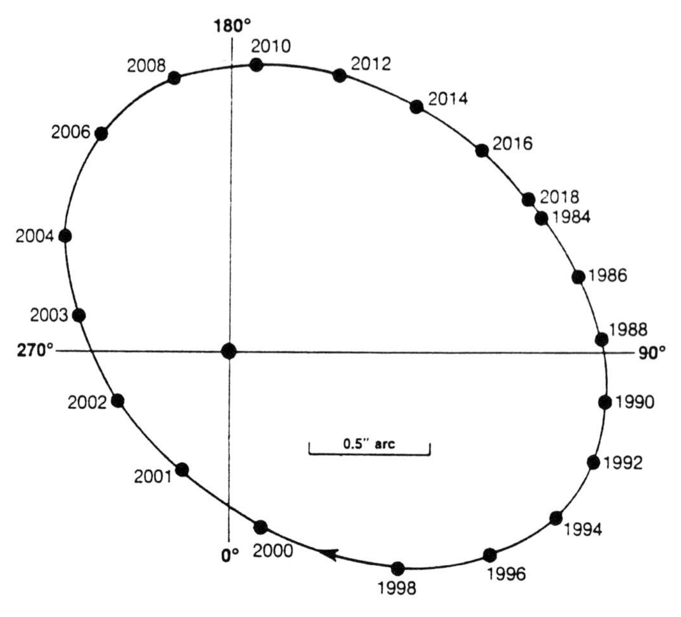

FIGURE 11.6
The orbit of ζ Her.

95 (Σ 2264)	18015n2136	5.0 + 5.1	6″.3	258°	1974

Interesting pair, due to the colours ascribed to it by various observers; Smyth and Webb called it 'apple green, cherry red', but most present-day observers see pale yellow and blue.

Hydra

ε (Σ 1273)	08468n0625	3.4 + 6.8	2″.7	294°	1990o

Bin- about 900Y; PA sl inc (2″.7, 318°, 2020); primary a close binary

β (h 4478)	11529s3354	4.7 + 5.5	0″.9	8°	1959

Sep dec, PA inc (2″.2, 340°, 1834)

54 (H III 97)	14460s2527	5.1 + 7.1	8″.6	126°	1954

Sep and PA dec (9″.6, 139°, 1783)

Indus

θ (h 5258)	21199s5327	4.5 + 7.0	6″.0	275°	1957

Sep inc, PA dec (3″.5, 305°, 1834); cpm

Leo

ω (Σ 1356) 09285n0903 5.9 + 6.5 0″.5 53° 1990o
Bin- 118Y; sep and PA inc (1″.0, 123°, 2030)

γ (Σ 1424) 10200n1951 2.2 + 3.5 4″.4 124° 1990o
Bin- about 600Y; sep and PA slowly inc (4″.5, 127°, 2020); grand object!

ι (Σ 1536) 11239n1032 4.0 + 6.7 1″.5 131° 1990o
Bin- 192Y; sep inc, PA dec (2″.7, 78°, 2050)

Lepus

h 3752 05218s2446 5.4 + 6.6 3″.2 97° 1953
+ 9.1 61″.2 105° 1898
AB PA sl dec (3″.2, 110°, 1837); pretty pair; M79 35′ to ENE

Libra

μ (β106) 14493s1409 5.8 + 6.7 1″.8 355° 1958
Sep and PA inc (1″.4, 335°, 1875)

Lupus

π (h 4728) 15051s4703 4.6 + 4.7 1″.4 73° 1956
Sep inc, PA dec (0″.8, 111°, 1835; elongated in 7.5-cm

μ (h 4753) 15185s4753 5.1 + 5.2 1″.2 142° 1955
+ 7.2 23″.7 130° 1955
AB sep and PA dec (2″.0, 174°, 1836); AC = Dun 180; all cpm

γ (h 4786) 15351s4110 3.5 + 3.6 0″.7 275° 1990o
Bin- 147Y; sep and PA dec (0″.0, 190°, 2030)

Lynx

12 (Σ 948) 06462n5927 5.4 + 6.0 1″.7 74° 1990o
+ 7.3 8″.7 308° 1959
AB bin- about 700Y; PA dec (1″.7, 44°, 2050) One of the finest triples in the northern sky

15 (OΣ 159) 06573n5825 4.8 + 5.9 0″.9 33° 1959
PA and sep inc (0″.5, 323°, 1844); primary yellowish

38 (Σ 1334) 09188n3648 3.9 + 6.6 2″.7 229° 1968
PA sl dec (2″.7, 240°, 1829)

Lyra

ε′ (Σ 2382)	18443n3940	5.0 + 6.1	2″.6	357°	1976
	Bin- about 1200Y; sep and PA dec (2″.5, 344°, 2020)				
ε^2 (Σ 2383)	18443n3937	5.2 + 5.5	2″.3	94°	1975
	Bin- about 580Y; sep sl inc, PA dec (2″.4, 73°, 2020); with ε^1, forms the celebrated 'double-double' (sep 3.5′)				

Musca

h 4432	11234s6457	5.4 + 6.6	2″.3	303°	1947
	PA sl inc (2″.3, 288°, 1836)				
β (R 207)	12463s6806	3.7 + 4.0	1″.3	36°	1990o
	Bin- about 400Y; PA inc (1″.3, 79°, 2050)				

Octans

λ (h 5278)	21509s8243	5.4 + 7.7	3″.1	70°	1946
	PA sl dec (3″.1, 83°, 1835)				

Ophiuchus

ρ (H II 19)	16256s2327	5.3 + 6.0	3″.1	344°	1959
	Sep and PA dec (4″.1, 3°, 1822)				
λ (Σ 2055)	16309n0159	4.2 + 5.2	1″.5	22°	1990o
	Bin- 130Y; PA inc (1″.5, 46°, 2020)				
36 (S,h 243)	17153s2636	5.1 + 5.1	4″.8	151°	1990o
		+ 6.6	732″.0	74°	1960
	AB bin- about 550Y; sep inc, PA dec (5″.7, 131°, 2050); all three cpm				
Σ 2173	17304s0104	6.0 + 6.1	1″.2	334°	1990o
	Bin- 46Y; (0″.2, 234°, 2003; 0″.9, 153°, 2011; 0″.2, 46°, 2022; 1″.2, 334°, 2036)				
τ (Σ 2262)	18031s0811	5.2 + 5.9	1″.8	280°	1990o
	Bin- 280Y; sep dec, PA inc (1″.0, 313°, 2050)				
70 (Σ 2272)	18055n0230	4.2 + 6.0	1″.5	224°	1990o
	Bin- 88Y; sep inc PA dec (6″.7, 122°, 2020); ideal fast-moving binary for small scopes				

Orion

14 (OΣ 98)	05079n0830	5.8 + 6.5	0″.7	349°	1990o
	Bin- about 200Y; sep inc, PA dec (1″.1, 271°, 2030)				

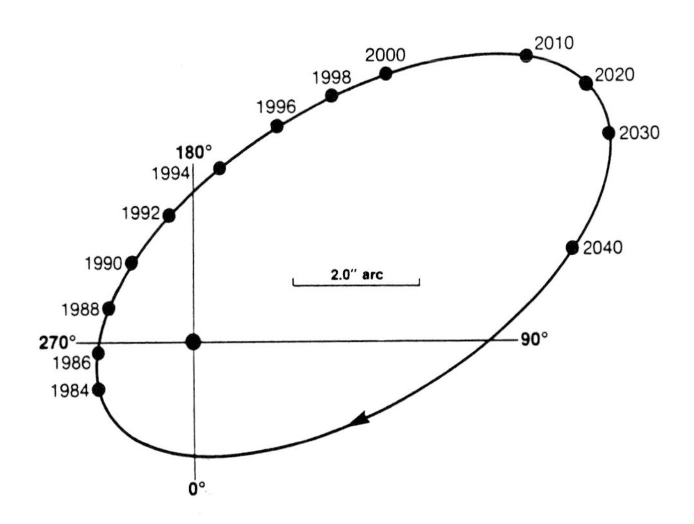

FIGURE 11.7
The orbit of 70 Oph.

β (Σ 668)	05145s0812	0.1 + 6.8	9″.5	202°	1954
	Rigel: light test for 7.5-cm scope.				
η (Da 5)	05245s0224	3.8 + 4.8	1″.5	80°	1959
	Sep inc, PA dec (1″.0, 87°, 1849)				
32 (Σ 728)	05308n0557	4.5 + 5.8	1″.0	42°	1990o
	Bin- about 600Y; PA dec (1″.1, 33°, 2040)				
ζ (Σ 774)	05408s0157	1.9 + 4.0	2″.3	164°	1990o
	Bin- about 1500Y; sep dec PA inc (2″.0, 171°, 2050)				
52 (Σ 795)	05480n0627	6.1 + 6.1	1″.6	210°	1959
	PA sl inc (1″.8, 200°, 1831); 7.5-cm test				
	Perseus				
θ (Σ 296)	02442n4914	4.1 + 9.9	19″.8	304°	1990o
	Bin- about 2700Y; sep and PA inc (20″.9, 306°, 2050)				
	Phoenix				
β (Slr 1)	01061s4643	4.0 + 4.2	1″.4	346°	1954
	Sep inc, PA dec (1″.0, 25°, 1891)				
	Pisces				
α (Σ 202)	02020n0246	4.2 + 5.2	1″.9	278°	1990o
	Bin- about 900Y; sep and PA dec (1″.7, 233°, 2050)				

	Puppis				
n (H N 19)	07343s2328	5.8 + 5.9	9″.6	114°	1952
	PA sl inc (9″.3, 105°, 1825)				
5 (Σ 1146)	07479s1212	5.6 + 7.7	2″.2	5°	1960
	Sep and PA dec (3″.3, 18°, 1831)				
	Sagittarius				
21 (Jc 6)	18253s2032	4.9 + 7.4	1″.8	289°	1959
	PA sl dec (1″.8, 297°, 1851)				
κ^2 (β763)	20239s4225	6.0 + 6.9	0″.8	234°	1952
	Sep dec, PA inc (1″.3, 204°, 1879)				
	Scorpius				
ξ (Σ 1998)	16044s1122	4.8 + 5.1	0″.8	39°	1989o
		+ 7.3	7″.6	51°	1975

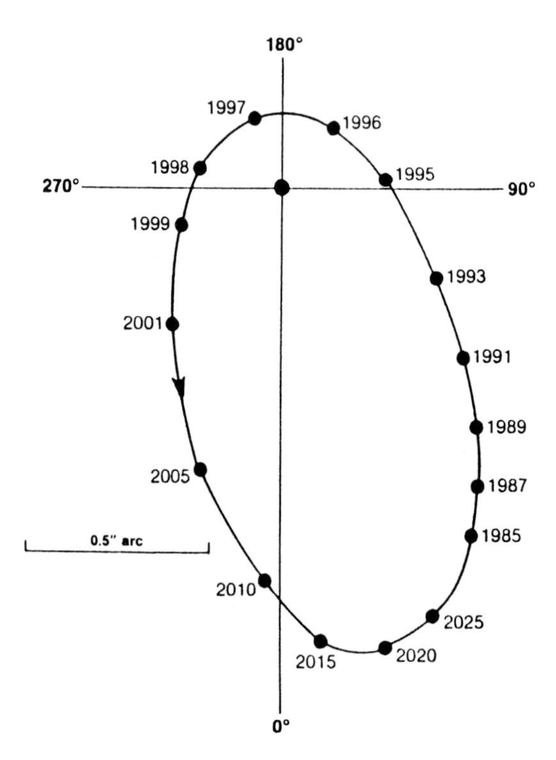

FIGURE 11.8
The orbit of ξ Sco.

	AB bin- 46Y; sep dec (0″.2, 203°, 1997), then inc (1″.2, 13°, 2120); AC sep inc, PA dec (6″.8, 79°, 1825); Σ 1999 (7.4 + 8.1, 11″.6, 99°, 1975) in same field				
ν (H V 6)	16120s1928	4.2 + 6.3	41″.1	337°	1955
	Primary is double (4.3 + 6.8, 0″.9, 3°, 1955), as is secondary (6.4 + 7.8, 2″.3, 51°, 1955); fine multiple system				
α	16294s2626	1.2v + 5.4	2″.7	274°	1990o
	Antares; Bin- about 900Y; sep dec (1″.9, 273°, 2050); edge-on orbit; red and green				
	Sculptor				
κ′ (β391)	00093s2759	6.1 + 6.2	1″.4	265°	1954
	Sep inc, PA dec (0″.9, 277°, 1876)				
τ (h 3447)	01361s2954	6.0 + 7.1	1″.9	335°	1990o
	Bin- about 2000Y; sep and PA inc (3″.2, 346°, 2050)				
	Serpens				
δ (Σ 1954)	15348n1032	4.2 + 5.2	4″.3	177°	1990o
	Bin- about 3000Y; sep inc, PA dec (4″.8, 173°, 2050)				
	Sextans				
γ (AC 5)	09525s0806	5.6 + 6.1	0″.6	67°	1990o
	Bin- 76Y; near widest sep through 2005 (0″.6, 51°), then closing (0″.1, 265°, 2033)				
	Taurus				
Σ 422	03368n0035	5.9 + 8.8	6″.7	268°	1990o
	Bin- 2000 + Y; PA inc (6″.8, 280°, 2050)				
80 (Σ 554)	04301n1538	5.7 + 8.0	1″.8	18°	1990o
	Bin- 190Y; Sep and PA sl dec (1″.1, 9°, 2050)				
	Triangulum				
ι (Σ 227)	02124n3018	5.3 + 6.9	3″.9	71°	1973
	PA sl dec (3″.8, 80°, 1836)				
	Triangulum Australe				
Rmk 20	15479s6527	6.3 + 6.3	1″.9	149°	1947
	Sep and PA dec (2″.4, 156°, 1835)				

Tucana

κ (h 3423)	01158s6853	5.1 + 7.3	5″.4	336°	1954
	Sep inc, PA dec (4″.7, 16°, 1836)				

Ursa Major

σ^2 (Σ 1306)	09104n6708	4.9 + 8.2	3″.6	358°	1990o
	Bin- 1000 + Y; sep inc, PA dec (5″.5, 339°, 2050)				
ξ (Σ 1523)	11182n3132	4.3 + 4.8	1″.3	60°	1990o
	Bin- 60Y; rapidly closing to 0″.9 (PA 0°, 1993), then again widening (3″.1, 115°, 2035)				
Wnc 4	12222n5805	9.0 + 9.3	50″.1	83°	1918
	A Messier object (M40)!				
78 (β 1082)	13007n5622	5.1 + 7.4	1″.5	57°	1990o
	Bin- 116Y; sep dec, PA inc (0″.6, 138°, 2030)				

Vela

I 67	08225s4829	5.2 + 6.2	0″.8	139°	1946
	Test for 15-cm				
δ (I 10)	08447s5443	2.1 + 5.1	2″.6	153°	1952
	PA dec (2″.6, 175°, 1894)				

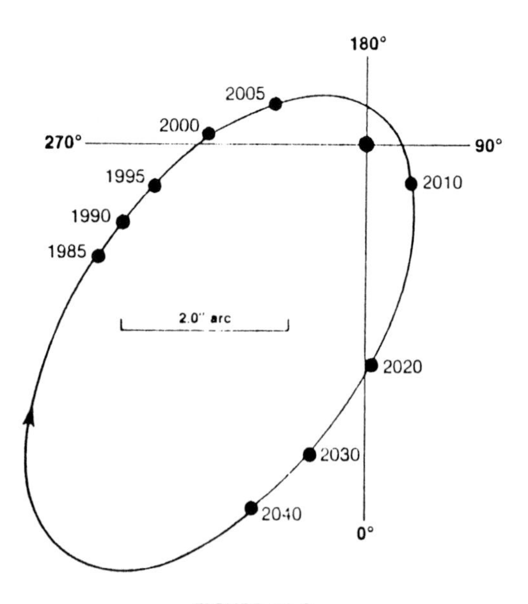

FIGURE 11.9
The orbit of γ Vir.

Virgo

γ (Σ 1670) 12417s0127 3.5 + 3.5 3″.0 287° 1990o
Porrima: bin-171Y; sep and PA dec through 2008 (0″.4, 126°), then sep inc (5″.1, 338°, 2050); beautiful twin pair

Vulpecula

16 (OΣ 395) 20020n2456 5.8 + 6.2 0″.8 115° 1959
PA inc (0″.6, 79°, 1844); resolution test for 15-cm

CHAPTER TWELVE · *Searching for Novae*

JAMES MUIRDEN
REWE, EXETER, ENGLAND

NOVA-HUNTING, AS A SERIOUS BRANCH OF AMATEUR ACTIVITY, IS of relatively recent origin. The reason for its sudden emergence is at first sight puzzling, since the discovery of a nova may well involve as many hours of fruitless searching as are required for a new comet, and the naming process does not 'immortalize' the discoverer in the same way. Who, for example, remembers the name of the observer who discovered the zero-magnitude Nova Persei in 1901, the brightest nova of the century? Nevertheless, a study of nova discoveries made over the past ten years, listed at the end of this chapter, shows beyond doubt that the contribution of a handful of amateurs overwhelms that of professional astronomers; there is also no doubt at all that discoverable novae are being missed and that there is scope for additional observers to make their mark. Furthermore, nova-hunting can produce tangible fringe benefits in the form of new variable stars and information about telescopic meteor activity, and sweeping across the Milky Way with binoculars is itself a delightful pastime.

THE NOVA PHENOMENON

All novae that have been examined in sufficient detail have proved to be binary pairs consisting of a red giant and a white dwarf star. Gas from the cool outer atmosphere of the red giant is swept up by its hot companion until a critical ignition point is reached. The brightness rise in a nova explosion is at least 10 magnitudes, and is usually achieved in less than a week; the return to its original brightness may take many years. Some novae are known to recur at intervals of a few decades, and it is possible that all novae are recurrent over very long time intervals.

The drastic nature of the phenomenon is part of its appeal: no matter how thoroughly a section of sky has been searched, a nova could have risen two or three magnitudes above the observer's search limit and be glaringly obvious on the very next night. By comparison, the comet-hunter can take a slightly more relaxed attitude, leaving with reasonable confidence a carefully-swept region for several nights before returning to it.

BRIGHTNESS AND DISTANCE

The maximum absolute magnitude of a typical nova is about −7 (although Nova Cygni 1975 reached the exceptional value of −10.2, second only to Nova Puppis of 1942, at magnitude −11.5). Making reasonable allowance for light absorption due to interstellar dust, Table 12.1 gives the approximate distance of standard novae of different apparent magnitude at maximum.

Nova Persei 1901, which reached magnitude 0.2 at maximum, has a distance of about 500 pc; Nova Cygni 1975 (magnitude 1.8) is about 1300 pc away. However, both of these novae were unusually luminous, and in any case distance estimates can only be made indirectly, based on assumptions about absolute magnitude, interstellar dimming, and the rate of expansion of any luminous shell following the outburst. The effect of interstellar dust is particularly important, since most novae are found near the plane of the Milky Way, where absorption can reach high values. In the case of typical novae, therefore, the zone containing objects peaking no fainter than magnitude 8 probably extends only a few thousand parsecs away from the Sun.

The brightness increase of a nova from the pre-outburst state to maximum brilliancy is at least 10 magnitudes, and may be much more. Nova Persei 1901 rose by 14 magnitudes, as did Nova Herculis 1934. Nova Cygni 1975, that most exceptional

TABLE 12.1
APPROXIMATE DISTANCES OF STANDARD NOVAE (ABSOLUTE MAGNITUDE -7), ACCORDING TO THEIR MAXIMUM APPARENT MAGNITUDE

Apparent mag.	Distance (pc)
0	150
2	400
4	1000
6	2500
8	6000

object, rose by at least 19 magnitudes. Few novae have been brighter than about magnitude 14 in their pre-outburst state.

Observations of many novae suggest that 15 days after maximum their absolute magnitude is about −5.2, which means that the more luminous the nova, the more rapidly it fades after maximum. The typical nova, of absolute magnitude −7, will therefore fade by two magnitudes in a fortnight, but a fast nova, such as Nova Aquilae 1982, fell by three magnitudes in only 10 days after maximum. Assuming that the rise to visibility occurs overnight, even a week of bad weather or neglect could mean that a faint but potentially discoverable nova rises and falls below the observer's threshold, unseen.

THE DISTRIBUTION OF NOVAE

Where novae occur, and where they are discovered, may be two quite different things. Except in the case of the bright naked-eye objects, novae will not be found where they are not being directly or indirectly sought.

The vast majority of novae have been discovered near the galactic equator, in the plane of the Milky Way and the Galaxy. Table 12.2 gives data for 119 novae brighter than magnitude 10.0 at maximum.

The concentration within the 20-degree band centred on the galactic equator is remarkable, 77% of the novae occurring inside it and 60% occurring in the 10-degree band south of the equator. Table 12.3 shows the distribution of the same novae in galactic longitude (0° marks the direction of the galactic centre in Sagittarius at 17 h 45 m, −29°, and longitude is measured eastwards).

TABLE 12.2

THE GALACTIC LATITUDE OF 119 NOVAE BRIGHTER THAN MAGNITUDE 10.0 AT MAXIMUM

No. of novae	Galactic latitude
0	> −39o
2	−39°– −30°
2	−29°– −20°
9	−19°– −10°
71	−9°–0°
21	+1°– +10°
7	+11°– +20°
2	+21°– +30°
3	+31°– +40°
2	>40°

TABLE 12.3

THE GALACTIC LONGITUDE OF 119 NOVAE BRIGHTER THAN MAGNITUDE 10.0 AT MAXIMUM

No. of novae	Galactic latitude
43	0°– 39°
18	40°– 79°
9	80°–119°
4	120°–159°
5	160°–199°
3	200°–239°
7	240°–279°
6	280°–319°
22	320°–359°

Again, a concentration is obvious, 70% of the discoveries being confined within an arc of 120 degrees around the direction of the galactic centre.

However, it does not follow that this observed concentration corresponds with the true distribution of novae. The great majority of these objects were probably fortuitous discoveries made during photographic patrol programmes for new variable stars or other objects of interest. These programmes concentrated upon the region of the galactic centre because it is particularly rich in such objects. Therefore, the number of novae discovered at other galactic latitudes and longitudes is almost certainly under-representative, assuming that a similar percentage of 'typical' stars in all observable regions of the Galaxy are likely to turn into novae.

This assumption seems reasonable. Novae are close binary stars in their early rather than later life; although one star is a red giant, it has evolved prematurely through mass loss to its companion. Novae are not observed in globular clusters, in old galaxies where star formation has ceased, or in old regions of spiral galaxies where the so-called Population 2 stars are found. Nor do they occur in clusters of very young stars. They are to be sought among the independent Population 1 stars in a galaxy's spiral arms, and with a minute exception all the stars detectable with amateur equipment fall into this category.

The following points are worth considering when planning a nova search programme.

1. Discoveries of novae brighter than magnitude 4.5 show less concentration around the galactic centre. Of 22 novae in this category, only 9 (40%) were within the galactic longitude range 320°–79°, compared with 70% of all novae.
2. However, the concentration towards the 20-degree band along the galactic equator is similar (68% compared with 77%).
3. Assuming that no novae falling within this brightness range were missed, this must represent the true distribution of the brightest (and on average nearest) novae.
4. It follows that the entire 20-degree band along the Milky Way must produce a far greater number of novae in the intermediate brightness range (say between magnitudes 5 and 8) than is suggested by the discovery figures, since these regions of the galactic arms are still up to several thousand parsecs away from the Sun (see Table 12.1).

THE FREQUENCY OF NOVA OCCURRENCES

A figure of approximately 20 novae per annum is generally accepted for galaxies like the Milky Way. This is based on theoretical work and on the numbers of novae observed in the few nearby galaxies capable of revealing them.

The discovery rate is about one-sixth of this, which is hardly surprising considering that the majority of events will be too distant to be made out using current search techniques. Even so, for the reasons stated above it is certain that many are being missed, and taking all the mitigating factors into account perhaps the true number is even greater.

To realize the number of novae that are being missed, consider the distribution of discoveries in the favoured Scorpius and Sagittarius regions since 1900, in 5-year periods (Table 12.4). These are all believed to have been made photographically. During the period up to the Second World War many of the discoveries were made by the distinguished ladies of the Harvard College Observatory. The bulge in the fifties was due principally to work by Haro and Zwicky. The discovery of novae was not their principal concern!

This bulge does not represent an extra contribution from fainter novae than ones that were being picked up earlier: all but two of them were recorded brighter than magnitude 10, and may have been brighter still at their true maxima. Four novae, ranging in brightness from magnitudes 4 to 11, were found in Sagittarius in 1936, and three in both of the years 1950 and 1952, the faintest of these being magnitude 9.8. Yet in the decade following 1954 not a single nova was recorded in Scorpius, and only three were noticed in Sagittarius! In the period 1980–90, thanks

TABLE 12.4
THE NUMBER OF NOVA DISCOVERIES MADE IN SCORPIUS AND SAGITTARIUS (THE REGION OF THE GALACTIC CENTRE), 1901–1965. THE VARIATION IS UNDOUBTEDLY DUE TO THE LEVEL OF PATROLLING CARRIED OUT

	Sco	Sgr	Total
1901–1905	1	2	3
1906–1910	1	1	2
1911–1915	0	3	3
1916–1920	0	2	2
1921–1925	2	2	4
1926–1930	1	5	6
1931–1935	0	3	3
1936–1940	0	5	5
1941–1945	2	4	6
1946–1950	3	4	7
1951–1955	4	8	12
1956–1960	0	1	1
1961–1965	1	1	2

to the efforts of a few amateurs, a total of 10 novae down to magnitude 10.5 were discovered, but almost certainly a greater number than this were missed.

SEARCH STRATEGIES

A summary of the above-mentioned facts and observations may be useful.

(1) Although novae may appear in any part of the sky, their far greater concentration near the plane of the Galaxy does not justify intensive searching elsewhere unless conditions demand it — for example, on a spring evening in the United Kingdom the Milky Way runs along the northern horizon.

(2) A general naked-eye survey of the whole sky to about 4th magnitude should be made on every possible occasion.

(3) Novae in the 5th–8th magnitude range may usefully be sought at all galactic longitudes, either visually or photographically.

(4) Observers south of about latitude 45°N could undertake intensive photographic surveys to at least magnitude 10 in the Scorpius–Sagittarius region, and the fainter the better. Historically the most productive region has extended from galactic longitude 350° to 010° and from latitude 0° to −10°. Based on the evidence cited above, a consistent programme could expect to net at least two novae a year if run from sites with a long season of visibility of this region — either in low northern latitudes or in the southern hemisphere.

EXTRAGALACTIC NOVAE

Amateur discoveries have recently been made in the Magellanic Clouds (located in about −60° Dec) and in the nearest spiral galaxy, M31 in Andromeda, Dec +40°. At the distance of the Magellanic Clouds, ordinary novae of absolute magnitude −7 would be expected to reach about magnitude 11 at maximum, and the discovery of novae somewhat brighter than this by McNaught and Garradd from Australia (see listing starting on page 272) proves that the Large Cloud in particular is a fruitful hunting-ground for the photographer. Bright novae in the Andromeda galaxy peak at magnitude 16 or 17, within reach of an exposure of just a few minutes using a 400-mm telescope, and this would be a particularly suitable project for observers in higher northern latitudes, where M31 passes near the zenith.

Having examined the more objective aspects of nova-hunting, it is time to examine the observing methods available to the amateur.

VISUAL METHODS

The sheer number of stars to be checked in a limited time restricts general visual searching to objects brighter than about magnitude 7.5. Since stars of this brightness can be made out with minimal optical aid, binoculars of about 50-mm aperture and a magnification of not more than ×10, to give a wide field of view, seem an obvious choice and are in fact popular with observers. Having said that, all four Eric Alcock's fine novae were found using 15 × 80 binoculars, but one reason for this choice was the possibility of picking up a faint, diffuse comet, and he was rewarded with the independent discovery of just such a comet, the 'earth-grazing' IRAS–Araki–Alcock, on 3 May 1983. Another advantage of the larger aperture is its ability to reveal the colour of fainter stars — near maximum a nova appears yellow or reddish owing to the strong line of Hα emission in the spectrum, a helpful clue if a suspicious star is spotted. Again referring to Alcock's work, he commented that 6th-magnitude Nova Delphini, discovered in 1967, stood out like a golden-coloured beacon in the large binoculars.

Charts – Sweeping may be done from memory, or by reference to charts. In practice, regular sweeping from charts will result in star patterns gradually being committed to memory without the observer being aware of the fact — just as a constellation may look 'wrong', without one being sure why, until a slow-moving satellite shifts against the stars. It is slow and awkward referring to a star atlas showing stars down to magnitude 7.5 — the sheets are too large (especially in a wind!) and contain confusing detail such as RA and Dec lines, nebulae, and so on. The writer made up a set of sweeping charts by tracing sections of the old *Atlas Coeli* on to A4 paper, omitting everything but the stars themselves. These were then slipped into a holder with a clean plastic front, illuminated by a red lamp, and arranged on an adjustable wooden arm attached to the observing chair so that the map could be twisted to the correct orientation.

Special holders to keep binoculars steady are available or have been made up by some amateurs. However, stand-mounted binoculars cannot be moved around the sky with the facility of a pair of hands. The writer found that acceptable steadiness could be achieved by adopting a redundant fireside chair, with wooden arms and a well-raked back, for the work. With the head supported against the chair back, the eyepieces against the cheekbones, the hands around the object-glass mounts (helpful in discouraging dew), and the elbows on the chair arms, an extremely rigid support for the binoculars is created. A wide range of altitudes can be covered comfortably, and the chart can be consulted without having to move the head at all.

Sweeping from memory – The American nova-hunter Peter Collins has described how he and his fellow-observers 'memorize binocular star fields by inventing small "constellations" — hundreds, even thousands of them — that resemble familiar people, places and things. I've found this task thoroughly enjoyable and easier than might be expected. It's gratifying how well these little asterisms linger in memory and appear like returning friends, just as naked-eye constellations do, with the start of each new observing season.' (*Sky & Telescope*, November 1988, p. 490.)

The observer who has become sufficiently familiar with the star patterns not to need a chart (except perhaps for occasional checks) has obvious advantages — the work will take less time, and can be carried out from anywhere. The nova patrol run by *The Astronomer* reported in the early days how participants were taking less time to sweep their allocated areas as they become familiar with the patterns. Another positive benefit, in generally cloudy sites such as in the United Kingdom, is that sweeping can be carried out when conditions are too poor for photographic work — if broken cloud is passing across the sky the experienced observer can select a gap and sweep in the star fields that are revealed during its passage. This would hardly be possible when sweeping from charts.

Potential for visual discoveries – A glance at the table of nova discoveries between 1980 and 1990 reveals that of 29 new (as opposed to recurrent) galactic novae, only three were discovered visually, and this includes Branchett's mystery object of 1981. What benefits, therefore, does visual sweeping have over photographic patrolling?

The first question is: could more of these novae have been discovered visually? Assuming that novae to a limiting magnitude of 7.5 would be identified, we then have the following list of potential novae:

Corona Austrinus 1981 (mag. 7.0)
Aquila 1982 (mag. 6.5)
Musca 1983 (mag. 7.5 when past max.)
Vulpecula 1984, No. 1 (reached mag. 6.5, but was discovered photographically on the rise at mag. 9.2)
Centaurus 1986 (mag. 5.6 when discovered, reaching mag. 4.6)
Andromeda 1986 (reached mag. 6.8 but was discovered photographically on the rise at mag. 8.0)
Hercules 1987 (reached mag. 7.5 but was discovered photographically on the rise at mag. 8.0)
Corona Austrinus 1987 (mag. 7.2)

Three of these eight novae were discovered photographically on the rise, when below magnitude 7.5. The others were photographed at or past maximum, but of course even the ones recorded on the rise may not have been identified until later, when the photographs were examined. So certainly most of these novae would have been announced first by visual observers, had the sweeps been made, and the immediacy of visual discoveries offers the chance of prompt follow-up by other amateurs and professionals.

A further advantage of visual work is that it can be carried out under poor conditions (passing cloud, or in moonlight or twilight) when photographic work is impossible or at least handicapped. A fast nova peaking at, say, magnitude 7 could be completely missed if photographic work were interrupted for ten days or so by a combination of moonlight and poor weather.

Telescopic meteors – Visual sweeping gathers up a useful harvest of telescopic meteors, which add interest to a watch and allow the observer to contribute to a practically defunct but potentially valuable area of amateur work if basic statistics are noted.

Telescopic meteors traverse the field so quickly that there is no chance of following them, but some or all of the following information can nevertheless be recorded:

1. Approximate location of field
2. Field diameter
3. Time
4. Speed: Some meteors appear as an instaneous line of light; others can be seen as a fast-moving point
5. Colour
6. Nature of train: Persistence of vision can leave a subjective train in the wake of a bright meteor, but occasionally a genuine expanding train like a luminous vapour-trail is seen
7. Magnitude and 'light-curve': In the case of a slow meteor, a change of brightness may occur during its passage
8. Particulars of path: The path may begin or end outside or inside the field, and by recording these details it is possible to analyse a large number of observations statistically and derive path lengths. A recommended code is to use 'O' for outside the field and 'a' for inside. Thus:
   OO = entered and left field
   Oa = entered field and ended inside field
   aO = began inside field and left field
   aa = began and ended inside field

PHOTOGRAPHIC METHODS

The author well remembers his amazement back in the sixties at the number of stars that could be recorded by an ordinary stationary 35-mm camera and an f/2.8 lens with a 20-second exposure, even in fairly poor suburban conditions. Using fast black and white film (T-max 400 is efficient and readily available), magnitude 9 stars will be visible on the negative. The slight elongation caused by the beginning of trailing helps distinguish true stars from dust-specks or emulsion flaws; further exposure is pointless, since no fresh stars are recorded and in fact sky fog begins to submerge the faintest ones.

Focal length and image scale – A standard lens of 50-mm focus will cover an area of about 28° × 41° on a 35-mm frame — equivalent to about 1100 square degrees. Such an area of the Milky Way will contain perhaps 5000 stars down to magnitude 9. The simple 20-second fixed camera method will not record stars as faint as this near the edge of the frame, since light losses caused by vignetting within the lens can approach a whole magnitude compared with the centre of the photograph, added to which the loss of definition at the margins will lose other faint stars. Even so, it is quite likely that a couple of thousand stars will await inspection on the negative.

Therefore the wholesale recording of star images is not a problem. Driving the camera on an equatorial mount means that much fainter stars can be photographed, although the star density becomes so great that the negatives are difficult to examine. The answer is to use a longer-focus lens and thereby increase the image scale. Robert McNaught searches down to magnitude 10 using an 85-mm lens, but the discoveries by Gordon Garradd of novae in the Magellanic Clouds, which typically peak at magnitude 11, have been made with a 300-mm telephoto lens, covering about 30 square degrees. Obviously it would be a major task to patrol large areas of the Milky Way with such a camera, but restricted areas, such as the Magellanic Clouds or the Galactic centre, may offer worthwhile returns in the form of fainter novae being discovered, or novae missed at maximum being caught on the decline.

To prevent the chance of a suspect object turning out to be a flaw or a flashing artificial satellite, exposures are always made in duplicate, one after the other. Even then, ghost images of bright stars, caused by reflection inside the lens, have caused false alarms.

Patrol fields – It may be worth considering the best way of mapping out a nova patrol. Accepting that most novae have been discovered within 10° north and south of the Galactic plane, the 'prime target' is this 20-degree strip of sky, which can be covered along the short dimension of, for example, a 50-mm focal length lens (field coverage 28° × 41°) or the long dimension of, say, an 80-mm lens (field coverage 18° × 26°). Set edge-to-edge, therefore, nine 50-mm frames or twenty 80-mm frames would girdle the entire Milky Way. The true number needed would be higher than this, according to the amount of overlap required, but of course only part of the Milky Way can be patrolled on any single night.

If this strategy is chosen, a problem immediately arises. Assuming that the camera is mounted equatorially, the frame will have a constant orientation with respect to RA and Dec. But since the median line of the Milky Way is inclined to the celestial equator, the camera must be readjusted between fields to keep the frame correctly orientated with respect to the Milky Way.

A 'galactic mounting' – Although the author has never constructed the following device, the suggestion may be worth considering. Instead of having a declination axis, the equatorial mounting carries a fixed circular plate graduated in perhaps 5-degree intervals, at the centre of which is an axis carrying the camera. This axis is inclined to the polar axis at an angle of 63°, which is the angle of inclination of the Milky Way to the celestial equator. When the mounting is suitably rotated in RA, the plane of the disk lies in the plane of the Galaxy, and by turning the camera a suitable number of degrees between each exposure it automatically tracks along the Milky Way (Fig. 12.1).

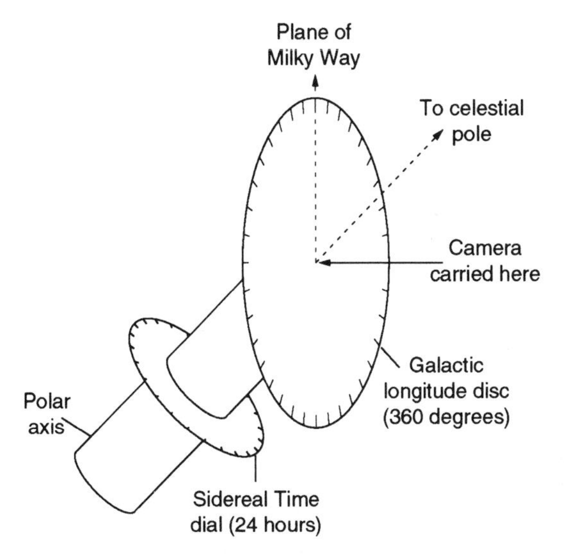

FIGURE 12.1
Principle of the 'galactic mounting'. The camera is mounted on an axis (not shown) at the centre of the galactic longitude disk, so that it turns in the plane of the disk. This axis is inclined at 63° to the polar axis. The sidereal dial on the polar axis is fixed so that when the pointer shows the Sidereal Time, the plane of the galactic longitude disk coincides with that of the Milky Way. The camera can then be pointed using the galactic longitude scale on the disk.

Processing – Prompt processing is a vital part of the operation; ideally a film should be developed the same night and checked as soon as it is dry. The difficulty of keeping up with a continuous stream of negatives is where most amateur photographic search programmes come to grief. Expense is not a minor matter, either, when several complete 35-mm films could be exposed during a week of clear nights: it is worth considering bulk purchase of film, which will work out considerably cheaper than buying individual cassettes. Anyone embarking on a serious photographic nova patrol must decide whether the time and the money are available to do the job properly. One advantage of living in the cloudier parts of the globe is the frequent occurrence of overcast nights, when negatives can be checked free from the temptation to go outside and expose some more!

Filing – Even if no novae are discovered, every pair of patrol photographs constitutes a permanently valuable resource. Years later there may be a sudden need to know whether a particular star, suddenly discovered to be variable, was bright or faint on that occasion. The record is useless if it cannot be located, and seriously devalued if it is not dated.

An observing book documenting each exposure should be kept, noting the field, the camera and film details, the exposure length, and the time when the exposure was made. Development details should also be noted, although if a standard procedure is followed (which it should be) the basic information need only be recorded once.

If negatives are separately mounted they can be filed by the coordinates of the field centre, but if they are preserved in uncut strips of film the sections will have to be given an identification, and a separate record kept of the field centres of the individual frames on each strip. A general index will also be needed, listing by date all the negatives which cover each field centre.

A comprehensive patrol will eventually accumulate thousands of negatives, and it is well worth giving careful thought to the cataloguing system to avoid later changes and re-thinks. Negatives must be kept in a dry environment.

Comparators – The time-honoured way of checking patrol negatives for intruders is to set the negative beside one taken earlier, and then to illuminate the negatives alternately so that each is viewed by the eye in turn. When this is done, any star that is on one and not the other will appear to blink on and off, while an object such as a minor planet, which has moved between the two exposures, will jump from side to side. The 'blinking' effect will also occur if a variable star has changed in brightness.

This device, known as a 'blink comparator' or 'blink microscope', would have to be specially built by someone with a workshop and some knowledge of optics. This difficulty is overcome in an alternative method based on standard equipment, the 'Problicom' technique. This is used by several observers in the United States, most notably William Liller in Chile, who has now discovered over a dozen novae photographically. The word stands for 'PROjection BLInk COMparator', since the negatives are superimposed on a screen using two identical projectors, with a rotating shutter cutting off each beam in turn — this shutter is the only item that has to be constructed. A rate of about five interruptions per second is satisfactory.

The 'stereo comparator', which sounds more sophisticated, is in fact very simple: it depends on the fact that if one negative is viewed with each eye, a star present on one and absent on the other produces a pseudo-stereoscopic effect. Although elaborate equipment can be built for the purpose, Gordon Garradd made his notable nova discoveries in the Large Magellanic Cloud by placing the two negatives on a strip of opalescent glass laid across two cans of sliced peaches, with a small fluorescent lamp beneath. The negatives were viewed through two ×6 magnifiers resting on the glass.

Superimposition – Whichever system is used, there must be some means of adjusting the two negatives so that the star images are superimposed. Ideally, both will be of exactly the same region of sky — all camera lenses distort, and it will be impossible to achieve good register if the same region is at the centre of one negative and near the border of the other. Using the Problicom method this is particularly important, since projector lenses add their own distortion. This really means that the camera should be directed at pre-arranged field centres using setting circles, allowing for adequate overlap so that every part of the patrolled region falls in an area of reasonable definition on at least one pair of negatives.

The 'galactic mount' mentioned above would ensure the desirable close registration of photographs taken on different nights.

REPORTING DISCOVERIES

Chapter 18 (Astronomical Communications) should be read by anyone likely to be involved in the discovery or confirmation of transient objects, but anyone embarking on nova-hunting should try to find two or three competent observers in different parts of the country who will be ready and willing to turn out at any hour of the night to check a suspect star. The reason for spreading the network as widely as possible is to increase the chances of one of them having good weather, just in case conditions at the discovery site are deteriorating.

In the case of a visual sighting, the immediate need is to obtain an approximate position and magnitude. Careful plotting on, for example, *Sky Atlas 2000.0* should allow an accuracy of a few minutes of arc, although the precision will be poorer if there are no nearby stars to act as guides. The epoch for which the atlas is drawn must also be quoted.

Estimate the magnitude with reference to convenient nearby stars using standard variable-star techniques (Chapter 10), identifying the stars used so that their precise values can be checked later and a more accurate magnitude derived. If no nearby star magnitudes are known, a 'guesstimate' to within half a magnitude will let other observers know what to expect, but as the star may be a rapidly-rising nova, care should be taken to make the real estimate as good as possible.

A photographic discovery will almost certainly be made at least some hours after the event, perhaps during the following day. By projecting the negative on to a screen, the star's position with respect to its neighbours can be drawn and then referred to an atlas. Alerting the local network will at least establish priority should the discovery be confirmed, while personal contact established with an observer on the opposite side of the world, where it is now night, could result in rapid checking.

In countries with well-established reporting networks, the national astronomical society should be able to introduce the observer into the system. The ultimate destination of all discovery messages is the IAU Central Bureau for Astronomical Telegrams, Smithsonian Astrophysical Observatory, Cambridge, MA 02138, USA, but a good network will 'weed' nova reports to eliminate known variable stars and minor planets before passing them on to Cambridge. If there is no national amateur–professional liaison, then investigate ways of tapping into another network, either by telephone or computer link. Make sure that the system is ready to handle a discovery *before* nova patrolling begins, or much time, and even discovery priority, may be pointlessly lost on that heart-stopping moment when a star stands out, whether through binoculars or on a negative, in a space between the well-known patterns of the night sky.

About the Author

James Muirden became interested in astronomy as a teenager, and owes much of his early encouragement to the writer and broadcaster Patrick Moore. Unable to afford a telescope, he ground and polished a 150-mm mirror in 1957 and began surveying the sky in the company of *Norton's Star Atlas* and *Webb's Celestial Ojbects for Common Telescopes*. In 1963 his first astronomical book was published, and he has written on the subject ever since. In 1964 he founded *The Astronomer*, the still-running monthly magazine devoted to rapid reporting of amateur observations, and edited it for a number of years. Urban dwellers for most of their lives, he and his wife now live in the country where they keep chickens and have planted a vineyard. Their children, Daniel and Emma, are both at university.

Contact address – Westfield, Rewe, Exeter, Devon EX5 4EU, UK.

Notes and Comments

Mystery object in Scutum – The following account is taken from *The Astronomer* for February 1981, and describes the still mysterious star seen by Dave Branchett (Eastleigh, Hampshire, England). It gives a vivid impression of the private 'agony and ecstasy' and need for mutual support felt by observers of such transient objects. — Ed.

> On the morning of 18 January 1981 I got up as usual to conduct some early-morning observations; this included Comet Panther and a few selected variable stars. Just before 6.15 UT cloud cover became evident from the south-west, so I decided prior to dawn and total cover to conduct a nova search sweep with binoculars through Scutum.
>
> At this time of the year, I conduct my sweeping from my bedroom window as this gives me higher elevation, permitting viewing above shrubs, trees and rooftops. So leaning out of my bedroom window I first located the star λ Aquilae, then swept in a westerly direction. On reaching the field of β and R Scuti, I noticed clearly an intruder star, placed to the south-west of β. My mind went blank, and I was dumbstruck. I checked my star charts and patrol print which goes down to about magnitude 10.5 but there was no star in this position.
>
> I then estimated the position of the star and compared its magnitude with a field star and estimated the suspect to be about magnitude 8.0. By this time cloud cover had reached the area and dawn was taking over. During the day I contracted various individuals whom I knew would be interested in this star. The following morning was one of total cloud cover, not only for myself but also for other fellow observers who had forsaken an hour or so of sleep to try and secure a second observation of the star. The morning of the 20th, which was Tuesday, was fine, cold and clear. The moon, nearly full, was over in the west in Gemini, its light drowning the background sky. Just after 6.00 UT I saw β Scuti rise. Soon it and the

surrounding star field, including R Scuti, were high enough to be examined, but to my horror there was no sign of the intruding star. My heart sank as I could not see the star which had been visible just 48 hours earlier.

I rushed outside and soon had the tripod for the Celestron 5 ready for use. I rapidly had the C5 locked on to R Scuti. At ×70 with a field of about 1 degree there was no sign of the star. I was disgusted and put the C5 away. I went to work, at least I could take it out on my work! I called Guy [Hurst] around 9.00 but he also saw nothing that morning. What should I say — I knew what I had seen but could not explain the sudden 'vanishing act' of this star! The young boy who works on my machine is a keen angler. I told him of the star and added that this must have been the 'one that got away'. I heard nothing until the afternoon when I received a call from Jim Muirden who also did not see the star. He told me that it could have been a fast nova but at the same time I felt that perhaps he was being kind and was giving me moral support.

Just after tea-time that evening I received another call from Guy; soon my disappointment rose to elation when he told me that the Royal Greenwich Observatory that very same morning had secured a stellar image of about magnitude 9.0 on a plate. I simply cannot find words with which to describe my feelings when I learnt of this. At this time the nature of the object is unknown, but my sincere thanks go to all who have been involved in looking for this object.

New variables on patrol photographs – In response to a request, patrol photographer Mike Collins (Sandy, Bedfordshire, England) sent in a list of confirmed variable stars discovered during nova patrol work. Not all are 'new' in the sense that they may have been previously suspected of variation, but the patrol photographs give substantial extra information.

For example, photographs taken on 10 January 1989 revealed a star that had risen to magnitude 9 from magnitude 11 two months earlier. This turned out to be a new Mira star: it had been suspected of variability in 1935, but had not been studied further.

Mike Collins' totals of 32 in 1989 and 18 in 1990 (with another three already in January 1991!) show that even if a nova is not detected, systematic patrol work is capable of producing many other discoveries if the negatives are scanned thoroughly and discrepancies carefully checked.

A Survey of Nova Discoveries, 1980–1992

This list has been compiled principally from reports in *The Astronomer*. The quoted positions are for epoch 1950.

1980 Sagittarius (18 h 16.5 m, −24° 45′)
Discovered by M. Honda (Japan) on 28 October at mag. 9.0.

1981 Scutum (18 h 44.2 m, −5° 00′)
A 'mystery' object seen at mag. 8.0 on 18 January by D. Branchett (England) and confirmed by one photograph taken on the 20th.

Corona Austrinus (18 h 38.6 m, −37° 34′)
Discovered at mag. 7.0 on a photograph taken on 2 April by M. Honda (Japan). By the 14th it had fallen to mag. 9.2.

LMC (05 h 32.7 m, −70° 24′)
Discovered on a professional photograph at about mag. 12 on 30 September by M. Wischnjewsky, but it could have been as bright as mag. 8 a fortnight before discovery.

1982 Aquila (19 h 20.8 m, +02° 24′)
Discovered by M. Honda (Japan) at mag. 6.5 on 27 January, using Tri-X film. A fast nova, it had fallen by three magnitudes only 10 days after discovery.

Sagittarius (18 h 31.5 m, −26° 28′)
Discovered on 4 October by M. Honda (Japan) at mag. 9. After a slight fade, it reached a second maximum of mag. 8.7 a month after discovery.

Nova in M31 (00 h 39 m 34 s, +40° 45′ 05″)
Discovered by J. Bryan and M. Brewster (Texas, USA) on 16 December at mag. 17.2.

1983 Nova in M31 (00 h 39 m 26 s, +40° 52′ 17″)
Discovered by J. Bryan (Texas, USA) on 8 January at mag. 15.1.

Musca (11 h 49.5 m, −66° 55′)
Discovered on 24 January by W. Liller (Chile) at mag. 7.5, when it was past maximum. A slow nova, it had only faded to mag. 12 after 15 months.

Sagittarius (18 h 04.7 m, −28° 50′)
This nova was recorded at mag. 10 on Tri-X patrol photographs by M. Wakuda (Japan) on 19 February, but he did not inspect them until October of the same year, and no announcement was made. It was also 'posthumously' discovered on a professional objective-prism plate examined a year after it was taken! Examination of patrol photographs showed that it peaked at mag. 9.5 on 13 February.

Serpens (17 h 53.2 m, −14° 01′)
M. Wakuda (Japan) discovered this nova on Tri-X patrol photographs on 22 February at mag. 7.7. It fell very swiftly, to mag. 12 by the end of the month.

Norma (16 h 09.8 m, −53° 12′)
Discovered on patrol photographs on 19 September by W. Liller (Chile) at mag. 9.4.

1984 Vulpecula No. 1 (19 h 24.1 m, +27° 16′)
M. Wakuda (Japan) discovered this nova on 27 July using a 200-mm telephoto lens and Tri-X film. It was then of mag. 9.2; patrol photographs taken three days earlier showed no object brighter than mag. 14 in the position, but a slow rise continued to its maximum of mag. 6.5 on 5 August.

Sagittarius (17 h 50.5 m, −29° 01′)
Discovered on patrol photographs taken by W. Liller (Chile) on 25 September at mag. 10.3, rising to mag. 9.7 on the following night. It fell by four magnitudes during the ensuing month.

Aquila (19 h 14.1 m, +03° 38′)
M. Honda (Japan) discovered this nova at mag. 10 on 2 December, fading to mag. 12 over the next five days.

Vulpecula No. 2 (20 h 24.7 m, +27° 41′)
Discovered on 22 December at mag. 6.8 by Peter Collins (California, USA). This nova was discovered on the rise, reaching about mag. 5.7 on the 24th.

1985 RS Ophiuchi (17 h 46.5 m, −06° 42′)
This famous recurrent nova was noticed bright by Eric Alcock (England) during a binocular nova sweep on 29 January. It was then at mag. 5.9. Observers in the United States also detected it on the same night.

Scorpius (17 h 53.3 m, −31° 49′)
Discovered on patrol photographs by W. Liller (Chile) on 24 September at mag. 10.5. It remained near maximum for about three weeks.

1986 Cygnus (19 h 52.7 m, +35° 34′)
This nova was discovered by M. Wakuda (Japan) on 4 August at mag. 9.4 using Tri-X film. A slow nova, it remained at around mag. 10 for a month after discovery.

SMC (00 h 34.9 m, −72° 21′)
Discovered by Robert McNaught (NSW, Australia) at mag. 10.2 on 25 October (85-mm lens, Tri-X film). This was only the sixth nova to be found in the Small Magnellanic Cloud, the previous discoveries being made in 1897, 1927, 1951, 1952 and 1974.

Sagittarius (18 h 00.5 m, −28° 00′)
A possible nova was discovered by R. McNaught (NSW, Australia) on patrol photographs (85-mm lens, Tri-X) taken on 25, 27 and 28 October, at mag. 10.4. The star was not noticed until an examination of the negatives almost a year later, and positive identification with any 'pre-nova' candidate has not been possible.

Centaurus (14 h 32.2 m, −57° 25′)
Discovered by R. McNaught (NSW, Australia) at mag. 5.6 on 22 November (85-mm lens, Tri-X film, exposure 1 minute), rising to mag. 4.6 on the 24th. [=V842 Cen]

Andromeda (23 h 09.8 m, +47° 12′)
On 5 December, after three years of work, M. Suzuki (Japan) discovered this nova on patrol photographs taken with a 200-mm telephoto lens. It was then at mag. 8.0, and reached a maximum of about 6.8 on 9 December. [=OS And]

1987 Hercules (18 h 41.5 m, +15° 16′)
Discovered by Sugano and Honda (Japan) on 27 January at mag. 8.0; it reached mag. 7.5 on the following night.

U Scorpii (16 h 19.7 m, −18° 47′)
This recurrent nova was first observed in 1863, and its latest outburst, to mag. 10.8, was discovered visually by D. Overbeek (Edenvale, South Africa) on 16 May. It is possible that the true maximum occurred about 13 May at mag. 9, but was lost because of moonlight. Ten days later it was down to mag. 14.

Sagittarius (17 h 56.3 m, −32° 16′)
This nova was discovered by R. McNaught (NSW, Australia) on May 18 at mag. 10.5, using an 85-mm lens and Tri-X film.

Nova Coronae Austrinae, 1949 (V394) (17 h 57.0 m, −39° 00′)
On 2 August, W. Liller (Chile) discovered a mag. 8.9 object which proved to be an outburst of this old nova. Patrol photographs showed that it reached mag. 7.2 on 29 July.

LMC (05 h 24.3 m, −70° 03′)
This nova was discovered by Gordon Garradd (NSW, Australia) on a patrol photograph taken on 21 September using hypered 2415 film (300-mm f/4.5 lens, 15 minutes exposure). It was then at mag. 12. A check of previous photographs showed it at maximum brightness on 17 September (mag. 9.5), and it fell by five magnitudes in the following fortnight.

Vulpecula (19 h 04.1 m, +21° 44′)
Discovered independently by two visual observers (Beckmann, Collins) in the United States on 15 November at mag. 7.1, and the following night on a patrol photograph taken from Japan (Sakurai). The nova was subsequently noted at mag. 8 on a patrol photograph taken on the 12th.

Novae in M31
Three probable novae were reported by J. Bryan (Texas, USA) on patrol photographs taken using a 410-mm reflector with Tri-X film.
Nova 1: 00 h 39 m 26 s, +40° 54.4′: mag. 17.5 on 21 November
Nova 2: 00 h 40 m 47 s, +40° 57.9′: mag. 17.5 on 21 November
Nova 3: 00 h 39 m 48 s, +40° 58.9′: mag. 16.8 on 20 December

1988 Andromeda (02 h 26.4 m, +39° 49′)
Discovered by Dave McAdam (England) on 21 March at about mag. 10 (300-mm f/4 telephoto lens). For a time its nature was in doubt: a spectrum was obtained at La Palma but was lost in the computer before it could be analysed. Not until July was a spectrum of the star (having by then faded to mag. 19) obtained at the Steward Observatory, Arizona, suggesting that it is a recurrent nova of the WZ Sge type.

LMC (05 h 36.0 m, −70° 23′)
Discovered by Gordon Garradd (NSW, Australia) on 21 March at mag. 11.4, using a 300-mm f/4.5 telephoto lens and 2415 film. This nova remained near maximum brightness until the end of the month.

Ophiuchus (17 h 08.8 m, −29° 34′)
Discovered by M. Wakuda (Japan) on 10 April at mag. 8.5 (200-mm telephoto lens, Tri-X). In two days this slow nova had fallen to mag. 10.0, but it declined by only a magnitude in the following month.

LMC (05 h 08.3 m, −68° 42′)
Discovered a day before maximum by Gordon Garradd (NSW, Australia) on 12 October at mag. 11.3 (300-mm f/4.5 telephoto lens, hypered 2415). It rose to mag. 10.4 on the following night, but had fallen to mag. 13 on the 19th.

1989 Scorpius No. 1 (17 h 52.1 m, −33° 13′)
Discovered by W. Liller (Chile) on 30 July at mag. 9.7, this object was identified as a further outburst of V745 Sco, a suspected nova recorded at mag. 11 in 1937.

Scorpius No. 2 (17 h 48.6 m, −32° 31′)
Discovered by W. Liller (Chile) on 17 August at mag. 10.0, this nova rose to mag. 9.4 on the following night. Liller took advantage of the dark sky during a total lunar eclipse to carry out patrol photography, and his efforts were rewarded.

Scutum (18 h 47.0 m, −06° 15′)
When discovered by a professional astronomer, P. Wild (Switzerland) at about mag. 10 on 20 September, this nova was already declining: patrol photographs suggest a maximum of about mag. 8.5 near the 17th.

1990 LMC (05 h 23.7 m, −69° 32′)
Discovered by Gordon Garradd (NSW, Australia) on 16 January at mag. 11.5 (300-mm telephoto lens, TP 2415).

LMC (05 h 10.7 m, −71° 43′)
Discovered by W. Liller (Chile) on 14 February at mag. 11.2. This nova appears to be a further outburst of one observed in 1968.

Sagittarius (17 h 56.1 m, −29° 10′)
Discovered by W. Liller (Chile) on 23 February at mag. 8.0, this is now classified as a new dwarf nova.

M31 novae
Discovered by J. Bryan (Texas, USA).
Nova 1: 00 h 40 m 20 s, +41° 00.6′: mag. 16.7 on 13 October
Nova 2: 00 h 39 m 52 s, +40° 55.4′: mag. 17.6 on 12 November

1991 Hercules (18 h 44.2 m, +12° 11′)
Discovered visually by M. Sugano (Japan) on 24 March at mag. 5.4, and independently by G.E.D. Alcock (England) on the same night.

Centaurus (13 h 46.6 m, −62° 54′)
Discovered by W. Liller (Chile) on 2 April at mag. 8.7.

Ophiuchus No. 1 (17 h 17.2 m, −26° 43′)
Discovered by P. Camilleri (Victoria, Australia) on 11 April at mag. 10 (135-mm lens, T-Max 400 film).

Ophiuchus No. 2 (17 h 40.1 m, −20° 06′)
Also discovered by Camilleri on a photograph taken on 11 April, at mag. 9.3, but not recognized until the negative was used for comparison purposes six months later!

LMC (05 h 04.2 m, −70° 22′)
Discovered by W. Liller (Chile) on 18 April at mag. 12.3. By the end of the month it had reached mag. 9, becoming the brightest ever observed in the LMC.

Sagittarius (18 h 11.0 m, −32° 13′)
Discovered by P. Camilleri (Victoria, Australia) on 29 July at mag. 7.0. Two days later it had faded by over a magnitude.

Scutum (18 h 44.3 m, −08° 24′)
Discovered by P. Camilleri (Victoria, Australia) on 30 August at mag. 10.5.

Puppis (08 h 09.7 m, −34° 58′)
Discovered by P. Camilleri (Victoria, Australia) on 27 December at mag. 6.4.

1992 Sagittarius (18 h 06.5 m, −25° 53′)
Discovered independently by W. Liller (Chile) and P. Camilleri (Victoria, Australia) on 13 February at about mag. 7.2.

Cygnus (20 h 29.1 m, +52° 38′)
Discovered at mag. 6.0 by P. Collins (USA) on February 19. At maximum, on 22 February it reached mag. 4.5.

Scorpius (17 h 03.9 m, −43° 12′)
Discovered at mag. 8.2 by P. Camilleri (Victoria, Australia) on 22 May, when it was still rising.

Sagittarius No. 2 (18 h 20.3 m, −28° 24′)
Discovered at mag. 8.5 on 9 July by W. Liller (Chile).

Sagittarius No. 3 (18 h 20.7 m, −23°)
Discovered at mag. 9.0 by P. Cammilleri (Victoria, Australia) on 13 October, three nights before maximum at mag. 8.0.

CHAPTER THIRTEEN · *The Search for Supernovae*

STEVE LUCAS
'SUNSEARCH'

THE EMERGENCE OF AMATEUR INVOLVEMENT IN THE VISUAL discovery of the supernova explosion phenomenon began in July 1968, with a serendipitous identification of an additional star in the galaxy M83, by Jack Bennett of South Africa, employing a 125-mm telescope. In 1979 another amateur astronomer, Gus Johnson of the United States, reported an intruder star in the galaxy M100, using a 200-mm telescope.*

These findings were perhaps chance encounters, but almost immediately provided an awareness that these minute points of light could be recognized with the use of the most modest of equipment.

During the early part of the 1980s, a mild-mannered minister and amateur astronomer from Australia has proved that these anomalies can be detected on a regular basis, provided the proper techniques, equipment, and facilities are coordinated to make this venture a reality. Through a systematic regimen of observation from 1980 to 1990 the Reverend Robert O. Evans discovered 18 supernova explosions, which have provided the professional community as a whole with some very important contributions. First of all, it is important to understand what happens during a supernova explosion. Table 13.1 gives a listing of all supernovae found visually since Bennett's discovery in 1968.

*In a letter, the author points out that photographic supernova discoveries were made by an amateur, G. Romano, in 1957, 1961 and 1970. — Ed.

TABLE 13.1

SUPERNOVAE DISCOVERED VISUALLY SINCE 1968

Name/year	NGC/IC	Mag.	Discoverer
1968L	5236 = M83	12.0B	Bennett
1979C	4321 = M100	11.6B	Johnson
1981A	1532	13.0V	Evans
1981D	1316	13.0B	Evans
1983G	4753	12.9B	Okazaki, Evans, Tsevtkov
1983N	5236	11.6B	Evans
1983S	1448	14.5V	Evans
1983V	1365	13.5V	Evans, Linblad, Grosbol
1984E	3169	15.2B	Okazaki, Metlova, Evans[a]
1984J	1559	13.2V	Evans
1984L	991	14.0B	Evans
1984N	7184	14.0V	Evans
1985P	1433	13.5B	Evans
1986A	3367	14.4B	Evans, Cameron, Leibundgut
1986G	5128	12.5B	Evans
1986L	1559	13.3V	Evans
1987A	LMC	4.5B	Shelton, Duhalde, Jones
1987B	5850	15.0B	Evans
1987L	2336	13.0V	Patchick
1987N	7606	13.4V	Evans
1988A	4579	13.5V	Ikeya, Evans, Pollas
1989B	3627 = M66	13.0V	Evans
1990K	150	14.0	Evans

[a]Discovered photographically, but first seen visually by Evans and reported by him.

TYPES OF SUPERNOVAE

All stars, be they large or small, eventually run out of fuel. What happens next in the evolutionary process depends to a large extent on the mass of the dying star. Massive stars, which are enormous producers of energy, will utilize more fuel than their smaller and more stable counterparts. In addition, as these massive stars consume their hydrogen fuel, enormous pressure is generated within their cores. Throughout the star's life, gravity resists this outward force and creates a balance between explosion and collapse: as primary fuels are exhausted the force weakens. The additional gravitational squeeze raises the temperature to the point where nuclear fusion succeeds in restoring stability by utilizing layers of carbon, oxygen and silicon around the core. However, a point is finally reached when even the energy from these fusion processes cannot withstand the gravitational force, and under the tremendous compression the collapsing star's matter undergoes fundamental atomic changes.

At this crucial moment, if a star's mass is at least 1.4 solar masses, a supernova explosion is possible. Energy in the form of neutrinos is expelled from the core, taking with it the star's last chance to resist any further gravitational collapse. Detonation occurs when the outer layers of the star rush in upon the core and are bounced back with such violence that the star explodes as a Type II supernova. The outer layers are blown into the interstellar medium, raising the star's apparent magnitude so drastically that it shines as brilliantly as an entire galaxy of stars. The resulting catastrophe may cause the core of the star to form a compact neutron star, or a rapidly-spinning pulsar. If the progenitor is large enough and the annihilation process surpasses certain criteria, the resulting event may cause a black hole.

In the alternative Type Ia event, a close binary star system may set the scene for the most powerful variety of supernova. Owing to usage of its fuel, one of the stars begins to expand to the red giant phase: as the star grows, gravitational instability causes mass transfer of accreting matter between it and its companion. The primary star, now stripped of matter, evolves into a white dwarf, with most of its carbon core exposed. The secondary star, now the red giant, begins transfer of mass back to the primary star. The white dwarf star cannot expand fast enough to compensate for the rising temperature that has occurred within its core, and the pressure there rises to enormous proportions. In a futile attempt to stabilize, fatal carbon burning is triggered once the star has exceeded the limit of 1.4 solar masses. This burning is short-lived, and the internal mechanism under so much pressure annihilates the entire star with an explosion four times that of the Type II supernova.

LIGHT CURVES OF SUPERNOVAE

The principal means of identifying a supernova's type is by spectroscopy. Secondarily, their identifying types may also be inferred through changes in the star's brightness, presented graphically as a light curve. While discussion of spectroscopy is beyond the scope of this chapter, a summary of light curves is appropriate for amateur observers.

As with other variable stars, any observer who diligently follows the progress of a bright supernova can prepare a useful and informative graph of changes in light over a period of time. The Type Ia supernova displays a fairly homogeneous and predictable light curve at any particular wavelength, while the Type II supernova has a variant in that two distinct sub-types exist — the 'linear' and the 'plateau' [1]. These light curves are illustrated in Fig. 13.1. It is theorized that both sub-types of the Type II supernova originate in the spiral arms of a particular galaxy, with the progenitor star thought to be more massive than eight solar masses. The main difference in light curves is that the 'linear' types lose most but not all of their

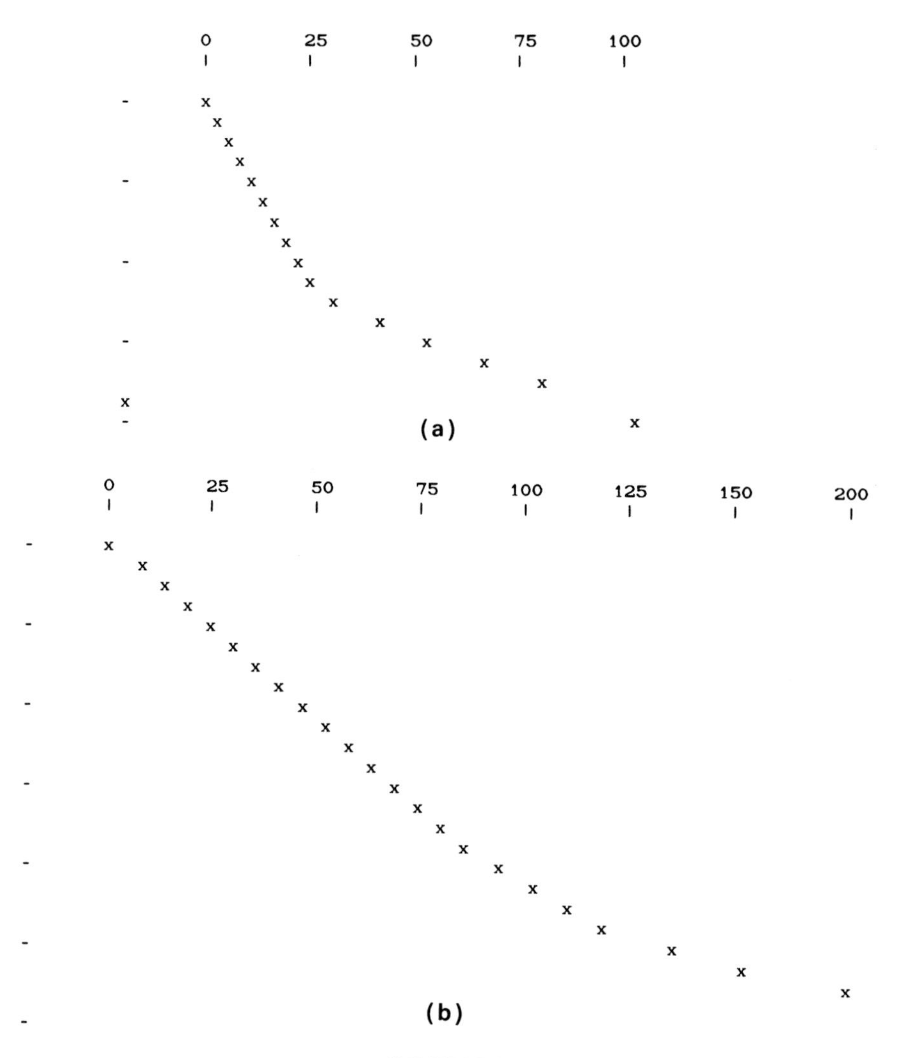

FIGURE 13.1
Typical light curves for supernovae. (a) Type I. (b) Type II 'linear'.

hydrogen envelopes before the explosion takes place, and therefore display a shorter length of visible linearity. In the 'plateau' type, however, a diffuse release of shock energy in the stellar envelope displays an immediate visible linearity in its light curve, which after a period of time drops off sharply. The kinetics which exist between these two sub-types vary with such irregularity that it is difficult to form a homogeneous sample of this type of supernova.

TYPES OF GALAXY AS PRIME SOURCES FOR SUPERNOVA

Supernovae may occur in any galaxy, but some galaxy types have been observed to produce more supernovae than others. Another influence upon the rate of

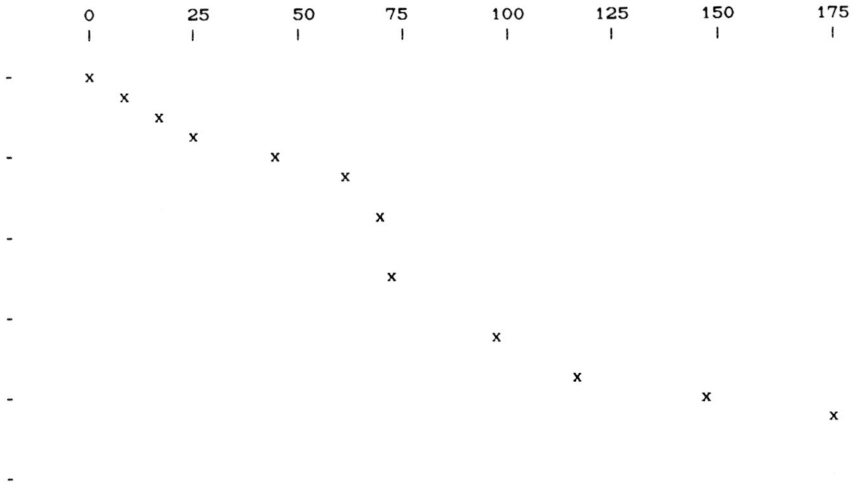

0 25 50 75 100 125 150 175

(c)

FIGURE 13.1 continued
(c) Type II 'plateau'.

supernova production is the size or luminosity of a galaxy: larger galaxies tend to be more productive than smaller ones. In elliptical galaxies (galaxies lacking in young stars) Type Ia supernovae are not likely to be observed. In the late-type spiral galaxies (galaxies which have predominantly young stars, with a mixture of older stars) both types of supernova are possible. So the most practical type of galaxy to produce a supernova is the latter. Table 13.2 displays a percentage factor ($\pm 0.05\%$) by morphological type of galaxy and associated supernovae of Types I and II, with

TABLE 13.2
MORPHOLOGICAL TYPES OF GALAXY AND PERCENTAGES (±0.05%) OF SUPERNOVAE

Type	%	Type	%
E	7	Sd	1
SO	5	Sdm	0.6
SOa	0.6	Sm	0.8
Sa	7	S	11
Sab	3	IO	1
Sb	13	Im	0.5
Sbc	8	I	4
Sc	23	nc[a]	10
Scd	4		

[a]Not classified.

all variations thereof that have been associated with them over the past 100 years ending in 1985 [2]. In that period of time only 661 events have been found that have been determined to be *bona fide* supernova explosions.

CONDUCTING A SEARCH

Observers who want the challenge of searching for supernovae should be prepared to make a long-term commitment, since years may pass between sightings. Some of the requirements include persistence to maintain the programme, proper equipment, appropriate charts, and reliable search techniques.

The observer should determine what personal limitations to set in terms of numbers of galaxies to be monitored for supernovae each month. Generally, they should be monitored at least twice per month for effective coverage. There is no sense in pushing oneself beyond tolerable limits. Trying to do too much may produce mistakes in judgement, but of paramount importance is the fact that the project may alienate the individual, and the programme may then become more of a chore than an enjoyable pastime.

EQUIPMENT

The main reason why amateurs have taken so long to become discoverers of supernovae is that the stars usually have very faint apparent magnitudes. Occasionally a supernova will achieve a 'bright' maximum of about 12th magnitude, but maxima generally occur below the 14th magnitude, which makes such stars easy to miss unless the observer is deliberately watching for them. Supernova type and galactic distance will determine the apparent magnitude. An estimated limiting magnitude for a successful visual study ought to be about 14.0, and a 200-mm or 250-mm telescope will meet this condition. A telescope of appropriate size, coupled with the observer's experience and good sky conditions, will make the observer more confident about identifying supernova candidates.

A smaller telescope may be used in the search for supernovae, but its limiting magnitude will become a serious constraint. While a bright supernova may be detected later in its apparition with a smaller telescope, it is the element of timing that is imperative in its detection. The event itself might have been observed two or three weeks earlier if a larger aperture had been used. Owners of small telescopes should consider only the brightest galaxies: this regimen will then generally yield the brightest supernovae.

CHARTS

Proper charts for galaxies with magnitude sequences of the surrounding field stars are of the utmost importance. Generally, observers either use published photographs of galaxies or attempt to draw the object at the telescope. Both methods have problems which can interfere with observations: highly-magnified images normally found in observatory photographs are not usually helpful to visual observers, while such images also tend to exclude distant regions of galaxies that represent good hunting grounds in a small telescope. Of course, in resolving a suspected sighting, all references which display the correct field star orientation are valuable.

Sketches at the telescope have the advantage of being matched to the artist's observing skill level. He or she will draw as much as is normally seen. However, the observer becomes a victim of personal equation, since drawings reflect errors that occur due to transparency, seeing and observing skill. Changes in limiting magnitude give rise to a 'now you see it, now you don't' situation. This may cause confusion when the observer spots a stellar image owing to better sky conditions and reports a possible supernova! Drawings always leave doubts concerning the angular size of a particular galaxy. Nobody should be surprised to learn that the visual appearance of a galaxy only represents a fraction of its true extent.

A third approach to chart-making combines drawings with photographs to achieve a practical compromise. Photographs with wide fields can be traced to create the underlying star-field for a chart. The appearance of the galaxy can be handled either by an artistic and careful drawing, or by simple line-drawings. How the galaxy is rendered is much less important than accurately presenting the star-field, since it is the stars that are being repeatedly observed and compared.

Figure 13.2, which is taken from J.T. Bryan's atlas of galaxies in Ursa Major [3], illustrates this technique. The author describes how the result was achieved:

> Each chart is based on a tracing of an enlarged portion of the Palomar Observatory Sky Survey ... The map is presented at an approximate scale of 5.5 arc seconds per millimetre. Simple line drawings are used to indicate the size, shape, and features within the galaxy. Recommended search boundaries are included to prevent observers from overlooking outlying regions of the galaxy. Associated with the boundaries are lines which converge on the centre of the galaxy and indicate the directions north, south, east and west. Sites of past supernovae are marked with small crosses. Visual observations were made with a 16-inch telescope to verify the position and approximate magnitude of field stars to approximately $M_v \sim 15.5$ (some measured). Fields are drawn within a limit of $M_v \sim 18.0$. Special attention was given to the starfield on the disk of the galaxy to search for stars hidden by photographic overexposure.

While it may not be possible for all amateur map-makers, it is important and desirable to locate any photoelectrically-measured stars immediately around galaxies, since these will serve as a basis for comparison if a supernova in the galaxy is suspected. Typically, it takes a search of professional literature to find these sequences, but without photometry observers are left to their individual skill in making light estimates.

Naturally, the chart should indicate a distance scale, which helps to orientate

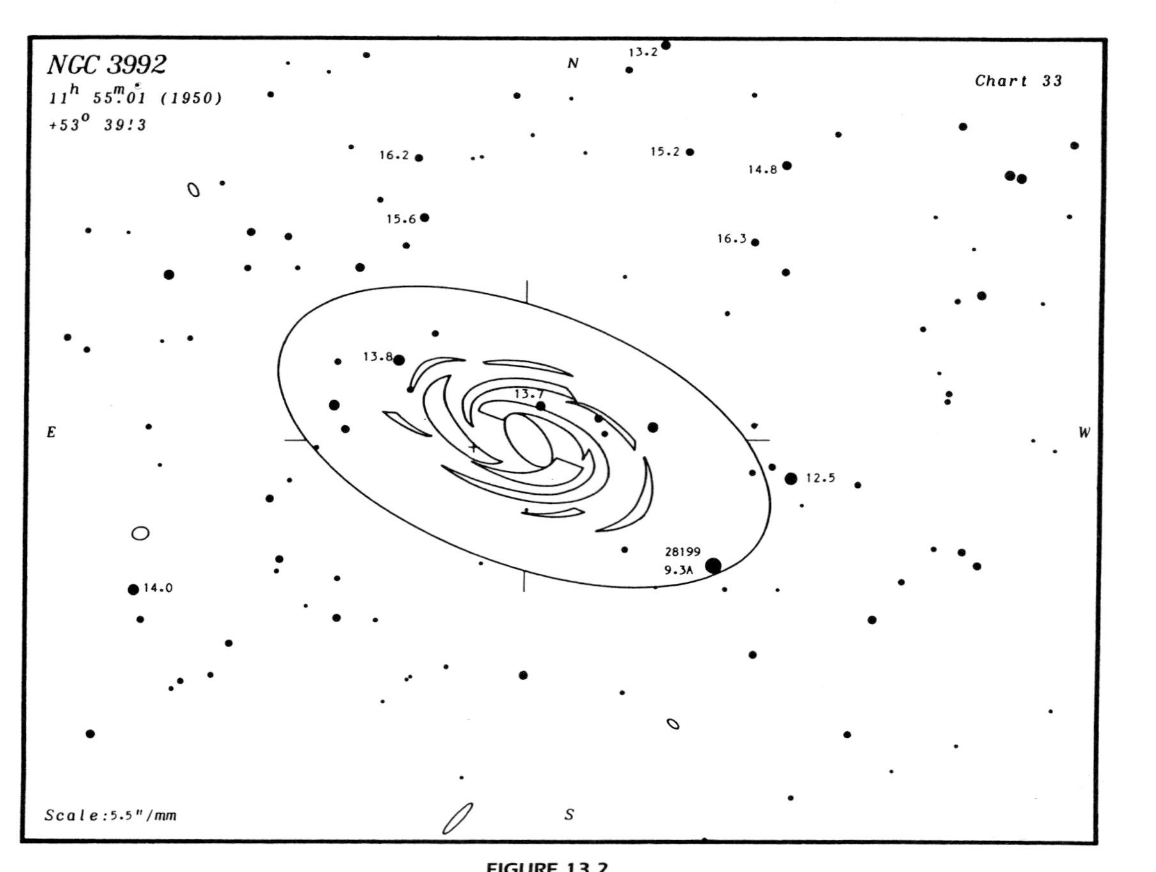

FIGURE 13.2
Supernova search chart for NGC 3992. Taken from reference [3].

the observer and to provide a means of specifying the location of a supernova. Stars are usually determined with reference to the galactic nucleus, in terms of distance in arc seconds and position angle or compass directions.

Finally, all charts must be checked and proved at the telescope. A traced star-field is almost always incomplete around and across the nuclear bulge of the galaxy, because the photograph from which it was taken is likely to have been over-exposed in this region and the stars lost. Careful visual observation will therefore disclose overlooked stars that must be added to the chart. This process is very important, since omitted stars in the nebulosity of the galaxy are certain to waste time later in the reporting of false alarms. Proof is most effectively achieved when several nights are used to examine a chart.

Regardless of the technique used to prepare maps, the final product should be able to help an observer answer three questions: Is a supernova present at or above the stated limiting magnitude? Where is it located with respect to the centre of the galaxy? How bright is the star?

Making charts is a long and tedious job. Those who want to spend more time under the sky and less time over a drawing table should look for sources of charts. The AAVSO and the BAA may be consulted in order to obtain these maps. Some individuals have privately published their work and offer it for sale through astronomy periodicals, and the Cambridge University Press has published *The Supernova Search Charts*, an example of which is shown in Fig. 13.3 [4].

SEARCH TECHNIQUES

At the next available opportunity, the observer should monitor the latest supernova to be reported. This is to gain an insight into the faintness of the type of object that is to be detected, and its appearance in the eyepiece. The searcher may then gain experience in making effective magnitude estimates and determining the distance of the supernova from the nucleus. Figure 13.4 shows the light curve of SN 1989B.

When checking a target galaxy, the area to be searched should be a distance equivalent to two or three times the long diameter of the galaxy: this will include the halo of the galaxy, plus any very faint spiral arms which are not apparent on any chart or photograph. Some supernovae have been detected as far as 10 arc minutes from the nucleus, as in the case of 1909A in NGC 5457 [2]. So keep in mind the size of the galaxy while monitoring for supernovae. While searching the galaxy, be aware of the nuclear area of the galaxy: many supernovae might be found around the brighter core regions of these target galaxies. Usually, higher magnification will distinguish the two objects if some unusual brightness has caused suspicion. Some galaxies, such as Seyferts, are brighter than others owing to their physical make-up. Careful examination of these objects is also part of the search process, as a supernova might be lurking somewhere *within* these brighter galaxies.

It is good practice to keep records of all galaxies monitored. The limiting

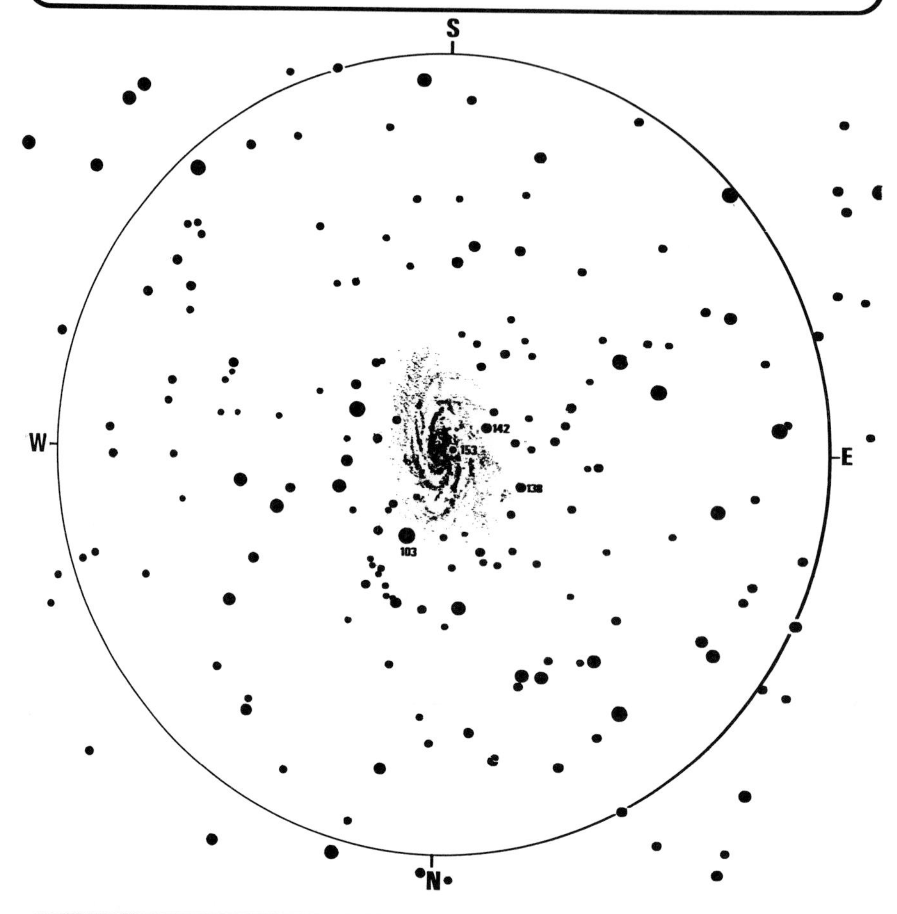

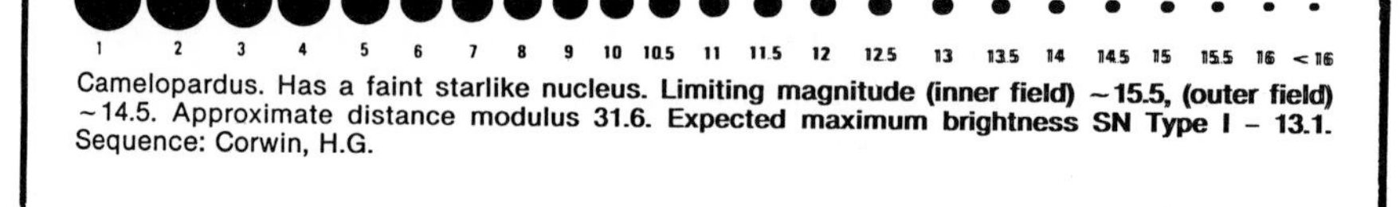

Camelopardus. Has a faint starlike nucleus. Limiting magnitude (inner field) ~15.5, (outer field) ~14.5. Approximate distance modulus 31.6. Expected maximum brightness SN Type I – 13.1. Sequence: Corwin, H.G.

FIGURE 13.3
Supernova search chart for NGC 2336. Taken from reference [4].

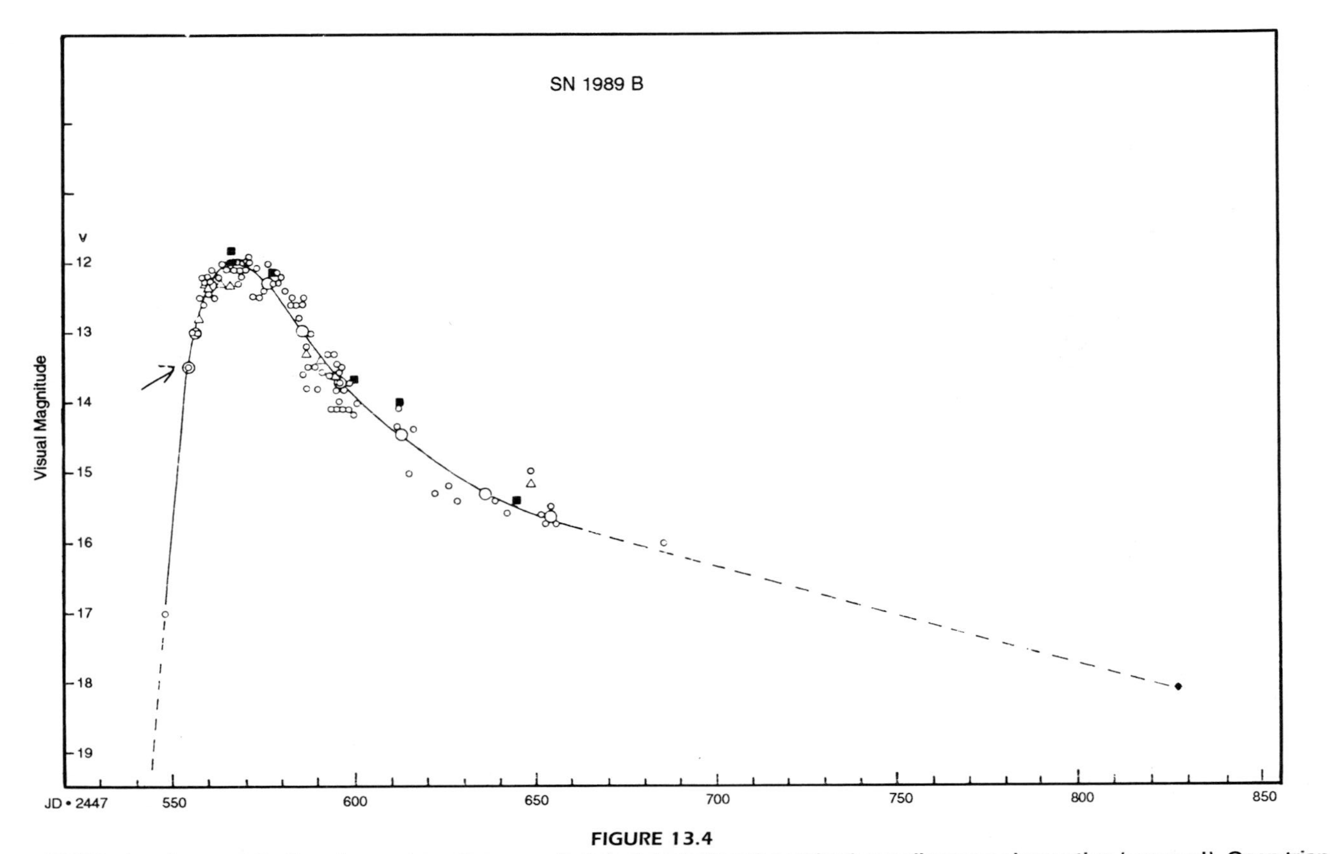

FIGURE 13.4
Supernova 1989B, showing contributions by amateurs (open small circles), including the author's pre-discovery observation (arrowed). Open triangles denote Asiager photographic observations. Squares are CCD magnitudes and large open circles locate mean points.

magnitude reached when viewing each galaxy should be recorded. This is important in two ways: first, the supernova rate may in the long run be determined by reports of negative and positive sightings; and secondly, the knowledge of having observed a galaxy to a particular magnitude will become important once a subsequent positive sighting has been made of a supernova in that galaxy. Such pre-discovery magnitude depth estimates, whether negative or positive, will assist astronomers in determining the particular characteristics of an event in its rise to maximum light.

Monitoring for extragalatic supernovae requires patience and careful observation. Repeated monitoring of target galaxies will enable the observer to learn star-fields so well that charts will become secondary in importance. This technique is used by the current foremost visual discoverer of supernovae, the Reverend Robert Evans, who has hundreds of galaxy star-fields tucked away in his memory. While his record is exceptional, other observers may with practice develop similar skills.

During the course of the search, observers may report sightings of objects other than supernovae. Chance encounters with asteroids occur regularly, especially near the ecliptic. Most of these objects have known orbits and appulses with various galaxies and are possible to forecast, but previously unknown variable stars can be troublesome since they are unpredictable. Sightings may be relayed to variable star associations, such as the AAVSO and the Variable Star Section of the BAA, for confirmation. Organized supernova search groups can also help in identifying these objects [5, 6].

VERIFICATION

The observer who has a supernova candidate should take special care in determining that the find is genuine. Observing the object an hour after the first sighting may reveal motion, showing it to be an asteroid. Obvious errors such as confusion with a field star or with a bright stellar nucleus of a galaxy should be checked. These careful determinations before an announcement will save much wasted time, expense and loss of credibility. On the other hand, being too careful might cause the supernova to be discovered by another. Eventually, observers learn to balance caution and confidence through knowledge of various galaxies and general skill.

What do you do when a suspect passes initial checks at the telescope? The obvious answer is 'Tell someone' — but whom? Direct contact with the Central Bureau for Astronomical Telegrams at the Smithsonian Astrophysical Observatory in Cambridge, Massachusetts is possible; however, owing to the frequency of reported false alarms, Bureau staff members may not give an unverified claim from an unknown observer much credence. Working with a recognized and experienced organization expedites the verification or refutation of a claim, and prospective searchers will do best to join such a group and work within its requirements [5, 6, 7].

To illustrate the importance of disciplined reporting, consider the following summary of the discovery of supernova 1987L in NGC 2336 by Dana Patchick, a member of Sunsearch.

On 26 July 1987, Mr Patchick observed NGC 2336 at dawn and noticed a magnitude 13.5 object near the galaxy. Before he could resolve his doubts, approaching daylight stopped observation. Bad weather prevented a second viewing until August 16, when he confirmed that a new star was indeed present in the field, although now fainter than on July 26. The next day Mr Patchick reported his claim to a verifier in the Sunsearch program, a verifier being a person with resources and contacts necessary to resolve most claims. An apparent supernova was visually confirmed at 40″ arc east and 75″ arc south of the nucleus of NGC 2336. After a final conversation with Mr Patchick, the verifier sent a telegram to the Central Bureau of Astronomical Telegrams to announce the claim. Eventually, a spectrum was obtained a Lick Observatory that confirmed a Type Ia supernova.

Hindsight suggests that the observer should have been more aggressive after suspecting the star on July 26. He was, however, unwilling to come forward until stronger facts were in hand. Since an amateur searcher's chief resource, after observational skill, is credibility, Mr Patchick's deliberate method was most understandable.

MONITORING SUPERNOVAE: THE ROLE OF THE AMATEUR

Once a supernova is discovered, regular monitoring is required to keep up with its progress. A bright star usually gets good attention so that its light curve is well recorded; however, a less prominent supernova is sometimes poorly observed, meaning that the light curve has gaps in coverage. Persons who search should also be ready to observe these objects — monitoring is less glamorous than discovering, but its practical value is high. Visual observations of supernovae are of sufficient interest that the Central Bureau for Astronomical Telegrams routinely publishes light estimates.

A recent professional article contains the following comment: '... magnitudes of many supernovae by repeated visual estimates with reference to close comparison stars can prove in some cases to be more accurate than those accomplished with certain instruments that measure magnitudes from Schmidt plates when the objects occurred on the bright body of the parent galaxy' [8]. In another article [9], this statement is made: 'The naked eye, (particularly when dark adapted) is quite sensitive to the 4600–4700 angstrom frequency, which seems to create a plausible explanation for the fact that the supernovae discovered by Tycho and Kepler, which were visual estimates, resemble very closely the typical photographic light curve of Type I supernovae.'

It has been further stated that a great deal of photometric data has been collected on Type Ia supernovae, giving the possibility of assembling a fairly large and homogeneous sample of light curves of this class of object. Owing to the importance of Type Ia supernovae as distance indicators, it seems worthwhile to attempt to

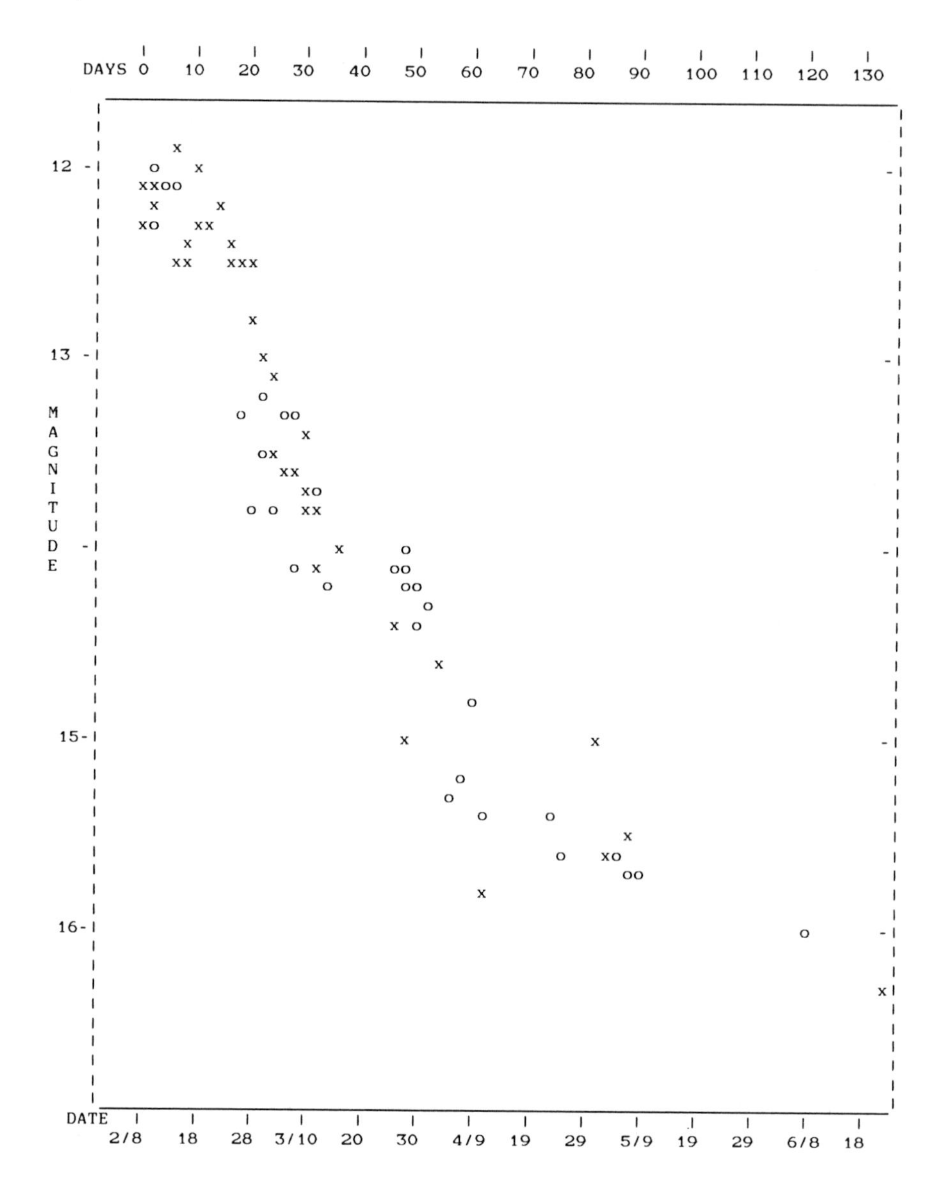

FIGURE 13.5
Observations of supernova 1989B in NGC 3627 (M66).

determine as accurately as possible the mean characteristics of their light curves. Here is an opportunity for amateurs to help with the appearance of each new supernova.

Figure 13.5 is a light curve of the bright supernova 1989B in NGC 3627. It contains observations announced in the IAU Circulars (×) and specifically identifies

Sunsearch (○) light curve estimates. Visual contributions appear throughout the $4\frac{1}{2}$-month apparition and agree well with observations made by other techniques. Similar graphs could be constructed for other supernovae if observers would respond with equal enthusiasm for each visible object.

PHOTOGRAPHIC SEARCHING FOR SUPERNOVAE

Visual detection of supernovae has been the theme of this chapter so far. Other methods of searching do exist, and while they are not technically outside the realm of amateur capabilities they might lie outside the average pocketbook — namely the initiation of an automatic search programme through the use of computers, CCD imagery and automatic large aperture scopes.

If photography is used as a amateur astronomer's tool for the detection of extragalactic supernovae, several criteria should be adhered to before this method will prove to be productive. Here are just a few of the considerations to be taken into account before such a programme is initiated:

The programme should ideally involve a team of equally inspired and dedicated astrophotographers, although one person may attempt such a venture. A base series of target galaxies should be established, and the exposure time for each object should be fixed so as to achieve a consistent limiting magnitude for comparison purposes. Photographs obtained in a night's work should be developed and analysed the same night with the use of a blink-comparator or other similar device to determine whether an intruder star is present. Only the brightest and closest galaxies should be considered for this venture.

Observers should carry out experiments on the galaxies in their programme to determine the most feasible limiting magnitude. Typically, using faster emulsions (for example, T-Max 3200), a 3-minute exposure on unhypered film, using a 200-mm f/5 telescope, will reach the 14th magnitude. In conclusion, it is imperative that the programme yield a continued and productive supply of photographs.

HOW BRIGHT WILL A SUPERNOVA APPEAR?

It is helpful to know the likely maximum magnitude of a supernova occurring in the programme galaxies. The average absolute magnitude of observed supernovae depends upon the assumed distance of the galaxies in which they occurred, which in turn depends upon the value taken for the Hubble constant. Using a Hubble constant of $H = 50$ km/sec per Mpc produces an absolute magnitude of -20.0 for a Type Ia event and -18.0 for a Type II event, while $H = 100$ km/sec gives an absolute magnitude of -18.5 for Type Ia and -16.5 for Type II events.

The estimated maximum magnitude for a supernova is computed by adding the absolute magnitude of the supernova type to the distance modulus of the galaxy. For example, if the distance modulus of M66 is 30.4 ($H = 50$ km/sec), then the theoretical apparent magnitude of a Type Ia supernova appearing in it would be magnitude 10.4 ($30.4 + (-20.0) = 10.4$). This simple computation gives a convenient indication whether the supernovae of a galaxy will reach maxima that observers can see. Arranging an observing programme by this criterion will save wasting time on galaxies that are too distant to show 'bright' supernovae. Unfortunately, distance moduli are not commonly available to amateur astronomers since they usually appear in professional literature.

A list of bright galaxies might be better accessed from *The Supernova Search Charts* [4], which provide a listing of 237 galaxies, while Table 13.3 gives a list of galaxies in Leo and Virgo and the theoretical maximum brightness of supernovae occurring within them.

TABLE 13.3
GALAXIES NORTH OF THE CELESTIAL EQUATOR IN THE LEO–VIRGO REGION. THE THEORETICAL MAXIMUM PHOTOGRAPHIC (BLUE) MAGNITUDES OF TYPE I AND TYPE II SUPERNOVAE ARE SHOWN

NGC	RA (1950)	Dec	Mag (Mb)	Size	Type (Mb)	Type II (Mb)
3351 = M95	10^{5} $41^{m}.2$	+11° 58′	10.5	7.4	10.5	12.5
3368 = M96	10 44.1	+12 05	10.1	7.1	10.9	12.9
3377	10 45.0	+14 15	11.1	4.4	10.4	12.4
3379	10 45.1	+12 51	10.3	4.5	10.9	12.9
3384	10 45.4	+12 54	10.7	5.9	10.5	12.5
3412	10 48.2	+13 40	11.5	3.6	10.8	12.8
3423	10 48.4	+06 06	11.6	3.9	11.1	13.1
3489	10 57.4	+14 10	11.1	3.7	10.1	12.1
3521	11 03.2	+00 14	9.6	9.5	10.5	12.5
3593	11 12.0	+13 05	11.7	5.8	10.0	12.0
3605	11 14.1	+18 18	13.1	0.7	10.4	12.4
3607	11 14.2	+18.20	11.1	3.7	11.1	13.1
3623 = M65	11 16.2	+13 22	10.2	10.0	10.6	12.6
3627 = M66	11 17.4	+13 16	9.7	8.7	10.4	12.4
3628	11 17.4	+13 52	10.3	14.8	10.8	12.8
3666	11 21.5	+11 37	12.4	3.6	11.3	13.3
3691	11 25.4	+17 12	13.5	0.9	11.4	13.4
3705	11 27.3	+09 33	11.8	5.0	11.2	13.2
3773	11 35.4	+12 23	13.1	0.5	11.2	13.2
3810	11 38.2	+11 45	11.4	4.3	11.2	13.2
4037	11 58.5	+13 41	12.5	1.5	11.7	13.7
4064	12 01.4	+18 43	12.3	2.5	11.7	13.7
4178	12 10.1	+11 09	11.9	5.0	11.7	13.7
4192 = M98	12 11.2	+15 11	10.9	9.5	11.7	13.7

TABLE 13.3 continued
GALAXIES NORTH OF THE CELESTIAL EQUATOR IN THE LEO–VIRGO REGION. THE THEORETICAL MAXIMUM PHOTOGRAPHIC (BLUE) MAGNITUDES OF TYPE I AND TYPE II SUPERNOVAE ARE SHOWN

NGC	RA (1950)	Dec	Mag (Mb)	Size	Type (Mb)	Type II (Mb)
4212	12 13.1	+14.11	11.8	3.0	11.7	13.7
4216	12 13.2	+13 25	11.0	8.3	11.7	13.7
4237	12 14.4	+15 36	12.5	1.4	11.7	13.7
4293	12 18.4	+18 40	11.2	6.0	11.7	13.7
4294	12 18.4	+11 47	12.6	2.4	11.7	13.7
4299	12 19.1	+11 47	12.9	1.1	11.7	13.7
4340	12 21.0	+17 00	11.9	2.2	11.7	13.7
4342	12 21.1	+07 21	13.5	2.0	11.7	13.7
4371	12 22.2	+11 59	11.7	3.9	11.7	13.7
4374 = M84	12 22.3	+13 10	10.2	5.0	11.7	13.7
4379	12 22.4	+15 53	12.3	0.7	11.7	13.7
4380	12 22.5	+10 18	12.4	3.0	11.7	13.7
4382 = M85	12 22.5	+18 28	10.1	7.1	11.7	13.7
4394	12 23.2	+18 29	11.8	3.9	11.7	13.7
4406 = M86	12 23.4	+13 13	10.0	7.4	11.7	13.7
4417	12 24.2	+09 52	12.1	2.2	11.7	13.7
4419	12 24.2	+15 19	12.1	2.3	11.7	13.7
4424	12 24.4	+09 42	12.3	2.5	11.7	13.7
4435	12 25.1	+13 21	11.7	3.0	11.7	13.7
4438	12 25.1	+13 17	10.9	9.3	11.7	13.7
4442	12 25.3	+10 05	11.3	4.6	11.7	13.7
4452	12 26.1	+12 02	13.3	1.3	11.7	13.7
4457	12 26.3	+03 51	11.7	3.0	11.7	13.7
4469	12 26.6	+09 02	12.3	3.0	11.7	13.7
4472 = M49	12 27.1	+08 17	9.3	8.9	11.7	13.7
4483	12 28.1	+09 17	13.4	0.8	11.7	13.7
4526	12 31.3	+07 58	10.6	1.2	11.7	13.7
4548	12 32.6	+14 46	11.0	5.4	11.7	13.7
4550	12 32.6	+12 30	12.3	2.0	11.7	13.7
4552 = M89	12 33.1	+12 50	10.8	4.2	11.7	13.7
4564	12 33.6	+11 43	11.9	1.8	11.7	13.7
4569 = M90	12 34.2	+13 26	10.2	9.5	11.7	13.7
4571	12 34.2	+14. 30	11.8	3.8	11.7	13.7
4586	12 35.6	+04 36	12.5	2.6	11.7	13.7
4595	12 37.2	+15 34	13.2	1.1	11.7	13.7

TABLE 13.3
GALAXIES NORTH OF THE CELESTIAL EQUATOR IN THE LEO–VIRGO REGION. THE THEORETICAL MAXIMUM PHOTOGRAPHIC (BLUE) MAGNITUDES OF TYPE I AND TYPE II SUPERNOVAE ARE SHOWN

NGC	RA (1950)	Dec	Mag (Mb)	Size	Type (Mb)	Type II (Mb)
4621 = M59	12 39.3	+11 55	10.7	5.1	11.7	13.7
4630	12 39.6	+04 14	13.0	1.1	11.7	13.7
4635	12 40.1	+20 13	12.9	1.5	11.7	13.7
4638	12 40.2	+11 43	12.0	1.1	11.7	13.7
4636	12 40.2	+02 58	10.5	6.2	11.7	13.7
4639	12 40.2	+13 32	12.2	2.0	11.7	13.7
4651	12 41.1	+16 40	11.4	3.4	11.7	13.7
4654	12 41.3	+13 24	11.1	4.7	11.7	13.7
4660	12 42.0	+11 28	11.9	2.8	11.7	13.7
4665	12 42.3	+03 20	11.4	4.2	11.7	13.7
4688	12 45.1	+04 37	12.4	2.2	11.1	13.1
4698	12 45.5	+08 46	11.5	4.3	11.7	13.7
4701	12 46.4	+03 40	12.8	1.5	10.3	12.3
4713	12 47.2	+05 35	12.2	2.3	11.7	13.7
4762	12 50.2	+11 30	11.3	8.7	11.7	13.7
4765	12 50.4	+04 44	13.3	0.7	10.5	12.5
4772	12 50.6	+02 26	12.4	2.7	11.2	13.2
4808	12 53.2	+04 34	12.6	2.2	10.5	12.5
4826 = M64	12 54.2	+21 57	9.4	9.3	9.2	11.2
4845	12 55.3	+01 51	12.2	4.0	11.3	13.3
4900	12 58.1	+02 46	12.1	1.7	11.1	13.1

It is not possible to use radial velocity values in conjunction with the Hubble constant to determine galactic distances or distance moduli, since each galaxy generally falls within some 'cloud', group, or cluster, each galaxy member of which is revolving around that group's centre of rotation with its own radial velocity. For example, in the Virgo Cluster (overall radial velocity 1150 ± 50 km/sec) some galaxies exhibit a radial velocity of 2200 km/sec, whilst others are travelling at −700 km/sec, despite the fact that they are all at the same distance. Supernova magnitudes in the Virgo Cluster are likely to be about 11.7 for Type Ia and 13.7 for Type II at peak brightness, although these values will be affected by the location of the supernova with respect to the galaxy, the type of galaxy, and how much absorption occurs either within the galaxy or along the line of sight within our own Galaxy.

CONCLUSION

The detection of these enormous stellar events represent one of the final bastions of scientific discovery left open to the amateur. Careful and repeated scrutiny of faint star-fields is the price that has to be paid for achieving such a goal. We observers are few in number, having dedicated ourselves to participate in a tedious yet exhilarating programme of amateur astronomy. Certainly it is not the easiest specialization, but it can be very rewarding to know that every time a galaxy is observed, the potential for a major discovery is present.

Extensive gratitude is extended to the Reverend Robert O. Evans and James T. Bryan for contributing their expertise and inspiration to this project — the former for giving me the insight to continue with the work, and the latter for guidance in protocol involving the Sunsearch project for so many years. Thanks also to Gregg Thompson for allowing Sunsearch to use a sample of pre-publication drafts of charts from *The Supernova Search Charts*, and to the many professional astronomers who have displayed confidence in our programme and have kept the channels of communication open to us. Good hunting to you all!

About the Author

Steve Lucas's study of the evening skies began, perhaps like most amateur astronomers, at a relatively early age. A $1\frac{1}{2}$-inch, 5-foot telescope of dismal quality (although not noticed at the time) penetrated the then un-light-polluted skies of Chicago, Illinois, where the Milky Way was seen with relative ease during the summer of 1952. Many visions of unseen — bright — astronomical objects passed like a parade before his young eyes in those days . . . fortified with pictorial literature that was received from some of the most prestigious observatories of the time.

Unfortunately his interest in the night skies did not follow his procession into manhood, until almost three decades later when it was once again kindled.

A more aggressive desire to contribute to astronomy inspired Lucas to want to undertake a programme that had never been attempted before on the amateur level. This programme was initiated by one of the giants of the astronomical community, on a professional level, Dr Fritz Zwicky. It was Dr Zwicky's intention to search the night sky for the existence of extragalactic supernovae. Why could the amateur astronomer not contribute to this programme? With the help of several professional and amateur astronomers Lucas was able to make this intention a reality. The culmination of this ideal was realized on 22 August 1987 when a supernova in the galaxy NGC 2336 was discovered by Dana Patchick of Culver City, California, and successfully verified by professional astronomers.

A desire to extend their aspirations has resulted in a cooperative effort with other amateurs from around the world. It is his intention to incorporate as many amateur astronomers as desire to seek the discovery of supernovae in other galaxies, and once having found them to monitor their behaviour so that important data may be amassed that might be of some use to the astronomical community at large.

Steve Lucas currently resides in a suburb of Chicago, Illinois, is 50 years old with a wife and two teenaged children. His current occupation involves participation in the trucking industry.

Contact address – 14400 Kolin Avenue, Midlothian, Illinois 60445, USA.

References

1. Doggett, J.B. and Branch, D. *Astrophysical Journal*, **90**, 11 (1985).
2. Evaluation compiled from *The Asiago Supernova Catalogue*. Osservatorio Astronomico, Dipartmento di Astronomia e Osservatorio Astrofisico dell'Universito, Padova-Asiago, Italy, 1989.
3. Bryan, J.T. and Corwin, H.G. *Atlass of the Ursa Major I North Cloud*, 1989. (See address under [7].)
4. Thompson, G.D. and Bryan, J.T. *The Supernova Search Charts*. Cambridge University Press, 1989.
5. 'Sunsearch' Supernova Search Network, Steve H. Lucas, 14400 Kolin Avenue, Midlothian, IL 60445, USA.
6. The UK Nova/Supernova Search Programme, Guy M. Hurst, 16 Westminster Close, Kempshott Rise, Basingstoke, Hants RG22 4PP, UK.
7. Supernova Search Committee of the AAVSO, James T. Bryan, 605 San Gabriel Overlook (east), Georgetown, TX 78628, USA [Northern Hemisphere]. Reverend Robert Evans, 57 Talbot Road, Hazelbrook, New South Wales 2779, Australia [Chairman].
8. Barbon, R. *et al.*, *Astronomy and Astrophysics*, **214**, 131 (1989).
9. Morrison, P. and Sartori, L. *Astrophysical Journal*, **158**, 541 (1969).

Notes and Comments

Despite the continued presence of SN 1987A in the southern sky, and two dozen faint supernovae found by professionals, 1988 was a very poor year for bright supernovae. The brightest was SN 1988A in M58 at mag. 14.5. This was the only new supernova I saw all year.

1989 started off differently, with two bright supernovae found during January.

SN 1989A was found in NGC 3687 by the Berkeley automatic search team, led by Dr Carl Pennypacker. At discovery, the supernova was mag.

FIGURE 13.6
Photograph of supernova 1989B in NGC 3627 (M66). Taken with a modified Celestron 11 at f/10 on 11 February 1989, 08.05–08.45 UT, using hypered TP 2415. (Photo credit Kim Zussman, Thousand Oaks, California.)

15.3, but later rose to 13.9, and proved to be of Type Ia. It was found on January 19, just before the full moon. This galaxy was not on my observing list and has not been observed by me.

SN 1989B appeared in NGC 3627 (M66).

Here is an account of the discovery of supernova 1989B in NGC 3627, by the Reverend Robert Evans. It was published in *The Astronomer*, March 1989. — Ed.

> My first sighting of the supernova was on January 30.5 UT, when it was around mag. 13.0. The supernova was immediately obvious with × 200 in the 16-inch [400-mm] Dobsonian. It could even be seen with low power, although the galaxy was still fairly low in the sky (from 34° south). After checking Gregg Thompson's chart, and estimating the offset of the supernova from the nucleus of the galaxy (15″ arc west and 50″ arch north) with a measuring rule, I rang Rob McNaught to see if there already existed any telegrams about it, and to start planning verification.
>
> Knowing that M66 had, at least in the past, been on the top priority observing list of the Berkeley automatic search, it was clear to me that verification would have to be very speedy if I was to have much chance of landing this discovery. Also, such a prominent northern galaxy would be observed by many other astronomers within a short time.
>
> Rob went to the nearby Uppsala Southern Schmidt, at Siding Spring, to expose a film of the galaxy, and develop it. While this was happening, I tried to find an asteroid–galaxy appulse list in my office — but no current one was found. So, I rang Dr Marsden at the Central Bureau in Washington to check their computer files for an asteroid near M66. No such asteroids were listed, and there had been no previous reports of the supernova.
>
> An hour later, Rob rang to say that his effort to photograph the supernova had failed. The photo had to be taken without him being able to see the object in question, and he had missed the target. Also, the sky had clouded over. He hoped that the sky might clear a little before dawn, about four hours away. All I could do was to ring the observers on the 2.3-metre telescope at Siding Spring, who were, of course, also clouded out, and I left a message on Carl Pennypacker's office answering machine in Berkeley, California, asking for help. Then I went to bed.
>
> Rob had notified Dr Marsden about his useless film and the cloud situation, so a few phone calls were made from Boston to alert interested people about the possible discovery.
>
> Just as dawn was breaking at Siding Spring, the sky cleared a little, and Rob managed to make a one-minute exposure with the Schmidt and a visual observation with his own telescope. With this second film, he measured an exact position for the supernova, and then he reported confirmation of the discovery to the Central Bureau.
>
> The first I heard of the successful confirmation was at breakfast time, when Carl rang in from Berkeley, having just heard of the confirmation directly from Boston. The automatic search had been busily at work, as usual, but had not observed M66 since January 21, before the supernova was obvious enough to be noticed.

An hour later, Rob rang in to tell the story of his success at making the verification, and how it had been possible at the very last moment.

The next day we heard that spectral studies had revealed the supernova to be of the classical Type Ia, and still approaching maximum light. In due course, we also heard that an Italian amateur had made an independent discovery of the supernova about six hours after my discovery. News of this Italian discovery had, however, been too late for the official discovery notice, and it would not be officially listed.

This report, showing what an experienced observer does when he suspects a discovery, emphasizes the following points:

1. There must be a confirmation network working on an 'all night' basis, with observers widely scattered to lessen the effect of bad weather.
2. The possibility of the supernova being a minor planet must be borne in mind, particularly when the galaxy is in the region of the ecliptic, as in the case of objects in Leo and Virgo.

The case of an American amateur who spotted a suspect supernova two days before discovery, waited a day or two to confirm his sighting and lost the discovery (and also lost astronomers an extra day of observing time) underlines how vital it is to ensure that any unusual-looking object is reported immediately to someone who is capable of deciding whether it is important or not. Lines of communication are absolutely vital, and with answering machines, fax, and electronic mail all widely available as extras to the telephone it is easier than it ever has been to reach the right people at the right time. — Ed.

CHAPTER FOURTEEN · *High-resolution Lunar and Planetary Photography*

GÉRARD THERIN
BLANC-MESNIL, FRANCE

MANY AMATEUR ASTRONOMERS WHO OWN A REFLECTING OR REfracting telescope must have thought about trying to record photographically the view seen through their instrument. Unfortunately, however, the ones who do make the attempt are soon put off by their poor results, not realizing that to succeed in astrophotography they must serve an 'apprenticeship'. In fact there are numerous problems (instrumental, technical, and even human) which must be overcome before it is possible to photograph details near the theoretical resolving power of the telescope being used, and persistent practice and progress is essential before the amateur can achieve the best possible results.

For high-resolution photography of the Moon and planets, an aperture of at least 100 mm is needed, and in general the optimum aperture is between 200 and 300 mm. An instrument of larger size than this will only prove its worth if the observing conditions are exceptionally good, or for photography of deep-sky objects. Atmospheric turbulence is the main obstacle to high-resolution photography, and unfortunately the choice of observing site is usually dictated by personal circumstances rather than astronomical factors!

TURBULENCE OR 'BAD SEEING'

The atmosphere contains air masses of different temperatures and therefore different densities, which act like converging or diverging optical systems. Light from a celestial object passing through the atmosphere therefore loses its original parallel nature, resulting in a loss of image quality which becomes more serious the greater the temperature difference between the air masses in its path. It is, therefore, essential to reduce as much as possible the principal controllable sources of turbulence, which may be defined as *local* and *instrumental*.

Local turbulence – This is caused by the immediate surroundings: within a hundred metres or so of the telescope there are certain to be heat sources which must be removed or at least reduced. Large areas of concrete exposed to the Sun for much of the day form a heat source which, like houses, should be as far away as possible. If you have a telescope mounted inside a dome, bear in mind that the air inside the dome is likely to be warmer than that outside, creating turbulence when it is opened up: for this reason, a shelter with a sliding roof is much to be preferred. If a dome is to be used, an opening of about three times the telescope aperture is essential, for this will allow the worst of the currents to escape on either side of the central part being used for observation. Similarly, the position of the observer, whose body and respiration are sources of heat, should be considered with respect to the light-path into the telescope.

Instrumental turbulence – This is often neglected by amateurs, although it can often reach an unacceptable level — 30″ arc or more. Some telescopes, despite having excellent optics, have never given good images simply because their tubes were made of poorly-conducting material, or were far too thick, thus greatly slowing down night-time cooling. At the start of a night's observation, the telescope as well as the observatory (especially if it is in a dome) is almost certain to be warmer than the surrounding air, and to minimize the period of poor seeing the whole telescope must cool to the night-time temperature as quickly as possible. Unfortunately, the most common materials used for telescope tubes — wood and fibreglass — are not the most suitable from this point of view: the tube is the site of major heat exchanges which can make the air swirl like heat up a chimney, and this effect will last longest if the tube is made of a poor conductor, or if it is excessively long.

At first sight, one solution for reflecting telescopes would be to close the upper end of the tube with an optical window. This does not work, however, since it actually slows down the cooling process; secondly, since glass allows infrared (heat) rays to pass through it, the resultant temperature gradient will deform the optical window. The same thing happens to the mirror in a reflecting telescope, whose optical surface, which faces the sky, tends to be a few tenths of a degree cooler than the rear surface. The result is that the reflecting surface is distorted into a slightly more concave shape, while the rear turns slightly convex. A thin mirror achieves thermal equilibrium more quickly than a thick one, and therefore has less chance of being affected by distortion, although, of course, it must be properly supported in its cell to avoid flexure.

In order to eliminate instrumental turbulence, the most effective answer is to have the kind of framework metal tube used in professional observatory instruments.

For relatively small apertures (200–250 mm) it is possible to use a rolled metal tube, provided the material is as thin as possible. In any case, it is vital to expose the instrument to the outside air long enough to achieve thermal equilibrium. If the night-time temperature drop is large, several hours may be needed!

COLLIMATION

The optical alignment of a reflecting telescope is never permanent, since mechanical factors, thermal expansion, and the effects of normal handling can slightly alter the adjustments. This means that there must be a way of realigning the components, and the careful observer will soon appreciate how important it is to check and refine the alignment regularly.

There are several ways of collimating a telescope. Despite its limitations, geometric collimation (making the mirror reflections appear concentric when viewed in daylight from the eyepiece position) is the method usually employed by amateurs. This assumes that the optical surfaces are perfect surfaces of revolution; also, the precision attainable with this method is not really acceptable in the case of primary mirrors working at f/5 or faster.

The only really effective method of collimation is to locate a star of the 3rd or 4th magnitude, using a magnification of about 1.5 times the aperture in millimetres (or more if the seeing is steady), and then to examine the intra- and extra-focal image of the star while moving the eyepiece holder slightly inside and outside focus. The appearance should be as shown in A and B of Fig. 14.1. If the rings are not perfectly concentric, as in C and D, it means that the instrument is not precisely collimated. To correct this, the mirror is adjusted so as to displace the star in the field of view in the direction indicated by the arrow. Once this has been done by the right amount, and the rings are concentric, the instrument can be considered to be in perfect collimation.

This is the only appropriate method for telescopes with very fast primary mirrors (and remember that collimation is vitally important with the popular Schmidt–Cassegrains, which have primary mirrors of f/2 or f/2.5). In fact, the slightest maladjustment produces a noticeable deterioration in their optical performance. But whatever the telescope, the most perfect optical components will never give good images if they are not adjusted correctly.

IMAGE ENLARGEMENT FOR ASTROPHOTOGRAPHY

The images of Mars, Jupiter, Saturn, and even the Moon formed at the primary focus of a telescope of normal focal ratio are far too small for a film to be able to record all the detail. Not even the well-known Kodak Technical Pan 2415, with a theoretical resolving power of 10 μm, could do this unless the focal ratio was about 20. In practice, however, it would still be necessary to increase the focal ratio by two or three times to obtain an image whose finest details were not lost in the emulsion grain.

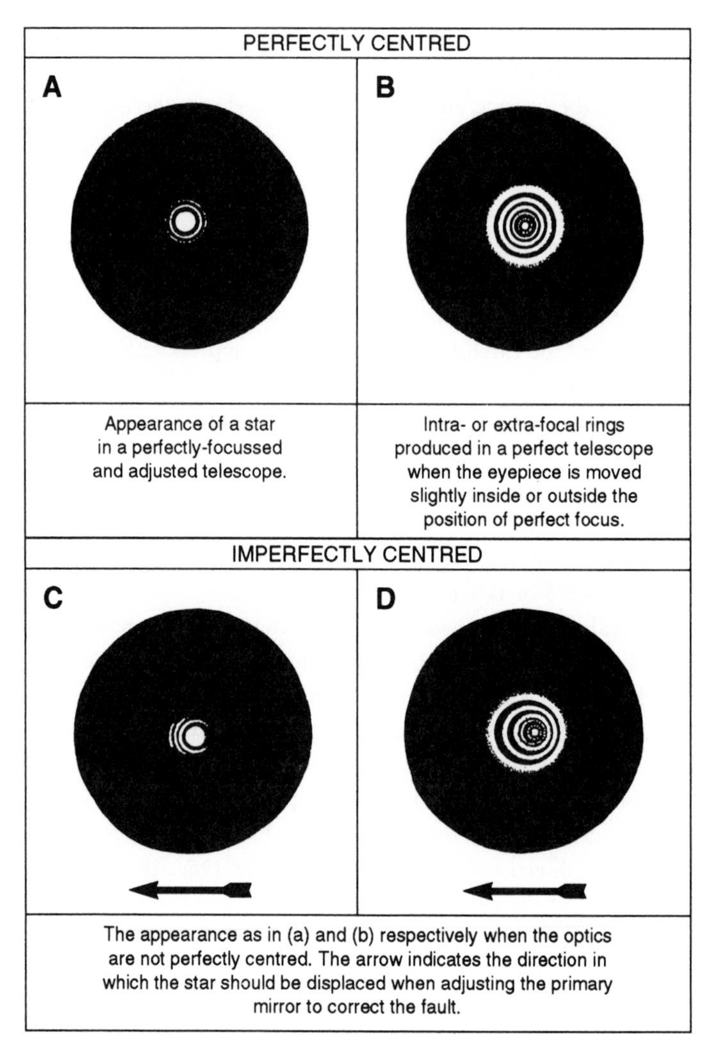

FIGURE 14.1
(a) Appearance of a star in a perfectly-focused and adjusted telescope. (b) Intra- or extra-focal rings produced in a perfect telescope when the eyepiece is moved slightly inside or outside the position of perfect focus. (Note that when a reflecting telescope is used, the central obstruction produces a dark spot at the centre of the rings.) (c) and (d) The appearance as in (a) and (b), respectively, when the optics are not perfectly centred. The arrow indicates the direction in which the star should be displaced when adjusting the primary mirror to correct the fault.

Fortunately, the amateur has several ways of optically enlarging the primary image available. Out of these we shall just recommend two: using an eyepiece, and using a Barlow lens.

Contrary to popular belief, these two methods are not optional alternatives: they complement each other, since each one is subject to fairly tight constraints.

Eyepiece Projection

An eyepiece is a kind of magnifying glass which allows the eye to examine in close-up the image formed by the telescope's lens or primary mirror. (For photographic work it is essential to use only well-corrected eyepieces such as the orthoscopic or Plossl type — simple designs like the Huyghenian, Ramsden, and Kellner are definitely not suitable.) To calculate the amplification A obtained in eyepiece projection (Fig. 14.2), when the distance between the eyepiece and film, or back focus, is B and the focal length of the eyepiece is F, the following formula is required:

$$A = B/F - 1$$

Therefore, for a given eyepiece, the greater the back focus the higher the magnification produced.

Since the observer's eye is, or should be, in a relaxed state, focused on infinity, eyepieces are also designed to form their image at infinity (Fig. 14.3a). The greater the departure from this condition, the more seriously the quality of the image formed by the eyepiece is affected, and when used for photography the eyepiece is not producing parallel light but rays converging to a focus on the emulsion, as in Fig. 14.3b. The longer the back focus (and therefore the greater the amplification) the more nearly parallel these rays become and the better the image quality will be. In practice, the amplification should be at least $\times 8$, and for lunar photography $\times 10$ is better. This is because one of the principal aberrations showing itself at low amplifications is field curvature, which makes it impossible to achieve sharp focus at the centre and edge of the field of view, which is particularly inconvenient for lunar photography. For planetary photography this effect is not so important, since higher amplifications are used and the sharp field of view need only be small.

A final piece of advice: for a given amplification it is better to use an eyepiece of moderate focal length with a considerable back focus than a short-focus eyepiece with a small back focus.

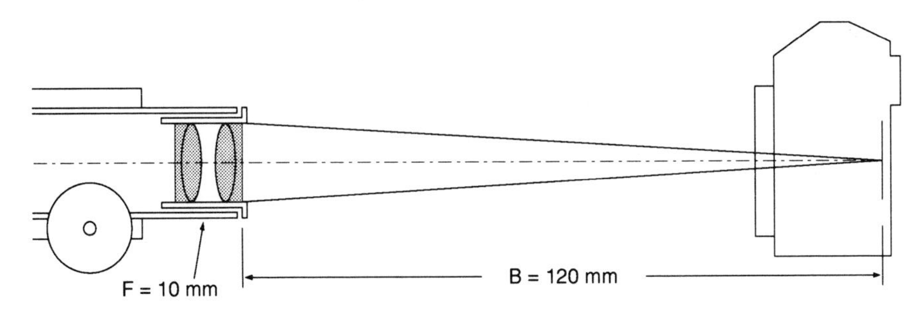

FIGURE 14.2
Using an eyepiece to amplify the focal length of the telescope. In this example a 10-mm eyepiece has a back focus of 120mm, giving an amplification of ×11.

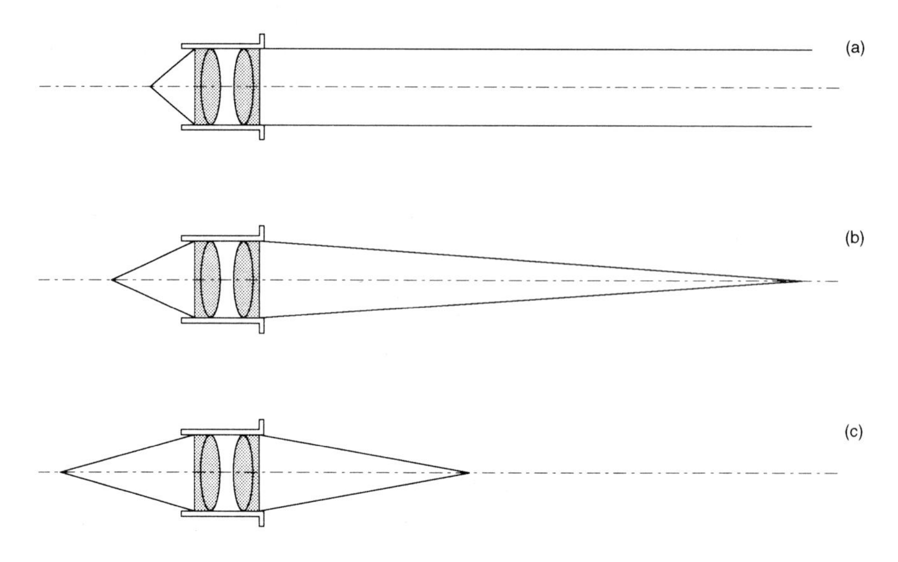

FIGURE 14.3
Using an eyepiece for visual use and for photographic projection. (a) Optimum performance is obtained when the image is at infinity. (b) Focusing the image on to a distant image plane (high amplification) results in slight image degradation. (c) Low-amplification projection (close image plane) gives a poor image, marred by field curvature.

The effective resultant focal length of the telescope is readily obtained by multiplying the original focal length by the amplification. For example, a telescope of 200-mm aperture and 1200-mm focal length (f/6), with an amplification of ×11, will have a resultant focal length of 13.2 m (1200 mm × 11 = 13 200 mm) and a corresponding focal ratio of f/66 (6 × 11 = 66, or 13 200/200 = 66).

Amplification using a Barlow Lens

The Barlow is a well-known accessory for visual observers, but is rarely used in astrophotography. It is a diverging lens which effectively increases the focal length of the telescope in the same way as the secondary mirror of a Cassegrain. Ensure that the Barlow being used is an achromatic combination consisting of two cemented lenses: single lenses (originating from bottom-of-the-range instruments) should be avoided, as they will have serious chromatic aberration and be generally of doubtful optical quality.

It is worth pointing out in passing that visually the Barlow–eyepiece combination is superior to use of an eyepiece on its own, since the f-ratio of the primary mirror is effectively increased — an f/5 mirror with a ×2 Barlow becomes an f/10 mirror, which noticeably improves the eyepiece performance with such short-focus instruments. Secondly, with a large focal ratio the focusing tolerance is increased.

Barlows are designed to work best at a specific amplification, which should be inscribed on the mount (usually ×2 or ×3). Theoretically, in order to preserve the designed performance, this amplification should be adhered to, although for photographic use a ×2 Barlow could be used at ×3, and a ×3 at ×4. When amplifications of between ×2 and ×8 are required, the use of one or two Barlows will give better results than an ordinary eyepiece, since they will be working at the amplifications for which they were designed. In particular, the field will be flatter, which as mentioned above is particularly important for lunar photography. By making slight adjustments, one or two Barlows can be made to produce a range of amplifications from ×2 to ×8.

The working of a Barlow is shown in Fig. 14.4 and the amplification may be calculated as follows:

$$A = B/F + 1$$

$$B = F(A - 1)$$

$$Y = \frac{F(A - 1)}{A}$$

where A = amplification, B = back focus, F = focal length of the Barlow, and Y = distance of the Barlow inside the focal point of the telescope.

As an example, take the case of a Clavé Barlow of 113 mm focal length. What will be the amplification with a back focus of 150 mm?

$$A = (150/113) + 1 = 2.32$$

Therefore, the intended amplification of ×2 is increased to 2.3 with this back focus.

If two Barlow lenses are used (Fig. 14.5), the calculation is as follows (this particular example assumes that both are identical Clavé Barlows). To begin with, the amplification given by Barlow no. 2 (the one nearest the film) is calculated. Using the dimensions in Fig. 14.5:

$$A = (120/113) + 1 = 2.06$$

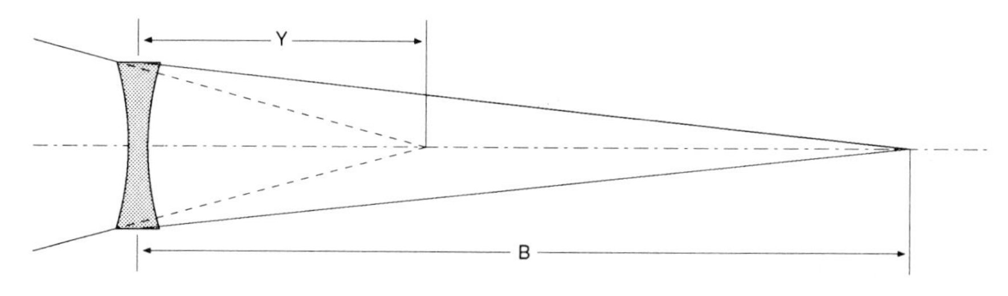

FIGURE 14.4
The principle of the Barlow lens. *Y*=original focus, *B*=back focus.

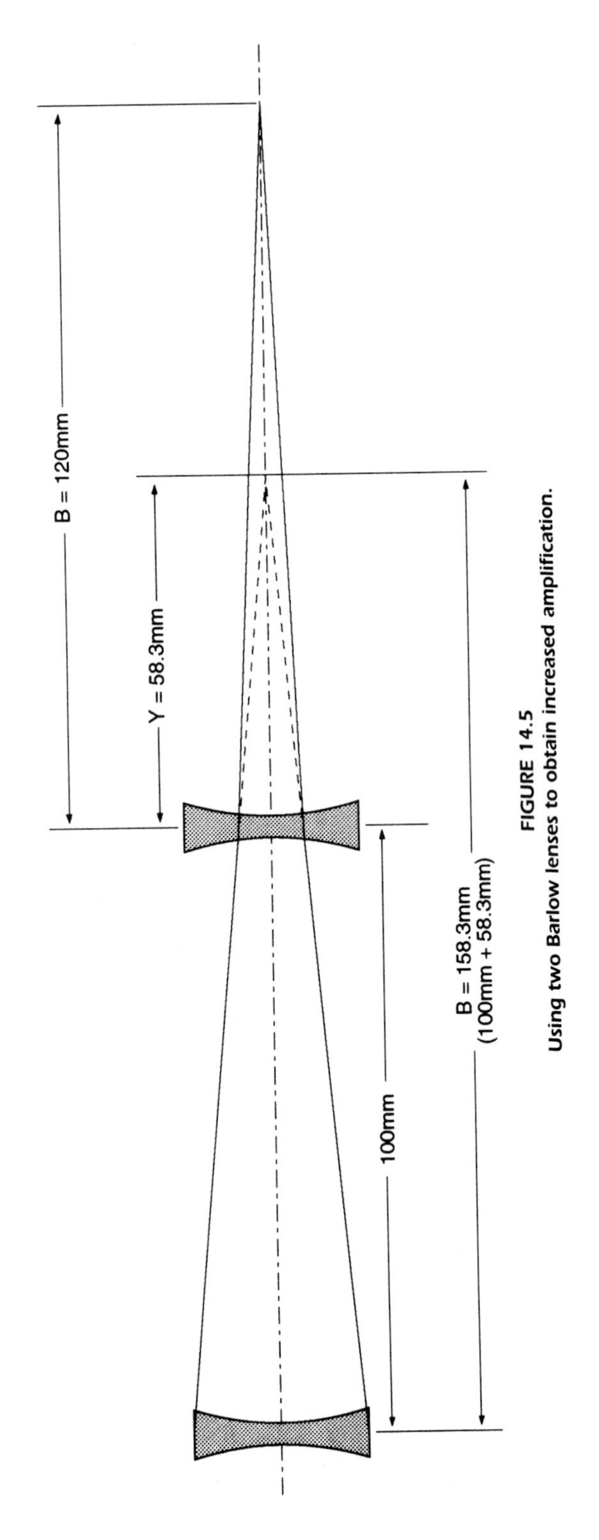

FIGURE 14.5
Using two Barlow lenses to obtain increased amplification.

Now calculate Y, which is the distance of Barlow no. 2 from the image formed by Barlow no. 1:

$$Y = [113 \times (2.06 - 1)]/2.06 = 58.3$$

The back focus of Barlow no. 1, which is situated 100 mm from no. 2, is therefore 158 mm, and the amplification must be

$$A = (158/113) + 1 = 2.40$$

Therefore the amplification of the whole system is $2.06 \times 2.40 = 4.94$. The effective focal length and focal ratio can be derived by multiplying the original values by the amplification.

STABILITY OF THE MOUNTING

In astrophotography, the quality of the mounting is at least as important as the quality of the optics. Too many amateurs have a tendency to forget this, but in fact the quality of the photographs obtained is very often limited by vibration of the mounting or by the accuracy of its drive. To realize the potential of a good optical system, the drive must be free from any periodic error amounting to more than the instrument's resolving power over several seconds of time, and in practice very few mountings are capable of such precision.

In addition, the mounting should be as massive and stable as possible to avoid vibration at every puff of air. From this point of view, short instruments such as Schmidt–Cassegrains are preferable to the Newtonian type of f/6 or f/7, assuming that the mountings in each case are of similar quality.

All sources of vibration affecting the quality of the photographs must be tracked down. Examples are the oscillation or swinging of certain types of flexible hand controls in right ascension and declination (which could with advantage be replaced by stiff rods with flexible joints); reflex mirror slam in the camera; and lack of balance around the two axes caused by incorrectly adjusted counterweights or even by the photographic equipment.

Most observers have their telescope mounted in the open air, and often confront a hostile foe — the wind. In windy weather, the feasibility of carrying out any work is dictated by the rigidity of the mounting. An examination through the eyepiece will indicate whether photography is possible or not, bearing in mind the type of object to be photographed and the length of exposure needed.

CHOICE OF FILM

The fairly modest aperture of amateur instruments, and the low level of image brightness when working with a large focal ratio (unavoidable in planetary photography), cause problems when deciding which film to use. To record on the negative the finest details visible through the eyepiece without losing them in the grain of the emulsion, it is necessary to use fairly large focal ratios, but atmospheric turbulence and residual drive errors prevent over-long exposures from being used — these should be between 0.5 second and 2 second, although those for Saturn have to be longer.

Planetary details are not sharp and well-defined, and moreover their contrast is low. Practical measurements on test scales have shown that photographic resolution is inversely proportional to the image contrast, and for this reason the Kodak T-Max 100 and 400 films, although often recommended, will never be able to record all the fine but low-contrast details on a planet, even at f/150. The search for a compromise between resolving power and sensitivity has always been a difficult one, usually resulting in the choice of a fairly slow film with good resolving power and improving its sensitivity and contrast by using an active developer such as HC110 or D19. However, the appearance of Kodak's Technical Pan 2415 some years ago has presented amateurs with the unique combination of high resolving power (about 320 lines/mm) and fast speed (about 200 ISO when developed for 12 minutes in HC110). Its speed, as well as its contrast, can be altered by varying the developing time or using a different developer. For these reasons, Technical Pan 2415 has a wide field of application, and a well-earned reputation as the answer to the astrophotographer's prayer.

CHOOSING THE FOCAL RATIO FOR DIFFERENT OBJECTS

Until recently there were two schools of thought with respect to image amplification: one group preferred to use fast films with a large focal ratio, while the other favoured shorter focal ratios and slower films. With the general acceptance of Technical Pan 2415 for high-resolution photography this debate seems to have died down, but it must still be possible to vary the focal ratio to suit the object being photographed.

The point to remember is that any increase in the focal ratio leads to a loss of image brightness, and therefore a compromise must be sought so that the focal ratio is sufficient to produce a large enough image while permitting a short enough exposure to minimize the effects of atmospheric turbulence, guiding errors, vibrations, and other interference.

In the following notes, the use of TP 2415 is assumed.

The Moon – (See Figs 14.6, 14.7 and 14.8.) Focal ratios of between 40 and 65 will be quite enough to produce a high-resolution photograph. The exposure time will vary from 0.5 second to 2 seconds, depending on the albedo of the area being photographed and the phase of the Moon, the exposure decreasing as Full Moon is approached. The film should be developed in Kodak HC 110, diluted 1 + 9 (using the pre-diluted developer available in a 500-ml bottle), for 8–12 minutes. This corresponds to the 'B' dilution (1 + 31) for HC 110 sold as 1 litre of concentrate. It is clearly better to use the 500-ml form, since the dilution can be measured and repeated more accurately and the developer, being used up more quickly, is less prone to oxidation.

FIGURE 14.6
Copernicus, 200-mm f/6 Newtonian at f/60, 1.5 seconds, TP 2415.

FIGURE 14.7
Clavius, 200-mm f/10 Schmidt–Cassegrain at f/60, 1 second, TP 2415.

FIGURE 14.8
Theophilus, 200-mm f/10 Schmidt–Cassegrain at f/60, 1 second, TP 2415.

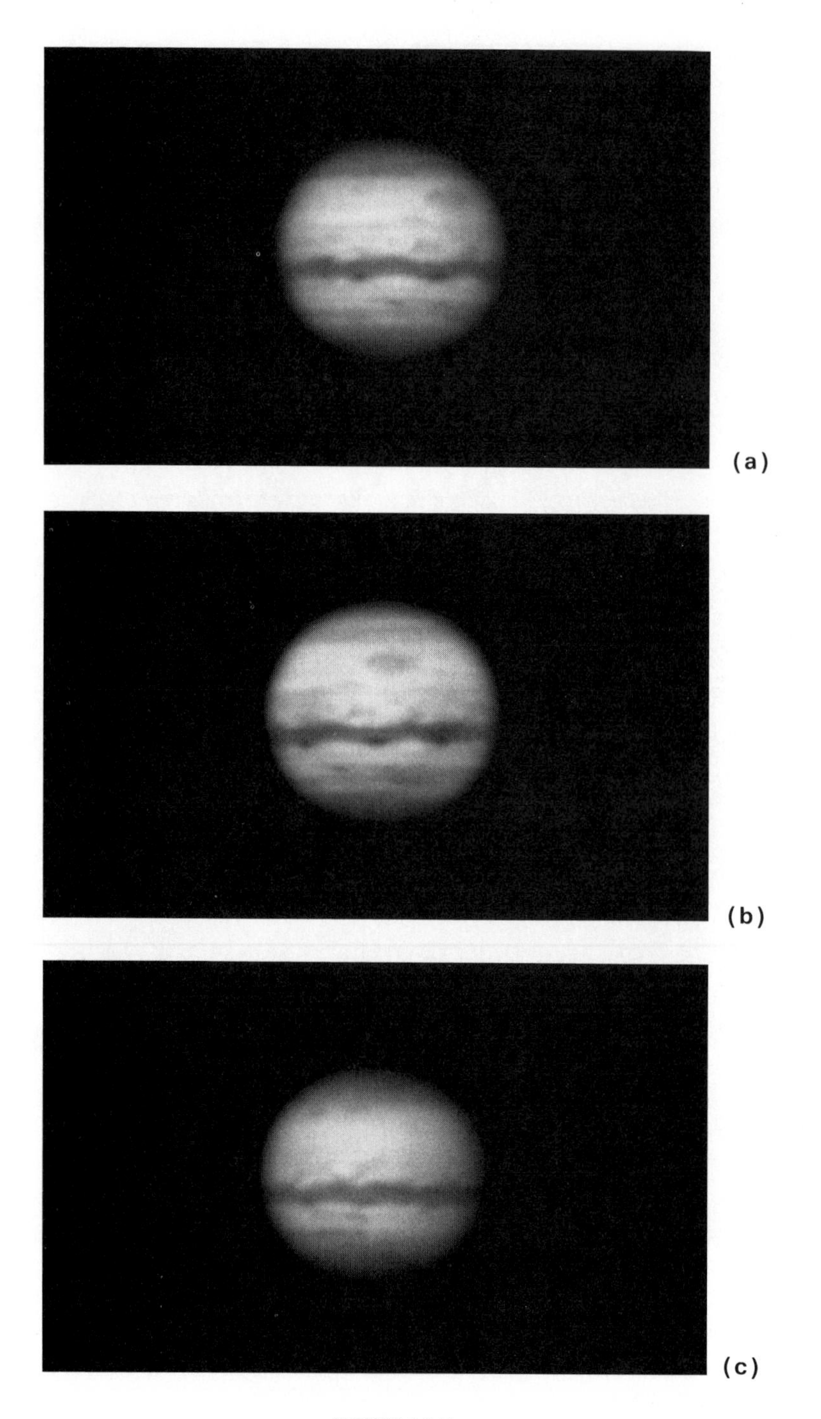

FIGURE 14.9
Jupiter, 200-mm f/10 Schmidt–Cassegrain at f/100, yellow filter, 3 seconds, TP 2415. (a) 18 January 1990, 22.02 UT. (b) 18 January 1990, 22.44 UT. (c) 22 February 1990, 20.02 UT.

Jupiter – (See Fig. 14.9.) To get photographs showing detail, it is essential to use an orange filter — a red one is better still, since it 'penetrates' more deeply into the Jovian atmosphere. Details at the edges of the belts will be far more numerous than if photographed in white light, although the contrast will be lower and the edge of the planet will have to be masked when printing to avoid losing it altogether. With an orange filter at f/80, the exposure time will be 3 seconds, or 4 seconds if a red filter is used. In either case the film should be developed for 12 minutes in HC 110 at 1 + 9. If you wish to take colour photographs, use reversal film (Fujichrome 50 or 100, Kodachrome 64 or 200), at f/55 or f/60.

Mars – The use of a yellow filter will give excellent results with Mars. Thanks to the high luminosity of the planet's surface, the exposure time is only 1 second at f/100. The film should be developed for 12 minutes in HC 110 at 1 + 9. For colour photographs use the films recommended above at f/80 to f/100.

Saturn – Because of its low surface brightness, Saturn is a difficult object to photograph. Furthermore, owing to its small size, a focal ratio of about 80 is necessary, which means that hypersensitized TP 2415 has to be used. Excellent results can be obtained by this means. An exposure time of the order of 5 seconds is needed, with development in HC 110 at 1 + 9 for from 12 to 14 minutes. Strangely, colour films give good results at about f/65.

FOCUSING TECHNIQUES

Every astrophotographer comes up against the problem of achieving a perfect focus. The trouble is that when the telescopic image is amplified, the brightness drops enormously and it becomes difficult to tell where the focus is. The standard camera focusing screen consists of a fine structure of microprisms which is a useful aid to sharp focusing in everyday photography, but unfortunately these just obscure the image at very low light levels. The screen must therefore be replaced either by ground-glass or a clear screen with lines on it. The clear screen is the more attractive of the two, since more planetary detail can be seen through the camera viewfinder, but despite the presence of the lines it is possible for the eye to focus a little way on either side of the focal plane. As a result, the focusing may not be accurate. This system should only be used at focal ratios larger than f/30; below this a ground-glass screen is the only way of ensuring a perfect focus.

Focusing must be very precise. This means that the focusing mechanism must be very smooth, and geared-down as much as possible to allow for fine adjustment. Even so, vagaries in the seeing make the focal point change constantly, and slight corrections will be needed to allow for this. It is best to refocus after every three or four exposures.

In this connection, a camera with an electrically controlled winding mechanism operated by remote control will be found particularly useful, since three or four consecutive exposures can be made without upsetting the focus, without accidentally moving the telescope, and in such rapid succession that the exposures may all be made during a period of steadier seeing.

USING A MANUAL SHUTTER

The shutter on a normal camera will make the telescope vibrate when it is operated. This means that a manual shutter must be used for making the exposure if the resolution limit is to be approached. A disk about 20% larger than the diameter of the telescope, painted matt black to avoid possible reflection, must be prepared. A handle makes it easier to manipulate. Set the camera shutter to the 'B' setting, hold the manual shutter in front of the telescope, and open the camera shutter, preferably using a pneumatic bulb attachment, which is preferable to the standard cable release. After waiting for a few seconds to allow any vibration to die away, remove the manual shutter for the required exposure time, replace it, and close the camera shutter.

Put a glove on the hand holding the shutter, as it is a source of heat currents across the light-path. This technique may seem primitive, but it is the only way of ensuring that none of the quality inherent in the telescopic image is lost.

SOLAR PHOTOGRAPHY

High-resolution solar photography is possible with an aperture of 100 mm — in fact, nothing is gained by using an aperture greater than 125 mm because of the limit set by daytime atmospheric turbulence. A 100 mm refractor is therefore especially suitable for this work. (See Fig. 14.10.)

However, solar photography can be risky unless certain precautions are taken. In particular, direct observation or photography must never be carried out using a simple dark filter placed near the eyepiece, since the concentrated heat could crack it. The ideal solution, for both types of observation, is to use an aluminized filter with a transmission of between 1/500 and 1/1000, placed in front of the objective. It is essential that the filtering should be effective right across the spectrum, including the infrared and ultraviolet, whose invisible rays can do great harm. [See also Chapter 1. — Ed.]

Having set the full-aperture filter in position, and during initial adjustment and focusing, there should be no sensation of glare around the Sun when it is viewed through the eyepiece. If there is, an additional neutral filter of sufficient density should be added at the eyepiece end of the telescope.

For photography, a coloured filter placed between the amplifying system and the film will improve the image contrast as well as lowering the intensity. A green filter gives very good results, although red is also worth trying since it slightly reduces the effects of bad seeing — but do not use it with Agfa Ortho 25, which is insensitive to red light.

Fine-grain films such as Agfa Ortho 25 or TP 2415, developed in HC 110 at

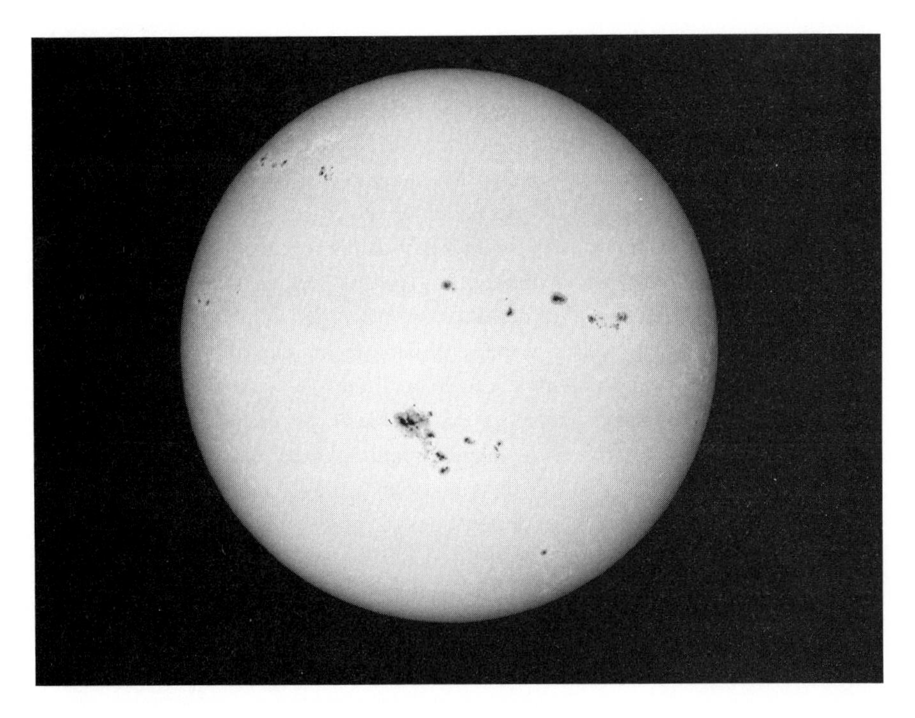

FIGURE 14.10
The Sun, 68-mm f/9 refractor at f/30, red filter, 1/1000 second, TP 2415.

1 + 9 for 8–10 minutes, give good results. The exposure time must be determined by experiment, since it depends so much on the altitude of the Sun, the atmospheric transparency, the degree of filtering and the film used, as well as the focal ratio. The best initial guide is to use an exposure meter. However, since the camera shutter rather than the manual shutter must be used, and the mirror-slam just before the exposure causes vibration, the exposure should be as short as possible, preferably not longer than 1/125 second.

HIGH-RESOLUTION STELLAR PHOTOGRAPHY

This may seem a curious heading when you consider that the best professional photographs taken from exceptional sites such as Mauna Kea in Hawaii only resolve to 0.5 or 0.6″ arc, which is approximately the resolving limit of a 200-mm telescope, although they were taken with apertures of several metres!

However, in stellar photography it is misleading to talk about *theoretical* resolution (unless you wish to indulge in your favourite pastime aboard the Space Shuttle). This is because there is no chance of choosing a period of good seeing to make the exposure, as with planetary photography. Atmospheric turbulence will degrade the star images in a random way whether the exposure is a minute or an hour: the resolution will not be very different in either case, and is certain to be far from the theoretical limit.

Nevertheless, deep-sky photography has really taken off with the appearance of hypersensitised TP 2415, which has succeeded in doubling the effective stellar resolving power and gaining almost two magnitudes, thanks to the excellent signal-to-noise ratio. It has totally outclassed for the purpose all existing emulsions, including Kodak 103a, for so long the professional astronomer's work-horse when it came to photography of faint objects. But there is another side to the coin: the quality of this film makes higher demands on mounting stability, telescope drive, guiding and focusing, as well as hypersensitizing technique, all of which need to be refined if the film's full potential is to be realized.

In stellar photography, remember that the basic parameter to be taken into account when seeking to achieve the highest quality photographs is the *focal length* of the telescope used, and that this determines both the resolution and the limiting magnitude. For most astronomical applications the power of a telescope is measured by its aperture, since this determines its theoretical limit of resolution, and therefore inreasing the aperture improves the resolving power. This argument is perfectly sound when applied to visual observation and planetary photography — but it does not apply to stellar photography. Two examples are given to explain why.

Similar aperture, different focal length – In this case, the resolving power of both instruments is theoretically similar, but suppose that one is a 200-mm Schmidt camera working at f/1.5 (focal length 300 mm) and the other a normal Newtonian with a focal length of a metre or more. The very small image scale of the Schmidt will be more seriously limited by the film's resolving power (even that of TP 2415), while its small focal ratio will reduce the maximum possible exposure time without sky fogging, and therefore the limiting photographic magnitude, compared with the Newtonian.

Similar focal length, different aperture – Both instruments will form images of the same scale on the film, but the larger aperture will reach the same magnitude more quickly; however, by increasing the exposure time, the smaller aperture will also be able to reach this magnitude, so the advantage of the larger aperture lies more in the more favourable exposure time than in its ability to photograph fainter stars. If an exposure time of $1\frac{1}{2}$ or 2 hours does not deter the photographer, then remarkable results can be achieved with telescopes of fairly large focal ratio. [In fact, because of the effect of sky-fogging at small focal ratios, the smaller aperture may even exceed the limiting magnitude of the larger because of the increased exposure time permitted. — Ed.]

Instruments of small focal ratio can of course have the focal length increased by using a Barlow or other accessory, therefore gaining the benefit of a larger focal ratio if required.

FIGURE 14.11
M42, 200-mm f/10 Schmidt–Cassegrain at f/6.2 (using a Lumicon focal reducer), deep-sky filter, 35 minutes, hypered TP 2415.

However, it is possible to have too large a focal ratio when photographing *extended objects* rather than stars. Even when photographing deep-sky objects of fairly high surface brightness, such as M27, M64, M82 and M104, anything slower than f/10 or f/12 may fail to record them at all, and therefore when photographing dim objects such as M33 or M101 a fairly small focal ratio is essential. It must also be remembered, that when photographing an extended object, doubling the focal ratio means that the exposure must be increased four times to obtain the same density in the photograph — for example, an object requiring 30 minutes at f/6 will need 2 hours at f/12.

To summarize, the resolving power and the limiting magnitude of a star photograph are determined by the focal length employed, but the focal ratio determines the exposure necessary to record an extended object.

Use of a long focal length accentuates faults which pass unnoticed in the traditional type of stellar photography using ordinary camera lenses. Attention should therefore be paid to the following points to ensure that the work is carried out with the best chance of success.

Some Technical Points

The mounting – This is of vital importance in high-resolution stellar photography, because its quality determines how good the guiding can be. It is useless trying to work with a long focal length on an unstable mounting. Points to consider are inherent structural rigidity, a firm base, and a means of attaching the tube without any residual play. Any vibrations must damp out as quickly as possible — for example, using a 200-mm telescope with a magnification of ×300, any vibrations caused by tapping the tube should die down in less than 4 seconds.

Long-focus photography requires such delicate guiding that motorized drives on both axes are recommended. In particular, adjustments in declination are too fine for manual compensation unless a geared-down system is used. The better the telescope drive, the easier guiding will be.

Aligning the polar axis – Careful alignment of the polar axis will help reduce the amount of guiding required during an exposure. If the telescope is not permanently mounted, a polar viewfinder is essential, as otherwise a precious hour of the observing session may be spent getting the alignment right. The best place for the viewfinder is on the polar axis, and it should be fixed rigidly to it, the displacement of the Pole Star from the optical axis of the viewfinder being marked by adjustable cross-wires. This arrangement is much more permanent than a system where the sighting telescope itself must be adjusted to the correct position. During the coming years the Pole Star will be approaching the celestial pole, and these small changes should be allowed for.

Focusing – This can be achieved in two ways. First, the image can be examined on a ground-glass screen in the camera. This is simple and straightforward, but it takes practice to get the focus exactly right, and people with pronounced astigmatism will have extra difficulty. The second method, using a Foucault knife-edge device, is more precise and repeatable, but it is important to make a proper attachment for the purpose — fixing a razor blade across the film guides in the camera with a piece of sticky tape is *not* the way to do it!

The knife-edge focuser consists essentially of a sharp blade with the chamfer towards the eye, attached to an extension which takes the place of the camera body so that the knife-edge is in exactly the same plane as the film. An empty aerosol can, with a 10-mm hole at one end so that the eye can be brought up to the knife-edge, could form the basis of the device.

The biggest problem lies in getting the knife-edge in exact register with the film. After initial measurement, visual comparison can be made by interchanging the focusing attachment and the camera body with a blade fixed to the film guides. The final checking will need to be done using trial photographs.

Guiding – Bearing in mind the possible exposure length, the telescope may have to be guided in right ascension through an angle of 30° or even more. Under these conditions the use of a separate guiding telescope can lead to problems of differential flexure and loss of mutual alignment between the two optical systems. The answer is to arrange a small totally-reflecting prism to intercept a portion of the light to one side of the area being photographed. The light is then focused by a guiding eyepiece, and a suitable guide star from the ones located in a narrow ring around the photographic field may be selected by rotating the attachment. This is not always easy, however, since the star must be sufficiently bright, it should be near the centre of the guiding field, and the definition so far from the centre of the telescopic field may be poor.

The guiding eyepiece must be equipped with cross-wires, but double cross-wires allow any movement of the guide star to be seen more easily. These wires must be illuminated, with the brightness adjustable to match the guide star. A system marketed by Vixen is now becoming available which projects a set of concentric circles into the eyepiece field: these can be moved anywhere from centre to edge, and enlarged or reduced to suit the focal length of the telescope and any swelling of the star image due to poor seeing.

Guiding for an hour or more is very demanding, but there is no reason for errors occurring, even when a long focal length is used, as long as the observer is seated comfortably in a position which allows the guiding eyepiece to be reached without strain during the whole of the exposure.

By using a long enough focal length any amateur can take amazingly good photographs with resolution rivalling the sort usually achieved with much larger apertures. Great care, discipline, patience, and of course experience are needed, but the best reward will come when you tell others who are convinced that your photograph of M51 was taken with a 400-mm telescope that in fact it was taken with half the aperture!

FIGURE 14.12
M51, 200-mm f/10 Schmidt–Cassegrain, 2 hours, hypered TP 2415.

FIGURE 14.13
M42 and the Horsehead Nebula (top), 300-mm telephoto lens at f/4.5, red filter, 60 minutes, TP 2415.

WIDE-FIELD PHOTOGRAPHY

Photography of the night sky using an ordinary camera with a wide-angle or even a telephoto lens is as exciting and spectacular as taking photographs at the focus of the main telescope. The reason is that large areas of sky (the constellations themselves, or diffused nebulosity such as the North America and California nebulae) cannot be photographed with the small field of view of most instruments.

Two important points need to be borne in mind. First, any misalignment of the polar axis will make stars near the edge of a wide field appear as short trials, centred on the guide star. Second, the camera must be fixed as rigidly as possible to the telescope tube via a small platform, bearing in mind how flexure can affect a long exposure. Additionally, the mounting must be adjusted for balance in right ascension and declination after the camera and platform have been fitted.

Always be prepared to stop down the lens if it does not give best results at full aperture. In fact a 50-mm f/1.8 lens will only give poor or average star images, but stopped down to f/4 the results should be very good. However, at apertures smaller than f/4 the necessary exposure time will become too long. Colour photography may, however, have to be carried out at full aperture.

CONCLUSION

I hope I have made it clear that the results achieved in astrophotography do not depend solely on the quality and power of the equipment used. Some people achieve remarkable results with modest equipment, while others, much better equipped, fall far short of its potential. In this field, as in many others, motivation and determination play an important part in the final result. You must always be prepared, both mentally and physically, to seize any suitable moment for undertaking high-resolution photography.

About the Author

Born in Paris in 1962, Gérard Therin describes himself as a dedicated high-resolution astrophotographer.

As a young child the 'little lights' in the sky always intrigued him, and nightfall often found him outside, contemplating the infinite! His true astronomical initiation, however, came with the acquisition of a 115-mm refractor.

His enthusiasm for the subject developed rapidly, and he decided to specialize in astrophotography, although he had to serve a long apprenticeship. Later on, in his quest for higher resolution, he obtained a Celestron 8 catadioptric telescope (aperture 200 mm) with which most of the photographs illustrating this chapter were obtained. His aim was then, and still is, to photograph the finest detail visible through the telescope in the moments of best seeing. In order to do this he has taken numerous lunar, planetary, and stellar photographs showing surprising detail considering the aperture used, and which demonstrate that in general amateurs underestimate the true potential of their instruments.

Nevertheless, his dream is to achieve a photograph that reveals every detail which the telescope is capable of showing under perfect conditions. Perhaps, one day, this dream will be realized?

Contact address – 7 Allée des Dahlias, 93150 Blanc-Mesnil, France.

Notes and Comments

The outstanding quality and wide range of Gérard Therin's photographs would do credit to any amateur-owned telescope — the fact that they were achieved with nothing larger than 200-mm aperture seems incredible. Could he or anyone else do better with a larger telescope? His stated belief that the optimum aperture for photography is 200–300 mm drew the comment from astrophotographer Martin Mobberley (Bury St Edmunds, Suffolk, England) that 'this may be true in typical seeing, but most amateurs are only interested in the rare moments of good seeing anyway. I think most amateurs would always opt for the largest possible aperture in order to minimize exposure times.'

Invited to comment generally, he made the following points:

> He says that TP 2415 has a resolution of 10 μm [100 lines/mm. — Ed.] In fact the issue of film resolution is far from simple. The official Kodak resolution is 320–400 lines/mm depending on the developer used, but this is only for very high-contrast black-and-white lines. In practice, amateurs may choose to project the image to allow for a resolution as low as 100 μm on a low-contrast object like Jupiter. In low-contrast cases the resolution is not solely the fault of the grain, but is also dependent on diffusion of light through the emulsion.
>
> I notice that he does not mention Rodinal in his choice of developer for TP 2415. Leading observers like Parker and Miyazaki use Rodinal diluted +25 or 1 + 50 and develop for 14 minutes for best results.
>
> He correctly comments that focusing at long focal ratios is easier with a Barlow, but this is only the case if the Barlow comes *before* the rackmount, not after!

TP 2415 film is not something you can normally purchase from a photographic shop. One supplier in the United Kingdom is 'Silver Print', 12 Valentine Place, London SE1 8QH. American astrophotographers can purchase hypered TP 2415 from Lumicon, 2111 Research Drive No. 55, Livermore, CA 94550. — Ed.

CHAPTER FIFTEEN · *Photoelectric Photometry*

ANDREW J. HOLLIS
DIRECTOR, ASTEROIDS AND REMOTE PLANETS SECTION, BAA

THE VISUAL OBSERVER IS THE BACKBONE OF AMATEUR ASTRONOMY, carrying out valuable research. The *BAA Journal*, the ALPO and AAVSO publications, all support visual observations which are abstracted into the professional reference manual, the *Astronomy and Astrophysics Abstracts*. The amateur has easy access to telescope time (in his own back yard), and he need only wait for favourable weather to obtain results.

Astronomical observation often consists of measuring light, or at least radiation within the electromagnetic spectrum. The measurement may not be made consciously, but, for example, drawing the Moon and planets or taking photographs involve types of light measurement. Visual observers of variable stars estimate their magnitudes by comparing their brightness with stars nearby. Under good conditions, estimates of white stars are accurate to ± 0.15 magnitude, although the accuracy of estimates of red stars is very much lower.

A visual observer can follow variables with an amplitude of at least 0.3 magnitude, but the observations may be biased. Preconceived ideas of what will be seen may influence observation (remember that Percival Lowell saw canals on Mars, Mercury, and Venus!).

Photography is useful for variable-star work if the telescope has a good drive. A photograph is a permanent record, magnitudes can be measured (and re-measured) to an accuracy of about ± 0.1 magnitude, and brightness can be referenced to standard magnitude scales if the right filters and film are used.

However, amateurs can also use modern techniques with microelectronics and

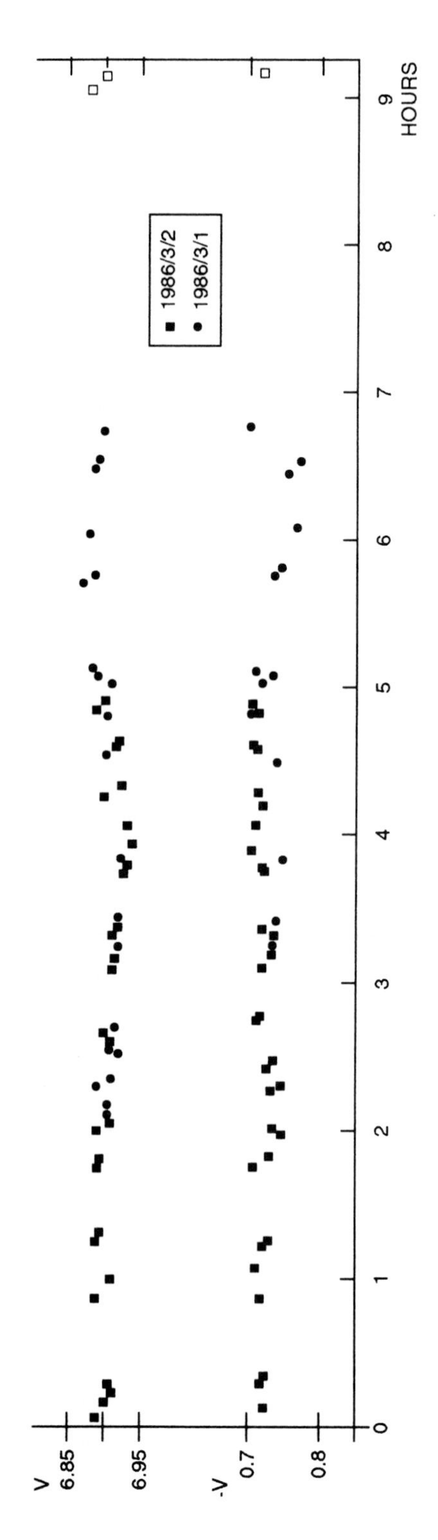

FIGURE 15.1
Composite V and B-V light curves for asteroid 1 Ceres in 1986. The range of variation in the V band is 0.05 magnitude. (A.J. Hollis, 135-mm reflector.) The open squares are repeated points.

computers. Photoelectric photometry (PEP) is one, and is encouraged by professionals, but it has advantages and disadvantages which must be considered carefully. For example, there is an additional cost, although the equipment can cost less than a good camera if it is home-constructed. A computer is not essential, but saves much hard work.

'Measuring light' is a fairly common activity — nearly everyone has used a photometer such as a camera light meter — but astronomy has more onerous requirements than photography. In particular, the sensitivity of an astronomical photometer must be at least a million times more than that for a camera.

A photoelectric photometer is a linear device (in other words, the signal generated is directly proportional to the brightness of the source). Comparing flux from different sources is easy, and measurements accurate to ± 0.01 magnitude are possible, although to achieve such accuracy demands rigorous observing and reduction procedures. Of course, the sky conditions must also be right.

Observations referred to standard magnitude scales can be made through standard filter sets. Special filters can also be used to observe in the ultraviolet or the infrared. The photometer calibration error is the only bias, but this error can be determined and corrected for. Results from several observatories (amateur or professional) can be combined with no loss in overall accuracy: in fact, many amateurs collaborate with professional astronomers.

Photoelectric photometry greatly increases the amateur's observing scope. For example, low-amplitude variables can be followed. The writer has recorded a light curve of the asteroid Ceres, with an amplitude of 0.04 magnitude and an error of ± 0.005 magnitude (Fig. 15.1).

THE MAGNITUDE SCALE

Even a casual glance shows that stars are not all the same brightness. The first star catalogue was compiled over 2000 years ago. In it, Hipparchus ranked stars into six ranges of brightness, from the 1st to the 6th magnitude.

The magnitude scale was not described mathematically until the last century. Herschel's experiments suggested that stars of the 1st magnitude were about 100 times brighter than those of the 6th. In 1856, Norman Pogson used this result to define the magnitude scale: a 5-magnitude difference in brightness corresponds exactly to a factor of 100 in flux. A 1-magnitude difference between stars is a difference in brightness of 2.512 (which is the fifth root of 100). The scale was related to that then existing by defining Aldebaran and Altair to be of magnitude 1.0.

In photoelectric photometry, this relationship is better considered in terms of logarithms:

$$F = 10^{(2/5)m} + \text{constant} \tag{1}$$

where F is the observed flux of intensity, and m is the magnitude. The constant is required to allow for the attenuation of the signal between the source and the measuring instrument.

The magnitude difference between two sources is given by:

$$F_1/F_2 = 10^{(2/5) \times (m_2 - m_1)} \tag{2}$$

Using logarithms to base 10, this becomes

$$\log(F_1/F_2) = \tfrac{2}{5} \times (m_2 - m_1) \tag{3}$$

hence

$$m_2 - m_1 = -2.5 \times \log(F_1/F_2) \tag{4}$$

which is the same as

$$m_2 - m_1 = -2.5 \times \log(F_1) - (-2.5 \times \log(F_2)) \tag{5}$$

The factor of 2.5 is exact, and not a rounding-off of the ratio 2.512: it is derived from $10^2 = 100$, and the fifth root of 100. These give $10^{2/5}$; $\tfrac{2}{5}$ is 0.4 and the reciprocal is 2.5. The negative sign comes since small numbers are brighter in magnitude terms than large ones.

Equation (5) is the most fundamental equation in photoelectric photometry. Light flux is measured directly by the photometer and the magnitude difference measured is thus directly calculable. However, this simplicity is a weakness, because there are many other factors that will be considered later. The weakness arises because observers may be tempted to ignore these other factors, and the main result of this is that the value of the observations is reduced.

WHAT TO OBSERVE

Before going into further detail about photoelectric photometry, something should be said about the fields of work for which it is suitable. In general, some experience must be gained before selecting any observing programme: most amateurs observe alone, and must be confident and capable both in operating the equipment and also in reducing the data.

There are many relatively bright stars of the W UMa type [see Chapter 10 — Ed.]. These stars have amplitudes of a few tenths of a magnitude, are continuously variable, and possess orbital periods of less than a day. Observing for two or three hours will show positive variations and may well cover a maximum or minimum of the light curve. They provide an ideal training ground for the beginner. There is also the satisfaction of seeing progress with each night's measurements (Fig. 15.2).

Many organizations encourage the use of photoelectric photometry and can give advice. The BAA and the AAVSO operate suitable observing programmes. They have members who can give advice on equipment and technique. The International Amateur–Professional Photoelectric Photometry Organization (IAPPP) provides a forum where professional astronomers publicize their requirements for specific observing projects.

Any observing project, to be capable of producing results, must be tailored to

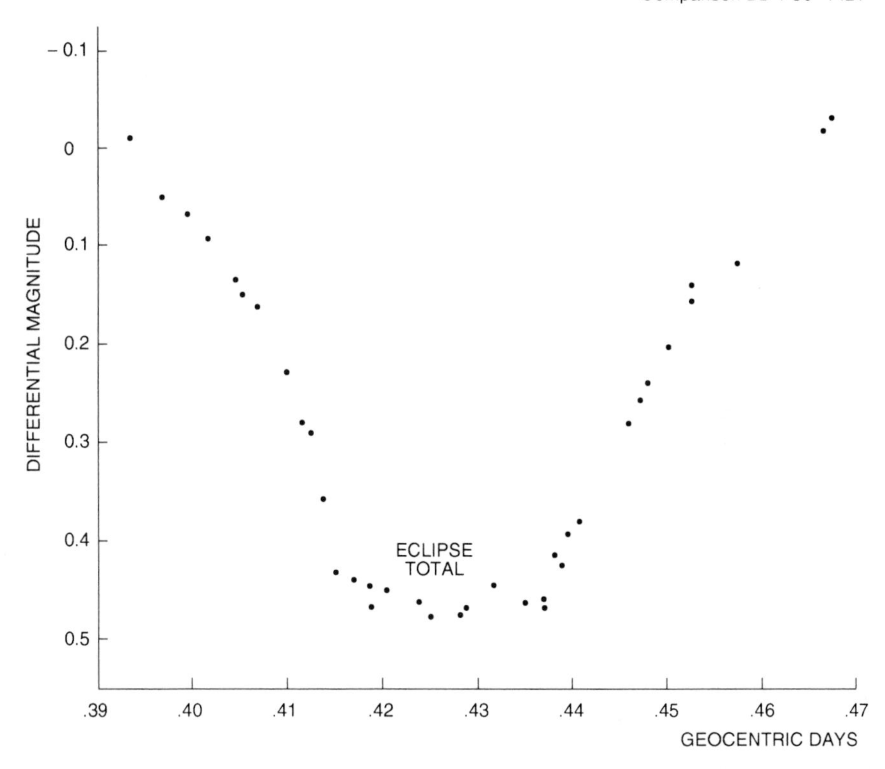

FIGURE 15.2
V-band photoelectric light curve of W UMa, showing a primary minimum observed on 1–2 April 1984. The eclipse is shown to be total by the flat minimum. (A.J. Hollis, 135-mm reflector.)

the likely observing conditions. In the United Kingdom, clear nights suitable for photometry usually come singly or in pairs every two or three weeks. The best projects for these conditions are therefore linked to objects that vary so rapidly that results can be obtained in a single night, or so slowly that measures every 10 to 20 days are sufficient to determine the variation. In Arizona, on the other hand, conditions are very different, and variables requiring monitoring for several consecutive nights can be followed.

Many short-period variables, and the majority of the asteroids, have periods less than a day. Sensible results can therefore be obtained during the course of a night. There are other forms of observing which can be undertaken: for instance, high-speed photometry of occultations of stars by the Moon or asteroids is possible. A prime example occurred on 3 July 1989 when Titan occulted 28 Sgr. The photoelectric trace obtained by Richard Miles and the writer is the most northerly obtained which showed the central flash. There are several other fields available, including photometry of the lunar surface, the planets, and planetary satellites and rings. Photometry of comets is also possible, although they pose special problems because they are extended objects, and special filters are required for normal observation. Suitable comets are rare, so there is not much support with amateur observers. Occasionally observing periods for flare stars or the cataclysmic variables are identified by professional astronomers — these are the periods when they are observing with X-ray or UV satellites, and they wish to know how the object is behaving at visible wavelengths. This is a fruitful avenue for amateur–professional cooperation.

The long-period Mira variables are probably adequately covered by visual observers, although they would be suitable for observing every week or two. The semiregular variables, many of which have low amplitudes, may well be suitable for the amateur with relatively few clear nights.

Finding new variables is a possible activity. From time to time lists of stars which are possibly variable are published, and many of these do subsequently turn out to be variable.

THE BASIS OF PHOTOELECTRIC PHOTOMETRY

The photoelectric effect was discovered by Hertz in 1887. Light falling on certain metal surfaces dislodges electrons. The number of electrons released per second is directly proportional to the intensity of the light. By collection and amplification of these electrons, it is possible to measure the incident light flux.

The first astronomical photoelectric photometry was carried out at Dublin in 1892. Since that time there has been a steady improvement in detectors and the supporting electronics. The early systems used bulky valves and galvanometers. These were replaced by transistors and pen-chart recorders. Modern photometers use integrated circuits and digital recording devices such as frequency counters or computers.

Photomultiplier tubes – The heart of any photoelectric photometer is the detector. In nearly all the photometers that amateurs use, this means the photomultiplier tube (PMT). The PMT is a kind of light-sensitive valve. Its early application in warfare was to generate random noise to jam radar reception — it is now used as a low-noise device to detect light! Photodiodes and CCDs are also in use, but they are about a factor of 1000 times less sensitive than the PMT. Also, the associated circuitry is noisy, and so in amateur hands these devices will not match the performance of the PMT. However, it is likely that this will change in the near future.

A PMT needs a high voltage, typically between -600 V and -1500 V, depending on the type of tube. The supply has to be regulated to within about 0.1% to allow the PMT to be sufficiently stable in operation. The effect of changing voltage on a PMT is proportional to the seventh power of the applied voltage, the significance of which can be investigated with a pocket calculator!

Common sense dictates that a PMT is a bad choice as a detector for photometry, and if there was any other choice, these tubes would already have become museum pieces: they are fragile, cannot be exposed to ambient light levels (especially with the high voltage connected), are sensitive to humidity and temperature changes, and respond to stray magnetic fields (is there a radio ham near you?). They also age by absorbing helium from the atmosphere, which degrades their performance.

Modern photometers use digital output. This output may be read by a frequency-counter or by direct input into a computer. Digital counting gives a high dynamic range (several magnitudes at a given amplification), and direct readout. Two approaches are possible — either counting photons or by conversion of the photon pulses to a direct current and amplifying this. In the amateur environment, either approach will yield similar accuracy so long as, in pulse mode, an end-window PMT is used, or, in DC mode, the PMT is dried with silica gel. Each method has its devotees, and this issue is probably the most contentious in photometry, with highly polarized views. My advice is to adopt whichever method seems the more suitable.

The photometer head consists of several items, beginning with a diaphragm aperture to exclude all but light from the area being measured. There should be an optical system to view the diaphragm and check that the star is central in the aperture. A Fabry lens is used to enlarge the image of the star on the sensitive surface of the detector: it serves to compensate for variations in the sensitivity of the surface of the detector, and for small errors in the telescope drive. Colour filters are used to restrict the wavelengths measured before the light reaches the detector.

The first stages of electrical amplification are usually also carried in the head. Photometer heads built in the past were massive, but the current generation are about the size and weight of a large camera.

Telescope and site – Telescope requirements for photoelectric photometry are unusual. First, the quality of the optics is not particularly important, since their sole function is to collect light and not to form good images — so if you have a poor mirror, rather than throwing it at the cat, build a telescope for photoelectric photometry. The only requirement is to have the optics in good alignment, and, if possible, clean. The telescope must also be capable of carrying the photometer head on its eyepiece mount.

Driving requirements are less onerous than for photography, the objective being to hold the star in a 60–90″ arc aperture for at least half a minute. For preference, the periodic error in the drive should be less than 15″ arc and it is best if the star can be held in the aperture for at least 15 minutes. The telescope should be balanced so that there is a slight positive bias: the drive then acts as an escapement, which will lead to less strain on it. The polar axis will also need to be well aligned.

One basic consideration is the selection of the best site. This would, in theory, be on top of a high mountain, above the main cloud layers. In practice, the advice

FIGURE 15.3
The author's observatory. (Above) The twin 350-mm and 250-mm Cassegrain reflectors. Part of the roof folds back, while the covered portion of the shed provides shelter for the electronics and the observer. (Below) The photometer attached to the 250-mm Cassegrain.

must simply be to find the darkest site, and for preference try not to observe as faint as the limiting magnitude. One good rule to remember is that if you can see the star in the viewing eyepiece, it can be measured. Put another way, the PMT is more sensitive than the eye.

OBSERVING PHOTOELECTRICALLY

Good equipment does not guarantee good results, but it is a valuable first step. In principle, photometry is easy: in practice it is not so straightforward. Frequently, results will have to be written off — when a good light curve is published, remember that it is unlikely to contain all the measurements made. Photoelectric photometry is capable of achieving high precision, to ± 0.01 magnitude or better, and so it is of great importance to operate the equipment properly and to be aware of potential sources of error.

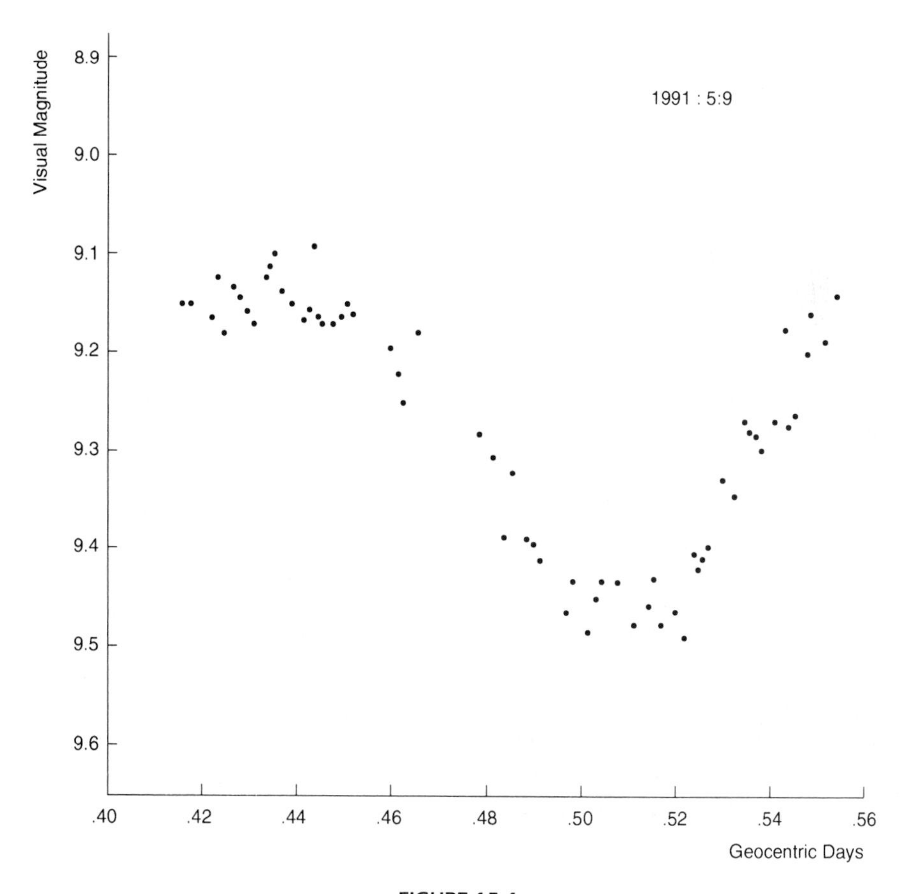

FIGURE 15.4
Minimum of AH Virginis. This trace was obtained using the twin telescopes on 9 May 1991, 21.50–25.11 UT.

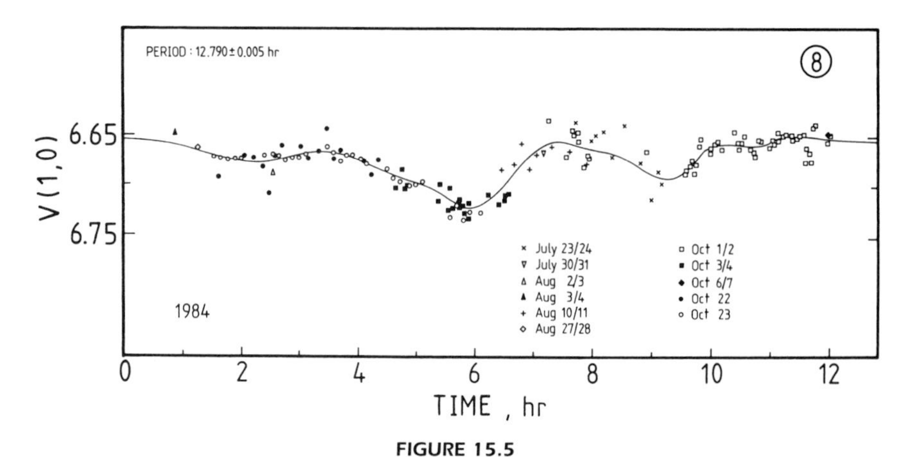

FIGURE 15.5
Composite V light curve of asteroid 8 Flora in 1984. Contributions came from Australia (C.S. Bembrick), France (M. Dumont), and the United Kingdom (A.J. Hollis and R. Miles). The extreme range of variation is 0.08 magnitude. The addition of several light curves is used to derive a rotation period.

Absolute and differential photometry – The first important consideration is the basic procedure. There are two primary types of observation — absolute photometry and differential photometry. Absolute photometry is used where the atmosphere is transparent and stable. Stars are observed in sequence to establish the atmospheric conditions and brightness of the object being investigated. It is an open-loop procedure and is probably best suited to professional observatories situated at a high altitude above the thickest parts of the Earth's atmosphere. Differential photometry is a closed-loop operation. Measures of the objects are referenced to the same standard stars continuously. The operator can check changes in the atmospheric conditions by noting the readings of the standard. This form of observing is to be recommended for observatories close to sea-level, because if conditions do deteriorate part of the night's measurements may still be usable.

The photometer's performance – Basically, observing photoelectrically is easy. It can lull the observer into a false sense of superiority. Initially it is worth spending some time in familiarization with the performance of the photometer. Once its performance is understood, when everything is working properly any deviations will indicate some problem either with the equipment or with the observing conditions. If the photometer is modified in any way, then its performance should be reassessed. However, once the photometer is performing satisfactorily, it should be left well alone — 'if it ain't broke, don't fix it'.

On a night when the sky is good, measure several standard stars in two filters — Visual and Blue, for example. Reducing these observations will indicate how close the equipment is to the standard Johnson system. it will not be an exact match because the standard system was set up using one of the reddest RCA 1P21 photomultipliers ever made — however, the photometer should be a fairly close match as long as the filters were carefully chosen. Correction factors for transformation of readings to the standard systems should be determined. Photometric readings are much more valuable if they are converted to a standard scale, because they will be usable by others.

The observing procedure must follow a defined sequence. There is some flexibility in approach if measures are logged manually, which means hard work in writing, but using a computer to log results gives less flexibility.

Diaphragm size – The first problem is to decide what size of diaphragm should be used for measurement. The aperture should be small enough to allow the light from the star to pass through while excluding light from nearby stars; it must be sufficiently large so that all the light from the star does indeed pass through, with minimal errors should the star be not quite centred in the diaphragm. This major problem tends to be glossed over in most manuals on photometry. If the sky is dark, with no moonlight, it is not easy to see where the edges of the diaphragm are, and so centring can be less than perfect. Periodic errors in the drive also cause trouble.

The basic rule for the amateur is to make the aperture as large as possible, because this will reduce the errors. It is unlikely that an aperture less than 60″ arc across will prove satisfactory, and conditions may well dictate the use of a 90″ arc aperture, or even larger.

Integration time – Another item which is not usually considered is the integration time. Photons arrive at a non-linear rate which can be described by Poisson statistics. Hence the rms error is given by the square root of the number of photons. To achieve an error of 1% (0.01 magnitude), 10 000 photons must be detected. For bright stars this is not a problem; 34 000 photons per second will arrive from a 5th-magnitude star using a 300-mm telescope, assuming a 15% efficiency (perhaps higher than is achieved in practice). In theory, an integration time of 0.3 second would be adequate. However, only 2000 photons per second arrive from an 8th-magnitude star, and about 325 from a 10th-magnitude one. As we wish for an overall error of 0.01 magnitude, the error from this source should be reduced to 0.3% or less, which corresponds to the detection of about 34 000 photons. Our 10th-magnitude star should be measured for over 100 seconds for each data point to give a photon noise error of 0.3%.

Comparison stars – The selection of the comparison star is important. It should be non-variable, near the star being observed, of similar colour index to the variable, and if possible of similar brightness. Ideally the comparison should already have been measured in the photometric system you use, and have been proved non-variable. Though these circumstances are unlikely, they do occasionally occur! If a field star has to be selected, then an additional check star of similar spectral class should be observed to confirm that the comparison star is non-variable. The comparison star should be within a degree of the variable if possible, to minimize the effect of air mass differences between the two. Similarly, if the colour index is the same as the variable then second-order colour effects will be negligible.

If you observe by differential photometry, always bracket observations of the variable by observations of the comparison. The observing procedure is thus:

CVV . . . VVCVV . . . VVCVV . . . VVC, etc.

C represents an observation of the comparison through each of the filters and a measure of the sky adjacent to the comparison through the same filters. V is the same sequence for the variable.

A similar sequence for the check star should be inserted, if one is used.

The number of integrations of the variable between each comparison set will depend on how constant the sky conditions are, and how rapidly the variable is changing. It is not possible to give advice on this matter — it is something that can be determined only by experience. If in doubt, bracket each set of measures of the variable by ones of the comparison — you then will not be wrong.

Heliocentric correction – If your interest is in observing stars, do not forget to apply the heliocentric correction to your times. Geocentric times of observation are not appropriate to observations of objects beyond the solar system. The light-time correction is given by

$$\text{LTC} = -0.005775\, R \cos\beta \cos(L - \lambda)$$

where R is the Earth's radius vector, λ and β are the ecliptic coordinates of the star, and L is the longitude of the Sun.

THE ATMOSPHERE

Climate is a controlling factor. In the United Kingdom the atmosphere is normally laden with water vapour, even if to a casual glance the sky appears to be clear. It is therefore necessary to think carefully about what you are doing, and to appreciate the ways of achieving the maximum accuracy in the photometer readings.

The atmosphere is not transparent: this is true with even the clearest of skies. Light is scattered and absorbed by components in the atmosphere. The main influence is water vapour, in the form of cloud and aerosols. Dust particles, either naturally occurring or from atmospheric pollution, can be a major influence, as can local inhomogeneities in atmospheric composition. These problems also exist for the visual observer, but they are not usually considered.

The first element is the atmospheric extinction. The amount of light loss depends on the thickness of the atmosphere through which the starlight passes. This is a minimum for an object in the zenith, and increases as the zenith distance (z) increases (or as the altitude decreases). The thickness of the atmosphere at the zenith is referred to as '1 airmass', which is not a constant value in absolute terms but is a valid way of assessing the relative thickness of atmosphere for the objects observed. The effect of extinction is colour-dependent. The Sun appears much redder near the horizon than the zenith because the extinction of the blue component is greater than that of the red as the zenith distance increases.

The common formula for the extinction relates it to the value of sec z. This is a simplification which does not really hold true. For air masses less than about 4, the true airmass (X) can be described by

$$X = \sec z[1 - 0.0012\ (\sec^2 z - 1)],$$

where z is the true zenith distance calculated from the observer's LST and the object's coordinates.

In visual wavelengths the primary extinction coefficient is about 0.3 magnitude per airmass, though this can vary from 0.2 to infinity (on a cloudy night). The normal range in the United Kingdom is from about 0.25 to 0.45. It has a higher value in Blue and Ultraviolet wavelengths, and a lower one in the Red and Infrared. There is also a colour-dependent second-order term which is not important if the objects observed are all of similar colour, but can become significant if they are of different colours. At visual wavelengths the correction for extinction is given by

$$m_{\mathrm{v}} = m_{\mathrm{v\ obs}} - (k'v + k''v\ (B - V)X)$$

where m_{v} is the corrected visual magnitude, $m_{\mathrm{v\ obs}}$ is the observed visual magnitude, $k'v$ is the primary visual extinction coefficient, $k''v$ is the secondary visual extinction coefficient, $B - V$ is the colour index of the star, and X is the true airmass.

Visual observers often make estimates of variable stars at low altitudes using comparison stars of greatly differing colour at a different airmass (altitude). This can be corrected for if the potential error is appreciated. For accurate photometry it is essential to correct for differential extinction. The problems are minimized if comparison stars can be found which are both of the same colour as the variable and also very close — a situation which it is not always possible to achieve!

Sky conditions invariably change. The obvious case is when dense clouds come over. More insidious are the bands of cirrus which are invisible on a dark night but which cause fluctuations in transparency of 10–20% during time scales of a few minutes. Usually these cannot be recognized until the data is reduced, and the night's observations have to be written off. It is not the presence of haze, fog, or cirrus which renders a night unsuitable for photometry so much as the changes in transparency that arise. The use of several photometers and telescopes allowing simultaneous photometry of variable, comparison, and sky is the only way of recovering data on nights when sky conditions are poor.

The seeing may be unsteady. Atmospheric turbulence can give rise to image wander. The stellar images may blur out to show an Airy disk 10″ arc across, or perhaps even larger. This image may appear to 'walk' by a few seconds of arc or more owing to changes in the refractive index of the Earth's atmosphere over time scales of a few seconds. It is important that the diaphragm aperture is large enough to contain the movement of the image. This effect is more significant as the airmass increases.

A final effect which is usually not considered is scintillation. Rapid fluctuations in the atmosphere cause changes in the total brightness of a star with periods of the order of 0.05 second. This scintillation can be approximated by

$$s = 0.09\, D^{-2/3}\, X^{-2/3}\, (1/2T)^{1/2}$$

where $s =$ the rms variation due to scintillation, $D =$ the telescope aperture in centimetres, $X =$ the airmass, and $T =$ the measurement time in seconds. For a 400 mm telescope this corresponds to 0.001 magnitude at an airmass of 1 and 0.01 magnitude for an airmass of 3 for a 10-second integration. Users of smaller instruments must be aware of this source of error, and allow for it when making their measurements by integrating for a long enough period. Visual observers should also be aware that this effect occurs, and be careful when making estimates. As an example, the occultation of 28 Sgr by Titan on 3 July 1989 occurred at an airmass of about 7 (an altitude of about 9°). With a 350-mm telescope, an rms variation due to scintillation of about 0.5 magnitude would be expected over a time scale of 0.05 second, which is a good approximation to the time scale for visual observation. The observed peak-to-peak effect due to scintillation was ± 0.75 magnitude, in line with the theoretical value. The scintillation is independent of the brightness of the source, and sets a lower limit on the accuracy obtainable.

For bright sources, scintillation is usually higher than photon noise, and is the major source of error in the measurements. It is possible that the additional fluctuations due to scintillation could overload and damage the PMT if the source is too bright. For fainter objects, photon noise is usually the most significant source of error.

Though not atmospheric in origin, the effect of moonlight should be considered. It is not always detrimental to photometry as long as the object under observation is not too faint or too close to the Moon. Moonlight increases the background sky brightness — as the sky is then faintly visible in the eyepiece, centring the star in the diaphragm is simplified and more precise. One advantage of moonlight is that changes in the sky transparency can be seen at a glance. Increasing cloud cover will lead to an increase in sky reading as it reflects more moonlight.

In general, sudden changes in the sky reading indicate either the drift of a star into the aperture or else a change in sky conditions. Any such change should be investigated immediately, and if sky conditions are changing rapidly observing should be suspended.

STANDARD SYSTEMS

For measures to have a lasting value it is important to transform them to one of the recognized photometric scales. The vast majority of amateurs work on the

Johnson UBV system which was established in the 1950s. Individual photometer–filter–telescope combinations will approximate to, but not equal, a standard. One of the first items an observatory should determine, either when new equipment is installed or when alterations have been made, are the transformation coefficients.

The major photometric systems have primary standard stars whose brightness in the different colours is precisely known and whose non-variability has been well demonstrated. Secondary standards have been referenced to these stars.

The most accurate way of determining the instrumental transforms is to observe reference stars of different colours which are close together. Three clusters are normally used for this — the Pleiades, Praesepe, or IC4665. A simplified technique also exists where close pairs of stars, one red and one blue, are observed. This procedure has to be done when the sky conditions are very good.

The corrected magnitude (considering visual magnitudes in this example) is given by

$$M_{v} = M_{v\,obs} + T_{v} X\ (B - V) + Z_{v}$$

where M_v is the standardized V magnitude, $M_{v\,obs}$ is the observed (instrumental) V magnitude, T_v is the transformation coefficient in V, $(B - V)$ is the colour index of the star, and Z_v is the zero-point coefficient. The zero-point needs to be determined each night, although it can change during the course of a night if sky conditions do not remain constant. This term disappears if differential photometry is used because the variable is continuously being referenced to a star of known brightness.

CONCLUSION

Photoelectric photometry is in use at many private observatories around the world. It is a practical proposition for amateurs who have an understanding of electronics.

The amateur can make a lasting contribution to astronomy using this technique. Long-term observing programmes can be carried out which would not be possible for their professional colleagues. At the present time there are a number of liaison organizations which can put the amateur with photoelectric equipment in touch with professionals requiring observations. The development of this can only assist in furthering astronomical research.

About the Author

Andrew Hollis is a chartered structural engineer living in Cheshire, England. He has been Director of the BAA Asteroids and Remote Planets Section since its formation in 1984. His interest in photoelectric photometry dates from a meeting with Richard Miles in 1982, which has developed into a close collaboration.

He is United Kingdom Wing coordinator for the IAPPP, and uses a Hollis/Miles design of equipment with two Cassegrain telescopes for his photometry.

Contact address – British Astronomical Association, Burlington House, Piccadilly, London W1V 0NL, UK.

Societies

The primary organization promoting photoelectric photometry amongst amateurs is the International Amateur–Professional Photoelectric Photometry Organization (IAPPP). Through its quarterly communications and by sponsoring meetings it serves to encourage the development of equipment and to provide a means of bringing amateurs with equipment into contact with those professionals requiring observations. There are local Wings in several countries and districts. Initial contact should be made via the IAPP at Dyer Observatory, Vanderbilt University, Nashville, TN 37235, USA.

Most national associations or societies encourage the use of PEP, and contact with them would be a good starting-point.

Both the Variable Star Section and the Asteroids and Remote Planets Section of the BAA support the use of PEP and the publication of results. Contact may be made via the BAA. The AAVSO runs a photoelectric photometry section.

Bibliography

As it is a primary tool of observation, nearly all books on techniques devote a section to photoelectric photometry. There are also more specialist books, and a selection of these is given below.

Cooper, W.A. and Walker, E.N., *Getting the Measure of the Stars.* Adam Hilger, Bristol, 1989.

Hall, D.S. and Genet, R.M., *Photoelectric Photometry of Variable Stars at the Smaller Observatory,* 2nd edn. Willmann-Bell, Richmond, VA, 1988. (A 'must' for observers. It covers the whole field from equipment to data acquisition and processing.)

Henden, A.A. and Kaitchuk, R., *Astronomical Photometry.* Willmann-Bell, Richmond, VA, 1989.

Percy, J.R. (ed.), *The Study of Variable Stars using Small Telescopes.* Cambridge University Press, Cambridge, 1986.

There are more specialist books:

Genet, R.M. (ed.), *Solar System Photometry Handbook.* Willmann-Bell, Richmond, VA, 1983.

Warner, B.S., *High Speed Astronomical Photometry.* Cambridge University Press, Cambridge, 1988.

The Fairborn Observatory Press (Mesa, Arizona) produces several books of value. The titles include:

Photoelectric Photometry Handbook (2 vols).
Advances in Photoelectric Photometry (2 vols).
Microcomputers in Astronomy (2 vols).

Notes and comments

Douglas S. Hall, a co-founder of the IAPPP with Russell Genet, observed in an article in *Sky & Telescope* for November 1988 that a complete set-up ready to fit on to a telescope could be purchased for $850. A 200-mm telescope could achieve 0.01-magnitude accuracy down to about the 8th magnitude, while a 350-mm telescope could achieve the same precision on stars a magnitude fainter than this. The Optec SSP-3 solid-state photometer is a small and simple model suitable for amateur use.

Hall comments that the owner of even a modest telescope can make photometric measurements of 100 000 stars, and research papers authored or co-authored by amateurs are appearing in professional journals in increasing numbers: 'Among these reports are discoveries of dozens of new variable stars, most of them naked-eye objects with very small amplitudes. Many types of variability are represented: pulsations, starspots or chemical peculiarities rotating in and out of view, eclipses, cataclysmic explosions, tidally-distorted shapes, and light reflected from one star off another.'

CHAPTER SIXTEEN · *Cometary and Asteroid Astrometry*

BRIAN MANNING
CHURCHILL, KIDDERMINSTER, ENGLAND

THE MEASUREMENT OF PRECISE POSITIONS OF COMETS, ASTEROIDS, novae and other objects by the amateur has, until recently, been exclusively a photographic process. However, the availability of commercial CCD cameras has now opened the way for the amateur to use CCDs for astrometry. The photographic method will be described first because the main principles apply to either method.

A large-scale, accurately timed photograph of the object and its field, including at least three and preferably four or more stars which are included in a catalogue of precise star positions, is required. Measurements are then made in x and y coordinates, of the positions of the object (or target, as it is called) and the catalogued field stars on the photograph. The final step is the reduction of the measured coordinates to the target's RA and Dec.

CHOICE OF TELESCOPE'CAMERA

For many amateurs the choice of photographic instrument will probably be dictated by the instrument already available, most probably a Newtonian reflector or Schmidt–Cassegrain. Another factor is the reference star catalogue which will be used, as the instrument must have a sufficiently large field of good definition to include several catalogue stars. *The Hubble Guide Star Catalog*, [2] with an average density of more than 400 stars per square degree, would allow the use of virtually any telescope. The SAO or PPM catalogues require a field of one or more square degrees to be reasonably certain of including enough stars. Schmidt–Cassegrains are therefore a borderline case. Newtonians with focal ratios of f/6 to f/8 are quite suitable. Field correctors should be viewed with caution as they may introduce distortion: the amount that would be insignificant pictorially might be a problem astrometrically and require correction in the reduction procedure. Ideally the focal length should not be less than one metre; however the late Dr R.L. Waterfield, who was a pioneer of amateur astrometry in Britain, produced 150 precise positions of comets with his 660-mm focal length f/4.5 Cooke triplet lens.

CHOICE OF MOUNTING AND GUIDING METHODS

Whatever the instrument, it must be well mounted and have smooth fine adjustments in both axes in order to produce first-class star images. Astrometric star images must be perfectly symmetrical — either nice round dots or smooth uniform trails if guiding for a comet or asteroid. The altazimuth mountings now being used by professionals would be expensive and difficult to make, and the equatorial is still the simplest and best; however, in recent years a number of equatorial platforms on which Dobsonian altazimuths can be mounted have been invented which appear to be capable of producing good photographs. Off-axis guiders are popular and certainly overcome the problem of relative movement between a guide telescope and the main instrument. For various reasons the author prefers to use a separate guide telescope: for example, there are no coma problems due to using an off-axis image, and it is easier to find a bright guide star.

A means of compensating for the motion of comets and asteroids is required, for two reasons. Firstly, fainter objects will not be recorded at all if allowed to trail; secondly, although a trailed star image is perfectly suitable for measurement, a trailed comet image is not. One method of compensation is to mount the film holder or camera on a slide propelled by a screw and stepper motor [1]. The slide is rotated to coincide with the direction of motion of the object in the focal plane of the telescope, and the motor steps the film along at the correct rate to keep the image stationary on the film. The more usual method, however, is to move the guiding eyepiece cross-hairs by means of a screw micrometer in the opposite direction to the motion of the target. It is also possible to use an eyepiece graticule and move the star along one division at appropriate time intervals. However, it is difficult by this method to achieve sufficiently small increments of motion.

FIELD SIZE AND CATALOGUES

The area of the field required is mainly determined by the star density of the available catalogues. From the beginning of 1992 all comet and asteroid positions have to be referred to the J2000 equinox, and the preferred catalogue is the *Positions and Proper Motions Star Catalogue* (PPM) [3]. It lists 360 000 stars at an approximate density of nine stars per square degree. The Hipparcos input catalogue with 118 000 stars is only suitable for amateurs with wide-field cameras. The actual density of stars in different parts of the sky does of course vary widely. Non-solar-system objects must still be referred to the 1950 equinox, and the SAO and AGK3 catalogues can be used, or a high-precision method of precession as described in current astronomical almanacs used to precess J2000 positions to B1950.

Based on experience with the old SAOC, and some experimental plots from the PPM near the Galactic pole, a field of one square degree will nearly always contain four or more PPM stars. A larger field is advised if possible. Occasionally the field may not contain the minimum of even three stars. When this happens the solution is to photograph overlapping fields and derive accurate positions for a number of non-catalogue stars common to both photographs. The position of the target can then be obtained from these stars.

PHOTOGRAPHIC MATERIAL

At one time it was believed that emulsion-on-glass plates were the only material with the necessary stability for astrometry, but in the 1970s it was found that film, particularly the polyester-based type, was quite good enough for cometary astrometry. Unfortunately, the most popular format is 35-mm which is rather small and not suitable for use with telescopes of greater than about 1.3 m focal length. The author, with a reflector of 1.9 m focal length, uses a true prime-focus attachment which takes a film size of 51 mm × 63.5 mm, ie. the standard 102 mm × 127 mm (4-inch × 5-inch) format cut into four pieces. In addition to a large field, this has the advantage of saving light normally lost by reflection at the flat, which, however must be easily removable. A further advantage is that the film plane can more easily be set perpendicular to the optical axis than it can at the Newtonian focus. Almost any emulsion can be used, but the very fast large-grained ones are not really suitable. Emulsions of 400 ASA are used quite successfully by many observers, but a finer-grained emulsion is desirable particularly if attempting to use focal lengths below 1 m. The writer uses hypered Kodak Technical Pan 2415 or 4415 exclusively.

PHOTOGRAPHIC EXPOSURE AND TIMING

Photographs of bright comets showing tail and coma detail are usually unsuitable for measurement, and for astrometry the exposure should be adjusted to show only the condensation when it exists. For example, Comet Halley when about 6th magnitude needed exposures of only 30–60 seconds with a 260-mm reflector. Some comets are very diffuse and are difficult to measure accurately. One should aim to record the start and finishing time of an exposure to an accuracy of one second, even though an error of a second or two will not be important for objects moving at less than 1″ arc per minute. For very fast movers such as comet Iras–Araki–Alcock 1983d even one second accuracy is not really good enough. The quoted time of the photograph is that of mid-exposure.

MEASURING MACHINES

The major problem encountered by anyone wishing to do astrometry is the measurement of the film or plate. The word ‘plate’ is used conventionally for either a true glass plate or ordinary film, and the professional device used for measurement is known as a plate-measuring machine or plate-measuring microscope. Such machines can accept plates as large as 330 mm square, and obviously are quite unnecessary for the amateur, who is most unlikely to use a format larger than 102 mm × 127 mm. The actual area over which measurement is required will be even less, and a device capable of a precision of one or two micrometres (thousandths of a millimetre) over an area of 50 mm × 75 mm will be adequate, except for very large telescopes.

Much engineering measuring equipment falls into this category. The toolmaker's microscope is one such machine, usually with a stage mounted on two precision slideways moving at 90° to one another (the *x* and *y* directions), and controlled by micrometer screws. The writer's machine is of this type (see Fig. 16.1). A variation has the stage on one slide with the viewing microscope on the other slide, as in M.J. Hendrie's machine (Fig. 16.2).

Both these machines are home-made, and this may well be the solution for the amateur telescope maker. Fortunately the components which are most difficult to make can be purchased: these are the micrometer screws and slideways. The writer

FIGURE 16.1
The author's plate-measuring machine. The viewing microscope is fixed, and the negative is mounted on two slides at right angles to each other. Note the digital readout.

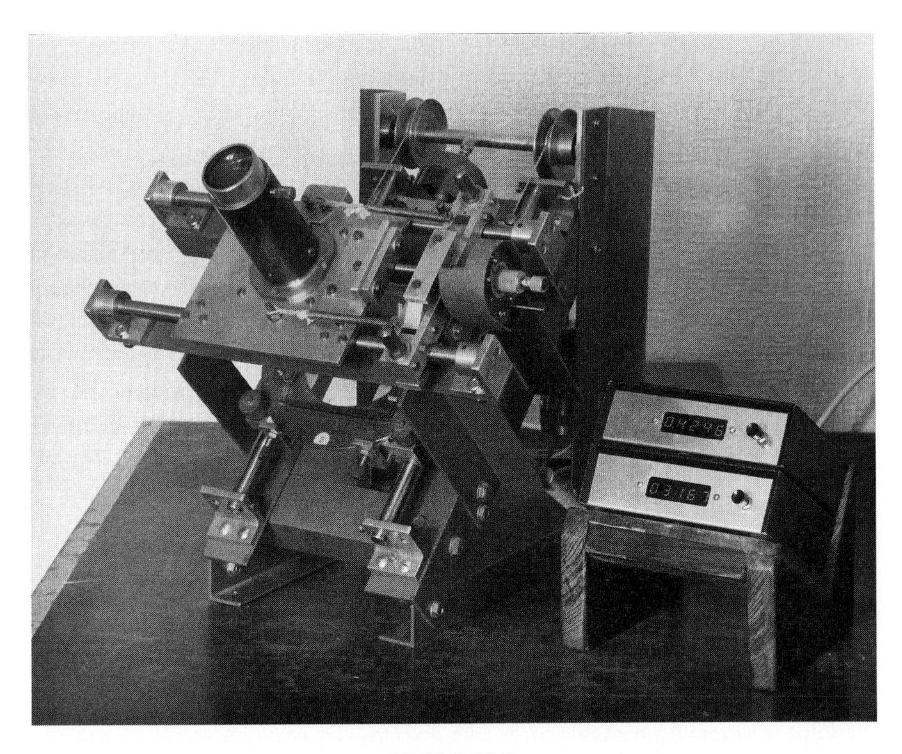

FIGURE 16.2
A plate-measuring machine constructed by M.J. Hendrie. The negative moves on a vertical slide, and the viewing microscope on a horizontal slide.

strongly recommends using linear ball bearing slides, because these almost entirely remove hysterisis and the need always to approach settings in one direction. The use of slideways is eliminated in Edgar Everhart's ingenious design, which uses large-diameter screws both to translate and to guide the plate stage [4]. Another possibility is Luigi Balbi's single-axis machine, which even dispenses with a micrometer screw [5].

The plate is viewed by transmitted light, and the stage on which the plate is placed must have an aperture slightly larger than the measuring area in order to allow for illumination. Usually the viewing microscope is a low-power compound type with a cross-wire in the eyepiece and a magnification of $\times 20$ to $\times 30$. For spectacle wearers a projection system such as used by the writer may be more convenient. This has a rotating screen to reduce the effect of grain. There is a graticule behind the screen and close to it, which, due to the large focal ratio of the projection beam casts a sharp shadow on to it. With projection a bright light source is required, but it is important that a heat filter is used, and that the power does not exceed about 10 watts.

For those constructors who enjoy electronics, the fitting of encoders (or, even better, linear encoders) to the screws for digital readout will be found a great boon. The tedium of having to make many micrometer readings, and the resultant occasional errors, are eliminated. It is also possible to transfer the readings direct to the computer at the touch of a switch.

At one time the purchase of even a small measuring microscope would have been out of the question, but today it is a possibility for some people. However, if construction and purchase are both ruled out, a method using photographic enlargement proposed by Charles Townsend [6] may be of interest. Strangely, the description calls for a linear measuring device several times more precise than is actually required, and I have suggested elsewhere [7] that an engineer's vernier caliper might be used instead. I have not actually used this method, but it was thought it should be included in order to provide every possible means of overcoming the main problem for anyone aspiring to take up astrometry.

MEASURING THE NEGATIVE

The first step in preparing to measure a negative is to identify the field and make a chart of the catalogue stars which appear on it. The use of the PPMC immediately raises a slight problem, it does not have a matching atlas as does the SAOC. *Uranometria 2000.0* [8] can be used but has fewer stars; also, many faint stars will not be common to both PPM and the atlas. A better solution is to have the PPM in computer-readable form (the diskette version is inexpensive) and plot and print to the correct plate scale the required field, as is done by the writer. All the stars on the negative can be used if desired, but six to eight is sufficient. When possible the stars should be evenly distributed around the target. Very bright and enlarged images should if possible be rejected, as well as badly comatic images or close double stars.

The chosen stars should be numbered and the negative sandwiched between two hinged glass plates. The selected stars and target can then be ringed with a fine overhead projector pen to aid identification under the measuring microscope. Exposures guided on a comet or asteroid will usually have trailed star images. Bisection of these images is more easily done if the negative is rotated to align the trails with the cross-wires in the measuring eyepiece. It is not recommended to try to measure to the trail ends. Several settings, for example five, should be made on the target, and the mean of the readings in each axis recorded. Similarly, more than one setting on each star may be made, but there is a danger that the task will become tedious, promoting mistakes and inaccuracies. It is good practice to make a second set of measurements with the plate or negative rotated 180°, that is, direct and reverse measurements [9].

REDUCTION

The task of converting the x and y measurements to celestial coordinates is now a simple one, with the aid of a computer. The old 'method of dependencies', well described by L.J. Comrie in the BAA Journal [10], although quite precise, is rarely used now. A method based on standard coordinates and described briefly by Marsden [11] is now generally used. The author uses his own adaptation to BASIC of the FORTRAN program written by Harrington of the US Naval Observatory [12] for the Comet Halley astrometric net. The rms values for the stars calculated by this program give a good indication if things have gone correctly. If they are larger than normally encountered, for example greater than 1″ arc, it may be that a star has been wrongly identified; the component of a double star for example. The author's program has a facility for excluding stars from the sequence, which is convenient for locating the offender, by noting which star restores the rms values to normal when excluded. If the target is remote from the only catalogue stars on the plate, its reduced position will be very sensitive to measuring errors and a note of this should be made. Instead of combining the direct and reverse measures as described by Gibson [9] and Harrington [12] the author finds the mean of the two reductions and submits this. It may be against the rules, but comparing the two sets of results readily reveals a systematic error in one's machine or eye.

Before submitting any positions it is essential to run a check on test data, to ensure that the reduction program is working correctly. Having done this, the data for a real negative can be entered, and a few seconds later we have a precise position — but is it really correct? A flaw in the film may have been mistaken for a faint comet image, an incorrect micrometer reading may have been entered, the time (or even the date!) may be wrong. It is essential to take every care to avoid submitting wrong positions. The practice of obtaining a pair of photographs on the same night is a good (although not hundred per cent!) check on film flaws, and is especially useful in the case of an object for which orbital elements are not available. When the elements are known, the measured positions can be compared with computed ones and the residuals used as a check. For this, a precise ephemeris program is required, with an additional section to correct for parallax. This is because the measured positions are topocentric (observer's position on Earth's surface), whereas an ephemeric is geocentric (for the centre of the Earth). Judgement must be used when considering residuals: for a well-established orbit they should be small, but for a preliminary or outdated one they may sometimes be several tens of seconds of arc or, rarely, even minutes of arc. A pair of observations taken on the same night and yielding very similar residuals, even though they may be large, is valuable in deciding whether to submit the positions.

Positions should be submitted to the Minor Planet Center, Smithsonian Astrophysical Observatory, Cambridge, MA 02138, USA. There is a preferred format for doing this, details of which should be obtained from the Minor Planet Center. Positions are quoted to 0.01 seconds of time in RA and 0.1″ arc in Dec, and the date in UT to 0.00001 day. The observatory longitude and latitude to as near 1″ arc as possible, and its altitude to 1 metre, are also required initially. Positions should not be corrected for parallax.

CCD ASTROMETRY

At the time of writing (March 1992) the author was aware of two observers in Japan and two in Italy using CCDs for astrometry, and although not having first-hand experience himself, was collaborating with H. Mikuz and B. Dintinjana of Slovenia in assessing the suitability of a flat-field f/4, 760-mm focus camera for astrometry. Exposure times with such an instrument are very short — one minute or less — and trailing due to the object's motion is generally insignificant.

If a long-focus telescope and a small CCD are to be used, then the use of the Hubble Guide Star Catalog for reference stars will be essential. However, Mikuz finds that fields acquired with a 574 × 384 CCD [13] on his 0.76-metre camera usually contain sufficient SAO or PPM reference stars for a reduction. One problem with using SAO or PPM is that, depending on the exposure, bright reference stars (e.g. magnitude 8 or brighter) may saturate the pixels and be unsuitable for measurement.

The acquisition technique for CCD images is fairly common knowledge already; however, the measuring process requires software which at present is not readily available, for example PCVISTA [14]. Instead of visually centring cross-wires on an image as for photography, the software computes the centroid, or 'centre of gravity' (CG) of the light, in the images of catalogued stars, and its *XY* coordinates in pixels. The reduction can then take place by the standard method, the focal length of the camera being expressed in pixel units.

The normal measuring accuracy of a measuring machine is of the order of 1–2 μm, whereas the pixel size of the usual CCD is around 20 μm. At first sight it would appear that a camera or telescope of very long focal length would be required to achieve high precision. Fortunately, although from rather limited information at the present time, it would appear that this is not so. Providing the reference star images and the target image are spread over several pixels, the centroids can be computed quite accurately, even from a camera of 760-mm focus. A trial reduction made for Mikuz using centroids supplied by him have yielded good residuals when submitted to the Minor Planet Center. Published results by T. Kobayashi indicate that he also obtains precise positions with a camera of the same flocal length, 760-mm.

There are, however, problems with such short focal lengths. Good seeing conditions can result in star images which illuminate only a small number of pixels. Provided the images are not too small, an adjustment of the star profile parameters in PCVISTA can be made. This still leaves a problem in the case of the image of a faint, highly-condensed comet, and particularly for an asteroid which may fall entirely within a pixel. The position would then be indeterminate to the extent of nearly one pixel.

A suggested solution would be to defocus slightly. The shadow of the secondary might, however, be a complication. The introduction of a small amount of spherical aberration or diffraction with an uncorrecting plate (!) or screen in front of the camera are suggested possibilities. Alternatively, a Barlow lens could be used to give a larger image scale, but would have the disadvantage of a smaller field for a given CCD, which would introduce the possibility of distortion, and necessitate the use of the *Hubble Guide Star Catalog*.

A problem may also arise with comets, owing to the lack of symmetry in their images. The CG of mass (usually, but not always, corresponding to the brightest part of the condensation), and not the CG of their light, is the required centroid for measurement. PCVISTA provides a means for dealing with this difficulty.

There is no doubt that as more experience is gained, CCD astrometry will be increasingly practised by the amateur; however, where wide-field work is required, photography and a measuring machine will presumably have a place for some years to come.

About the Author

Born in 1926 in Handsworth, Birmingham, Brian Manning is married to a very tolerant wife and a 260-mm reflecting telescope. He first remembers an astronomical interest as a child when walking home with his parents from a grandparent's home when skies were really dark and stars could hardly be missed. He borrowed a ship's telescope from his maternal grandfather, who had helped to survey the Canadian Pacific Railway route as a young man, and looked at the Moon with the telescope resting on the bars of a gate.

In 1939, his parents bought him a 25-mm 3-draw telescope (×12) for 35 shillings, the limit of the family budget, but it was good and he was very excited to see the moons of Jupiter with it. In 1947 he took a photograph of asteroid Iris with this telescope on a mounting driven by an alarm clock. His father was a toolmaker and taught him to use a lathe, etc., which was useful when eventually coming upon a book on mirror making by George McHardie. A 55-mm mirror was made from glass blown out of a roof by a German bomb, in 1940. In 1948 he constructed a 180-mm equatorially-mounted reflector, and in 1950 he constructed the 260-mm 1.9-m focus reflector which is still the main instrument.

In the middle 1950s onwards he become interested in the ruling of diffraction gratings, and constructed one of the first interferometrically controlled ruling engines, which eventually produced 75-mm gratings of near theoretical resolution in the higher orders. In 1975 he subscribed to *The Astronomer* magazine and submitted prime-focus photographs which led to a request to photograph Comet Chernykh 1977l for astrometric measurement. In 1982 he received an invitation from Donald Yeomans of JPL to join the Halley Astrometric Network.

In 1985 he took early retirement from the University of Birmingham and constructed a measuring engine for home use. In October 1989 he made an asteroid discovery with sufficient observations for an orbit to be calculated — the first British discovery for 80 years.

Contact address – Moonrakers, Stakenbridge, Churchill, Kidderminster, Worcs DY10 3LS, UK.

References

1. Arbour, R.W., An amateur's computerised camera for the automatic tracking of comets. *BAA Journal,* **96**, 1 (1985).

2. Space Telescope Science Institute, *Hubble Guide Star Catalogue.* Distributed by the Astronomical Society of the Pacific, 390 Ashton Avenue, San Francisco, CA 94112, USA.

3. Roser, S. and Bastian, U. *Positions and Proper Motions Star Catalogue.* Astronomisches, Rechem-Institut, Heidelberg, Germany.

4. Everhart, E. Constructing a measuring engine. *Sky & Telescope*, 289 (Sept 1982).

5. Balbi, L., A wire micrometer for photographs. *Sky & Telescope*, 310 (Sept 1987).

6. Townsend, C., Photographic astrometry with a single axis micrometer. *Proc. RTMC*, 1988. (Orange County Astronomers, 2215 Martha Avenue, CA 92667, USA.)

7. Manning, B.G.W., Astrometry. *The Astronomer*, **26**, 305 (1989).

8. Tirion, W. *et al., Uranometria 2000.0*, in 2 vols. Willmann-Bell, 1987–88.

9. *Cometary astrometry.* Proceedings of the Halley workshop, Munich, 1984. Issued by Jet Propulsion Laboratory, California Institute of Technology, 4800 Oak Grove Drive, Pasadena, CA 91109, USA.

10. Comrie, L.J., Note on the reduction of photographic plates. *BAA Journal*, **39**, 6 (1929).

11. Marsden, B.G., How to reduce plate measurements. *Sky & Telescope* (Sept 1982).

12. Harrington, R., *Astrometric Plate Reduction Procedure.* Issued by Jet Propulsion Laboratory, California Institute of Technology, 4800 Oak Grove Drive, Pasadena, CA 91109, USA.

13. (H. Mikuz Crni vrh private observatory, alt 700 m). The 574 × 384 CCD is the property of the University of Ljubljana and is used at Crni vrh for observation of comets when not being used for the University programme on variables. B. Dintinja, staff member of the University, is a computing and instrument construction expert.

14. Image processing program PCVISTA distributed by University of California of Berkeley, available for PCs, computes centroids.

Notes and Comments

Brian Manning's report of his first confirmed asteroid discoveries, made in October 1989, was published in the November 1989 issue of *The Astronomer*, from which this following account is extracted. — Ed.

> The search started earlier this year with the discovery of what appeared to be an asteroid trail on a photograph of 1645 Waterfield. Although follow-up photographs disproved it, I was very interested when the search that Guy Hurst [Editor of *The Astronomer* — Ed.] obtained for me from Brian Marsden's computer [At the Bureau for Astronomical Telegrams, Washington — Ed.] revealed that there were three additional asteroids, all

below 16th magnitude, in the field. Each one was recorded as a star-like object with no perceptible trailing, and it raised the possibility of recording new ones by guiding at the average rate of asteroids near opposition. Not a new idea, of course. Communications from Robert McNaught and Brian Marsden revealed that the Japanese were making many discoveries, and they urged me to 'have a go'.

Nothing was done during the summer due to the low elevation of the ecliptic, and then Robert's latest contribution to *The Astronomer* [see below — Ed.] provided the necessary push to get me going, although, to be honest, I thought my chances would be pretty slim.

The first attempt was on the field of 2664 Everhart, but neither Everhart nor any other asteroid was recorded. I then learned that Comet Schwassmann–Wachmann (1) had brightened, and as its motion near the ecliptic was similar to that of an asteroid it made a suitable subject for an asteroid search negative. I would get something for my efforts, even if no asteroids! The evening of the 4th of October being clear I made two exposures, one of 20 minutes and guided at the comet rate, the other 25 minutes at twice the rate.

These were developed next morning, and I first examined the negative with the longer trails under a low-power binocular microscope (× 9). Quite soon a nice 16th-magnitude dot was found and confirmed in a slightly different position on the other negative (now asteroid 1989TE). The second asteroid (1989TF) was found by a fluke. I suddenly noticed a short north–south trail which checked out as a 17th-magnitude dot on the other negative, although there should not really have been a trail. Serendipity in asteroid discoveries, I suppose. Next came a spell on the measuring machine to obtain the four positions. Guy Hurst was informed and answered the phone with a cup of tea in hand. Soon after I received a call from him to say that he had successfully contacted Brian Marsden's computer at the Smithsonian Astrophysical Observatory, and that there was no numbered asteroid on my negative. Great jubilation.

The next night (5th October) was exceptionally clear and I was able to obtain two confirming exposures, but not without a calamity: one negative was badly out of focus at one end because the film did not seat properly, but luck was on my side as well, as 1989TE was very nearly coincident with a fat star image. The only image I could find on the sharp negative for the other asteroid was quite obviously a flaw, as I thought, as it was much too sharp for a true image. However, I decided I might as well check the out-of-focus negative, and to my delight there was just the faintest circle of a few grains displaced by the correct distance.

A few days later a slightly disappointed Guy phoned me to say that Brian Marsden had identified 1989TF with 1960F, for which five observations and an orbit were obtained in that year. However, the rediscovery after 39 years, and the establishment of a good orbit, is still very nice. A number of single observations of 1989TE have been located which are very useful for an orbit. It may soon be numbered and looks fairly certain to be a discovery.

About ten days later I made another search of the October 4th negatives, and incredibly came up with two more asteroids. The third one was, unfortunately, just off the edge of the negatives taken on the following night and may be lost. The fourth is designated 1989TN1, and so far rates as a discovery. Brian Marsden has burned out the satellite link to Guy, who has in turn posted me on all the latest news, and I am grateful to them for their interest and effort in following up the discoveries. Comet Schwassmann–Wachmann (1) was also recorded!

The interest has been amazing. For two days the telephone never stopped ringing, and my wife and I were besieged by photographers, local radio, and television.

In the inspirational article to which Brian Manning referred (*The Astronomer*, September 1989), Robert McNaught points out that during a period of almost ten years (1978–1987) amateur astronomers discovered 197 asteroids — 160 from Japan,

FIGURE 16.3
A photograph of asteroid 3698 Manning, taken by the author using his 260-mm reflector. The asteroid is the faint spot at the centre of the photograph; the telescope was driven to follow its predicted motion during the exposure.

and 37 from Italy. It was the discovery of an asteroid by the Japanese observer Urata in 1978 which opened up this field of amateur discovery.

There are two ways in which an amateur can play a part in an asteroid discovery:

1. To be the first ever to record a new asteroid, and to make at least two observations on different nights. This will allow its location and further measurement during the following dark-Moon period. Once its position has been measured over 30 days (preferably 60 days) its recovery at the following opposition is assured.
2. To record an asteroid on at least two nights, which can then be identified with previously-obtained images that were not good enough to provide a reliable orbit.

An asteroid is never given a number and name until it has been recorded at two apparitions. In practice, with the huge computerized archive of single-image asteroid positions, a Type 1 discovery is unlikely. It would probably have to be an Earth-grazer, appearing unusually bright because of its closeness, and then its apparent motion would be so much faster than the normal tracking speed to which a telescope is set that it would not be recorded anyway!

Since this chapter was written, Brian Manning has continued with his labour of identifying and measuring asteroidal images. His asteroid 1990FJ, a magnitude 17 object, was confirmed by pre-discovery positions dating back to 1980, and it has now been given the permanent number 4506 and named by Brian 'Hendrie' in honour of M.J. Hendrie, for 10 years director of the BAA Comet Section. Thereby achieving official status as a new asteroid. Another discovery by Brian has been numbered and named (4751) Alicemanning, in honour of his wife. — Ed.

CHAPTER SEVENTEEN · *Starting in Video Astronomy*

TERRY PLATT
BINFIELD, BERKSHIRE, ENGLAND

THE CCD, OR CHARGE-COUPLED DEVICE, HAS HAD AN ENORMOUS impact on professional astronomy in recent years, and it is in the tradition of amateur astronomers to try to keep up with such advances whenever possible. These hopes have been frustrated until recently by high prices and lack of availability, but all this has changed with the growth of the home video market. Manufacturers are now producing high-grade CCDs at UK prices of less than £100, and complete camcorders are in the shops for under £500, while in the United States CCD imaging systems designed for use at the telescope are available from prices of about $500 upwards. It would be misleading to suggest that you can record faint deep-sky objects with nothing more than a home video system, but the brighter planets, the Moon, and the Sun are within reach of simple equipment, and the electronically able amateur can build a system to reach the limiting magnitude of the telescope available. Although space is limited, I hope to give enough advice for you to make a start in the field while avoiding the worst pitfalls and to suggest ways of advancing towards a truly high-performance system.

First, I will briefly describe the CCD and try to show how it can best be used in astronomy.

HOW A CCD WORKS

The CCD is the solid-state equivalent of the vidicon camera tube, used almost exclusively in small TV cameras for many years, and has now almost totally replaced it, owing to its very small size and low power consumption. Its great advantages, from the astronomer's point of view, are the excellent sensitivity to faint light and precise dimensional stability, combined with the ability to integrate light over long periods as does photographic film. This last feature is most important if a CCD is to be used for deep-sky work, but is not available on home video equipment and is the great limiting factor for such systems. The time between readouts of a normal video camera is only 40 milliseconds (ms), or 33.3 ms in the United States, and this amounts to an 'exposure time' of only about 1/25 s — hardly long enough for sensing a faint comet or nebula! To use long exposures it is necessary to buy or construct special drivers for the CCD and to store the image in some form of memory so that it can be continuously displayed. An outline of these techniques will be given later.

At this point it should be mentioned that two alternative CCD structures are at present in common use, which differ in the way in which the charge is handled as it exits from the array. Most home video equipment uses CCDs of the 'interline transfer' variety, whereas industrial and scientific CCDs are generally of the 'frame transfer' type. Performance is not greatly affected by these differences, although the frame transfer CCD is a little more sensitive: the amateur will not find this of great concern in most cases, especially as the interline type is quite a lot cheaper at the present time.

The interline CCD (Figs 17.1 and 17.2) is based upon a thin rectangular wafer of very pure silicon, about 10 mm along each side, on the surface of which an array of about 250 000 light-sensitive diodes has been constructed by a photochemical process. Alongside each sensor is a charge storage region or 'well' which can hold up to about 100 000 electrons (produced by light falling on the diode) before overflow or 'saturation' occurs. During the integration period mentioned above, this charge

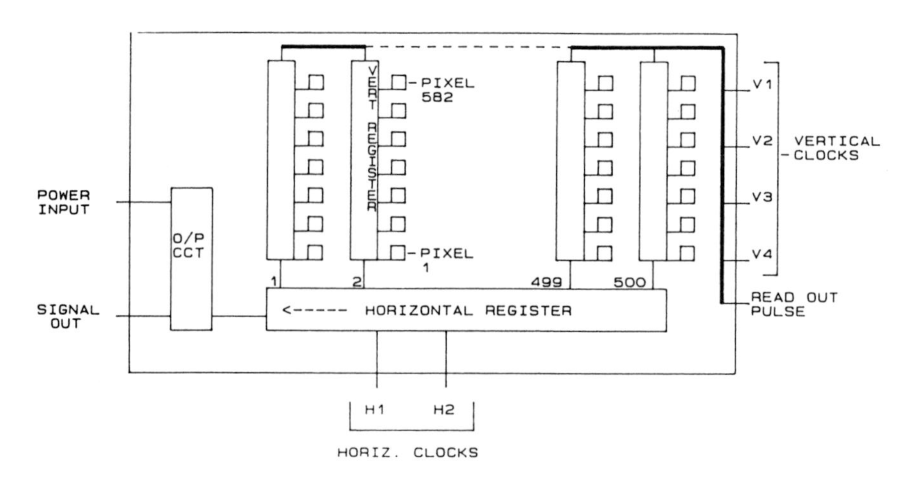

FIGURE 17.1
Typical CCD structure (interline transfer).

FIGURE 17.2
A home-built CCD camera using a Sony ICX 021L CCD (seen in box at left). The cooling fins are to dissipate heat from the Peltier cooler.

accumulates at a rate proportional to the light intensity on the diode, and at the end of the integration time it is read out by a voltage pulse into a 'charge-coupled shift register' which stretches from top to bottom of the diode column in question. Each diode column has its own shift register alongside, and each register is driven by the same 'clock' signal as all the others. This means that, once the diode charge has been transferred to the registers by the readout pulse, the entire pattern of charge can be moved down the array by applying clock pulses to the register circuit.

In the interline type of CCD, the charge packets being pumped out of the ends of the registers are deposited directly into a horizontal register which lies across the entire diode array. This register is clocked by the CCD driver at such a speed that its contents are ejected from its output during one line scan of the TV image. A typical CCD will have 500 diodes or 'pixels' per line, and the European (CCIR) line scan time is 64 microseconds (μs), about 52 μs of which is video data. The American EIA equivalents are very similar. A clock frequency of about 9.5 MHz is thus used to drive the horizontal register, and each pixel voltage is present at the output for about 0.1 μs, during which time it is sensed by a field-effect transistor and delivered to the video output pin. This 52-μs string of voltage pulses represents one TV line of the picture, and, by clocking the vertical registers once after each line, a sequence of scan lines can be output in the manner of a normal TV camera. A European standard TV picture has 625 lines, 588 of which carry picture data, and the American standard is 525 lines with 502 used. The Sony ICX 027 BL is a typical interline CCD for the European standard, or ICX 026 BL in America.

In the frame transfer type, the vertical registers move the signal charges down into a storage region during the frame blanking period, and the video signal is clocked out of this memory area while the next field is integrating on the photodiodes. The Philips NXA 1011 and EEV CCD02 are of this type.

PERFORMANCE LIMITATIONS

All CCDs have a high sensitivity to light, but the ultimate performance under poor illumination is limited by spurious charge generation and electronic random currents ('noise'). Such random signals are responsible for the granular appearance of pictures taken at low light levels, and have a similar effect to the coarse grain of a high-speed photographic film, where random density variations constitute the noise signal. An image of a faint, low-contrast object will be most severely affected by noise, and this makes good pictures of planets with a lower surface brightness than Mars difficult to obtain with standard video cameras — I will cover the subject in more detail later.

Dark current is always present, but can be greatly reduced by cooling the device to as low a temperature as possible. A drop of 10°C will reduce the current by a factor of about 2.5, and professional equipment uses liquid nitrogen to allow exposures of hours without serious image degradation. 'Peltier effect' coolers, using electric current in a grid of thermoelectric junctions to 'pump' heat out of the CCD, can reduce the temperature to −50°C or so and are much more practical for the amateur. However, this is available only to the person who constructs the camera from components, and is not practical or necessary with the short integration time used in home video equipment.

Electronic noise is mainly a function of circuit design and CCD quality, and is not open to improvement by simple methods — however, some possibilities will be discussed later.

ASTRONOMY WITH A HOME VIDEO

I will now give some advice about using simple video equipment to the best effect on astronomical subjects where the user does not wish to get involved with building complex electronics.

The reader will be familiar with the structure of a TV picture in which a sequence of several hundred horizontal scan lines is displayed 25 or 30 times per second to build up a rectangular 'raster'. The resolution of detail in such a picture is obviously limited by the number of lines in the vertical direction and by the number of CCD pixels horizontally. If the purpose of using the video camera is to record useful data on the Moon, Sun, or planets, it is important to avoid losing detail owing to the coarse image structure of the camera. Equally, an excessive amount of noise due to too much magnification (and hence low brightness of the image projected on to the CCD) will also ruin a good picture. A compromise is essential if good results are to be obtained, and some method of adjusting the image scale is required to set the camera resolution as close to the telescope resolution as possible, compatible with adequate image brightness. One method is to mount the camera to view the field through a good-quality eyepiece, such as a 25-mm orthoscopic, and then to use the zoom facility of the camera to set the image scale. However, this is undesirable from the point of view of the amount of glass involved, and if the camera lens can be removed it is better to use a Barlow lens to project the image directly on to the CCD.

Most modern video cameras are highly automated to make them easy to use for their intended purpose of recording Aunt Mabel on holiday, but this can cause many problems when used on a telescope. Autofocus systems must be set to manual, colour balance set for daylight, and the auto iris disabled by ensuring that the image brightness is not excessive (the Moon and Sun can cause the iris to stop down if too much light falls on to the CCD). Most errors are readily seen in the viewfinder picture, but care is needed to avoid mistakes which cannot be corrected later.

The Moon

Using a camera–telescope combination as described above, excellent-quality recordings of the Moon can be made in which detail close to the resolution limit will be discernible during moments of good seeing (Fig. 17.3). The eye is very

FIGURE 17.3
Clavius, photographed on 17 January 1989 with the author's 318-mm tri-Schiefspiegler, using the camera shown in Fig. 17.2.

efficient at averaging out rapid changes in a TV picture, and this visually suppresses some of the noise which might be unacceptable in a single still image. This will be apparent if the recording is first viewed normally and then in the 'freeze frame' mode, but unless still photographs are to be taken from the tape, the noise will not be a major problem. Occultations of bright stars by the Moon, lunar eclipses, and searches for Lunar Transient Phenomena (LTPs) are all within easy reach of a home video camera with a medium-sized telescope, and offer a rich field for useful work.

The Sun

The comments made above with respect to lunar recordings are also generally applicable to the Sun, but of course a full-aperture solar filter will be essential unless a reflecting telescope with uncoated optics is used. The CCD is very resistant to damage by intense sunlight (far more so than the human eye), but using a dark shade glass on the eyepiece can never be recommended for solar observing, if only because of the serious optical distortion which is bound to affect the unprotected lenses and mirrors.

The use of an Hα filter with the camera can give excellent results on the solar surface and prominences, and this can be a rewarding field for the video experimenter.

The Planets

Planetary video is not difficult if the image scale is kept small, but attempting to match the camera resolution to the telescope will generally give a disappointing amount of noise in the picture, obscuring fine detail. Venus will allow a large image to be recorded, as there is plenty of light, but Jupiter does tend to be a 'noisy' subject unless the aperture is large (at least 300 mm). Mars is fairly good as the surface brightness is quite high, but the need for high magnification does offset this to some degree. There is no doubt that very useful videos can be made of the above planets for verifying drawings, timing occultations, and timing transits of features on Jupiter, but a conventional CCD camera and video recorder will not give 'photographic' quality pictures, and refinements are necessary.

Some improvement can be made by investing in a 'video enhancer', which is a readily available piece of equipment sold by specialist video retailers for improving the clarity and colour rendering of tape copies. Many types are on the market, but a low-cost unit will be adequate, as the sharpness enhancing and colour balance features are all that you need for astronomy. The device is connected in line between the camera and recorder, or between recorders when copying, and can be adjusted for best results. Sharpness boosting should not be overdone, as noise will also be increased and 'outlining' of the images becomes objectionable. Colour balance improvement is often useful for emphasizing tints of planetary features, and, combined with sharpness boosting, can give a valuable result (Fig. 17.4).

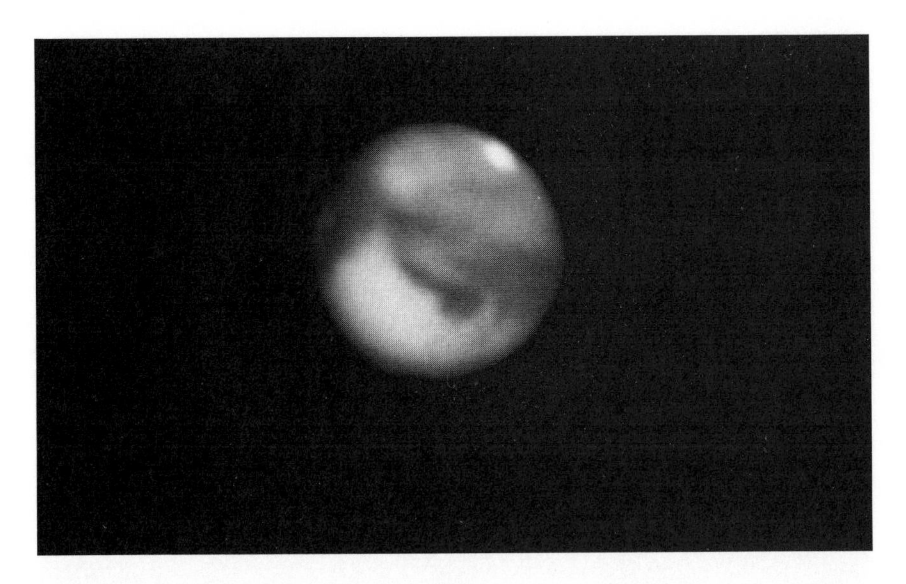

(a)

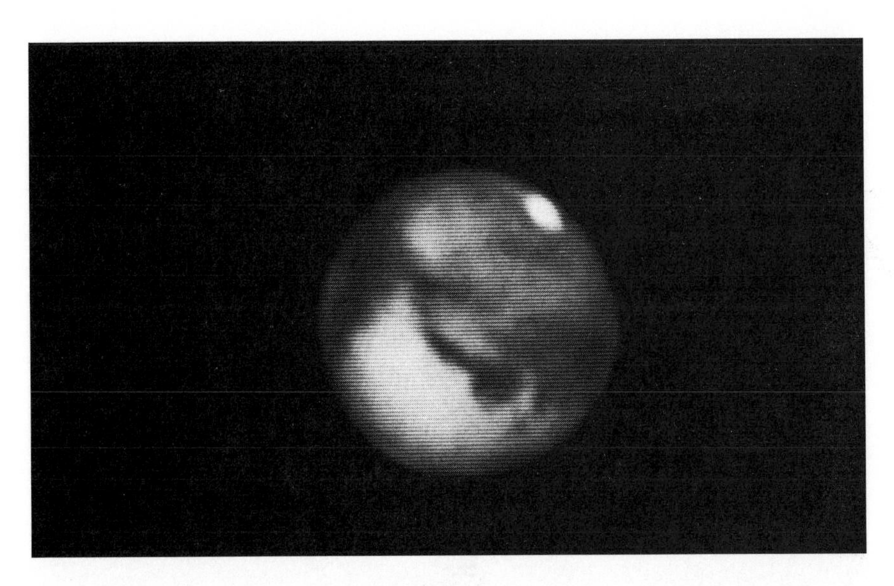

(b)

FIGURE 17.4
Planetary photographs taken with a 300-mm Newtonian and the camera shown in Fig. 17.2. (a) Mars, 19 September 1988 (unprocessed image). (b) The same, after processing through high-pass (sharpness-boosting) and median (noise-smoothing) filters. (c) Mars, 12 October 1988, showing the Solis Lacus region. (d) Mars, 6 November 1990, showing dust storm over Aurorae Sinus. (e) Jupiter and Ganymede, 28 December 1988. (f) Jupiter, 1 March 1991.

(c)

(d)

(e)

(f)

IMAGE INTENSIFICATION

If you are keen to get the most out of your video system, it may be worthwhile purchasing an image intensifier to give a large increase of sensitivity. The cost is rather high, but it will bring much fainter objects into range so long as the pale green image is acceptable. Efficient coupling of the camera and intensifier is difficult but important: the optical gain of the intensifier may be 20 000 times, but loss from poor light-transfer to the CCD can easily amount to 100 times, reducing the improvement to 200 times (approximately six magnitudes) or less. If the camera is used on 'Macro' it can be focused onto the output window of the intensifier from a distance of 25 mm or so, when light transfer will be fair. The addition of a short-focus magnifier (loupe) to the optical train may improve transfer by reducing the effective focal length of the camera lens, and is worth a try.

The intensifier will allow recording of many relatively dim phenomena, such as occultations of faint stars by the Moon and planets, meteor shower monitoring (using a wide-angle camera lens on the intensifier), as well as permitting better planetary images with less noise. The writer obtained a very useful recording of the occultation of 28 Sagittarii by Titan on 3 July 1989 by coupling a Sony 'Handycam' plus intensifier to a 300-mm reflector. Titan, at magnitude 8.2, was clearly visible on the recording, and the internal camera clock, displayed on the image, allowed precise timings to be derived.

It should be noted that older intensifiers ('first generation') have massive pincushion distortion of the image, and a second-generation type will be found much more satisfactory. Intensifiers are also very easy to damage with bright light, so be careful with torches and other light sources when using one!

FRAME GRABBING

This text is much too brief to cover more advanced techniques in any detail, but I am sure that some guidance would be useful to those with electronic knowledge and determination to succeed. It is obvious from the above that standard video equipment cannot approach the performance which professional CCD systems can achieve. Much of this is due not to the CCD but to the lack of integration of the light for a long time before readout. Since such a device would produce only one frame every few seconds or minutes, it is necessary to have some form of picture 'memory' to hold the result of readout and display it in a still image. Such a memory is often called a 'frame grabber', and is where the complication of a CCD system tends to lie (Figs. 17.5, 17.6).

A large digital memory is needed if a full-resolution TV picture is to be stored: this will typically be 262 144 bytes of 8 bits each to give a grey scale with 256 (2^8) levels. Such a store will hold an array of 512 × 512 pixels, or a similar rectangular raster to match the CCD, and can be written to, or read out, in real time (video speed) if constructed with modern 'dynamic' or 'static' RAM chips.

The camera output signal is an analogue voltage and must be digitized by an

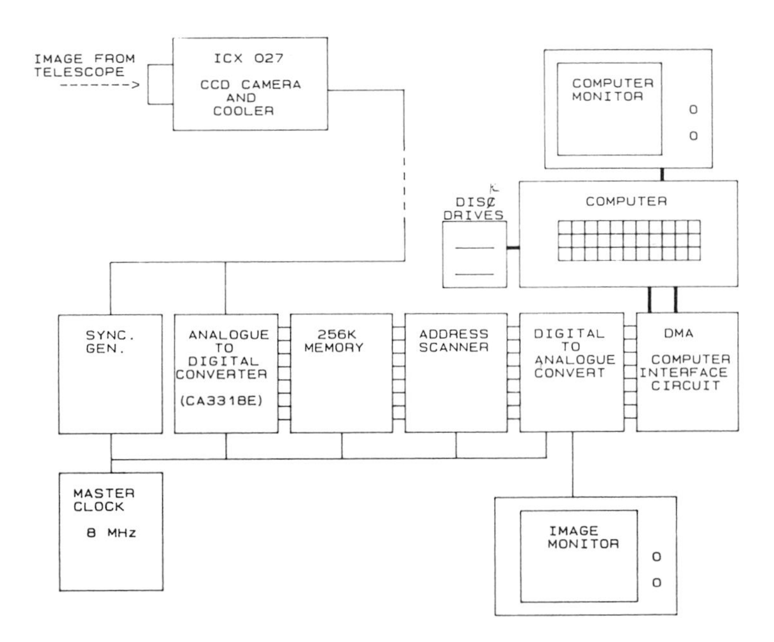

FIGURE 17.5
A CCD camera and digital memory as used by the author.

FIGURE 17.6
The author's frame grabber, monitors and computer.

'analogue to digital' converter, preferably at video speed. This is extremely fast, but many suitable chips are now available at quite low cost. An example is the RCA type CA3318E, which can digitize at up to about 15 million pixels per second for a UK price of less than £40.

Additional components include a scanning address generator to read the memory in synchronism with the picture scan, a camera controller to select integration periods and synchronize it to the memory, and a computer interface board for reading data into a machine for processing. Such a device will cost only a few hundred pounds to construct, but needs much electronic know-how to design and debug. A possible way around most of the problems is to purchase a commercially-available frame grabber designed for use with a PC or similar machine, and to adapt a camera module to allow integration. A search in the advertisements of a computer magazine will usually reveal a frame grabber or two at a reasonable price, but be sure to read the specification very carefully. It is essential to buy a unit that will grab frames in real time (in the time taken to read out one frame), as only one frame will be delivered at the CCD readout. Some low-cost grabbers read many sequential frames at low speed to build up the image, and clearly are useless for astronomy. Another feature to check is the grey scale resolution or number of levels between black and white. Most cheaper units give 64 grey levels (6 bits), which is just about adequate, but 256 levels (8 bits) are much better and almost essential if computer enhancement is to be used. A 6-bit picture of a low-contrast object tends to show obvious 'contouring' across the surface, and this becomes much worse when contrast is boosted by the computer. Colour image grabbers are fairly common, but grey-scale resolution is often sacrificed to store colour data and a lot of data needs to be handled if colour image processing is done. For these reasons I recommended a monochrome camera and grabber, especially as monochrome cameras are somewhat more sensitive than colour types, the internal colour filters always absorbing some light. The equipment will also be cheaper!

The most readily available camera module is probably the Philips 56740 (USA standard is 56471), which has a frame transfer NXA1011 CCD fitted, and is very compact. Unfortunately, at the time of writing, this is available only in a 'normal' TV standard form, and does not have an integration facility. However, if you are adept with a soldering iron it is possible to allow integration by preventing the vertical clock signals generated by the SAD 1019T from reaching the CCD until readout is desired (Fig. 17.7). The module will need to be operated in its non-interlaced mode (294 active lines) to avoid having to integrate twice for each image, but this will not degrade the picture too seriously. I suggest that the interested reader contact the manufacturers, Philips, for more information (see sources section, below).

If the CCD is not cooled, exposure times will be limited to about ten seconds before the leakage becomes severe. This, however, is more than enough for the bright planets, and excellent detail can be recorded on to computer disk for later

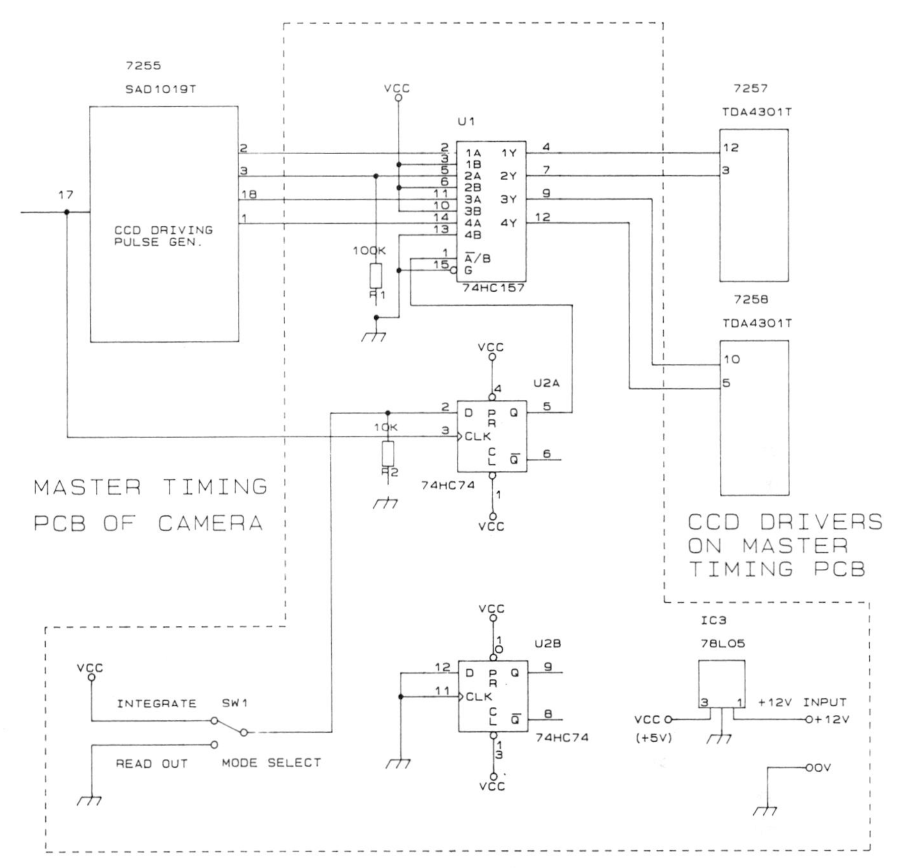

FIGURE 17.7

Modification to Philips 56470 series camera to allow integration of image on CCD. Circuit within dotted line is added to camera between SAD1019 and TDA4301 drivers. The pins of the SAD1019 and TDA4301 devices shown are normally connected and these tracks are cut to insert the above circuit. The camera must be set to operate in the progressive scan mode by grounding pin 3 of the mini DIN input socket on the camera. When SW1 is on 'Integrate' the CCD will store photoelectrons. When set to 'Read' the single field is dumped to the output (this will require some delicate surgery on the camera to cut tracks and the reader is reminded that Philips will not replace a damaged camera unit if he makes a mistake!).

processing. My own method is to cause an image-capture cycle to occur by pressing a button, and to assess the quality of the result on the monitor before either saving the image on disk or pressing the button again. In this way, an observing session will provide about six good disk recordings, and these can be subjected to enhancing programs at leisure the following day.

COOLED CAMERAS

If a Peltier cooler is added to the camera, the CCD temperature can be reduced to about −20°C in a few minutes, or a two-stage cooler will give perhaps −50°C in a similar time. A thermal break, such as a Perspex plug, can prevent condensation from forming on the CCD window, but my own preference is to house the camera head in a sealed box with silica gel and to use an exposed area of metal on the cooler as a 'cold finger' to trap moisture. A 'skylight' camera filter seals the box front and is well spaced off from the CCD to avoid dewing of the filter surface.

A few minutes' deep-sky exposure can be used at −20°C, and 10 or 15 minutes at −50°C. As the quantum efficiency of a CCD is around 60%, compared with 1–10% for hypered film, a 15-minute exposure will almost certainly reach the sky-fogging limit of the telescope system, unless your sky is exceptionally dark (Fig. 7.9).

IMAGE ENHANCEMENT

Once an image has been stored, it is usually beneficial to enhance the result by applying suitable computer programs. It would be impossible to go into detail here, but the simplest techniques that are often used involve so-called 'point operations', where individual pixels are modified in value according to a simple set of rules. Some software for image processing is commercially available, but many of the routines are quite easy to develop for oneself. Changing contrast according to a suitable mathematical formula, for example a square law or logarithmic law, is a typical case. The simplest way of doing this is to set up a table of pixel values in the memory at addresses which can be called by the present value of a pixel. The table would have 256 values, from 0 to 255, which follow the required law (for example, the value in location 129 will be 128 if the law is linear, but 64 for a square law), and are stored in sequential addresses from, say, 64 000 to 64 255. A pixel from the image has its value of between 0 and 255 read, and this is added to 64 000. The result is used to call the corresponding address, and the value in that address is substituted for the pixel in the image. This is repeated for all pixels until the complete picture is modified.

Such processing is best done in machine code to give a fast run time, but BASIC can be used if time is not critical (compiler BASICS, such as Microsoft 'Quick Basic' are quite fast). Many other routines for increasing definition and smoothing noise are possible, but involve more complex programs and must use machine code if the time is not to run into hours. The essence of these programs is to examine an area centred on the pixel to be modified, and to derive the change to the central value from the result of this review (Fig. 17.8). If the average of the surrounding pixels is subtracted from the central pixel, the result represents a high-frequency component in the picture. Adding the difference to the central pixel will increase the high frequency, and hence fine detail, in the image.

The 'median filter' is used to smooth out noise, and can be performed by scaling the central and surrounding pixels in such a way as to find the pixel value which is midway (the median value) between the largest and smallest present. If the central pixel is not of this value, it is modified to this value, otherwise it is left

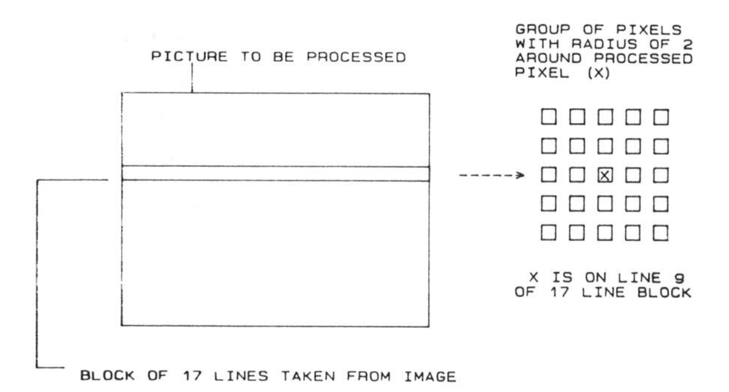

FIGURE 17.8
Performing high-pass (sharpness-boosting) and median (noise-smoothing) filters. To execute a high-pass filter the following calculation is performed: (1) All pixel values in group are totalled. (2) Total is divided by 25 to give average. (3) Average is subtracted from value of pixel *X*. (4) Remainder is added to value of pixel *X*. (5) Pixel *X* in new image is given value found in (4). (6) Process is repeated for all pixels on line 8. (7) New line is selected by moving block down 1 line. (8) Repeat (1) to (7) until whole image is processed. (This is a simplified example of a high-pass filter — most such procedures use 'weighting' of pixel values according to their distance from the central pixel.) To execute a median filter the following calculation is performed: (1) Read values of all pixels in block and store in file. (2) Sort values into ascending order (bubble sort). (3) Find median value (in file location 13 for this radius). (4) Set pixel *X* in new image to this median value. (5) Process whole line as in (1) to (4). (6) Process rest of image by moving block down 1 line and repeating until all done.

FIGURE 17.9
M27, photographed in 160 seconds with a Sony ICX 027BL CCD on a 200-mm f/5 reflector.

alone. This filter will suppress noise without serious loss of definition, and is often applied after the use of the sharpness-enhancing filter, which emphasizes noise.

This concludes my brief look at video techniques. I am sure that such methods will become ever more popular amongst amateurs, and the benefits to lunar and planetary astronomy, in particular, will be considerable. I conclude with some suggested reading and sources of components — if you try video, I am sure you will not be disappointed.

About the Author

Terry Platt is an electronic design engineer and has the position of technical director of a small company in Maidenhead. He is married with two children of 10 and 16 who think that Dad is a man who lives in an observatory at the bottom of the garden and calls in at the house now and then for a meal! He is now 45 years old and has had a keen interest in science in general since junior-school days, when astronomy and electronics became his hobbies.

He is a native of Gargrave, a small village in the Yorkshire Dales where the scenery is marvellous and light pollution slight, but there is no work in electronics! He has, therefore, lived in Berkshire for the last 20 years and has concentrated on the planets because of the light skies.

Much dismantling and rebuilding of old radios and televisions, combined with the grinding of mirror blanks, has led to a career in electronics and a home observatory containing a 318-mm off-axis planetary reflector with a CCD camera attached.

He is a long-term member of the BAA and hopes that others will be encouraged to apply modern electronics to astronomy by reading of his experiments. He would like to thank Mr R.K. Payne of Memco for his help and encouragement in developing CCD equipment, Mr D.C. Gaskin for assistance with computer programs, etc., and last but not least his wife Christine for tolerating 20 years of solder splashes, pitch fumes, and astronomy magazines!

Contact address – September Cottage, Murrell Hill Lane, Binfield, Berkshire RG12 5DA, UK.

Sources and References

Technical information on CCDs and camera modules may be obtained from the following.

Philips Technical Handbook: *CCD monochrome imaging modules* (*types 56470 to 56475*). From Industrial Electronics Division of Philips Components Ltd, Mullard House, Torrington Place, London, WC1E 7HD, UK (071-580 6633).

Sony Data Book: *CCD Camera and Peripheral.* From Hakuto International UK Ltd, Eleanor House, 33–35 Eleanor Cross Road, Waltham Cross, Herts, EN8 7LF, UK (0992 769090).

Suppliers of components (CCDs, driver ICs, etc.) include the following.

Hakuto International (see above) for Sony devices, as well as Pronto Electronic Systems Ltd, City Gate House, 399–425 Eastern Avenue, Gants Hill, Ilford, Essex, IG2 6LR, UK (081-554 6222).

Norbain Imaging Ltd, 37 Boulton Road, Reading, Berks, RG2 0LT, UK (0734 864411) for Philips devices.

Real-time frame grabbers: Imaging Technology Inc., 13841 SW 106 Street, Miami, FL 33186, USA ((305) 385-7092).

CCD camera heads: Santa Barbara Instrument Group, 1482 East Valley Road, Suite 601, Santa Barbara, CA 93108, USA ((805) 969-1851).

Astronomical CCD systems: Abracadabra Ltd, Briar House, Foxley Green Farm, Holyport, Maidenhead, Berks, SL6 3LA, UK (0628 21571).

Works on image processing, etc.

Gonzalez, R.C. and Wintz, P. *Digital Image Processing*, 2nd edn. Addison-Wesley, Reading, Mass., 1987.

Green, W. *Digital Image Processing, A Systems Approach*, 2nd edn. Van Nostrand, New York, 1989.

Buil, C. *CCD Astronomy*, Willmann-Bell Inc., PO Box 35025, Richmond, Va, 1991.

Notes and Comments

Alan Macfarlane (Seattle, Washington), who has pioneered the use of astronomical video recording for educational purposes, read this chapter and agreed that 'video is the medium of the future, and has much potential, especially for solar system object recording'. The detail visible in Terry Platt's Mars and Jupiter photographs, taken with a relatively modest telescope, from low altitude and in a built-up area, exceeds the finest that were taken anywhere in the world, regardless of aperture, using orthodox (or should we say old-fashioned?) photographic techniques.

An article by Macfarlane published in *Sky & Telescope* for February 1990 is useful additional reading for anyone starting on this exciting venture. Some selected points are presented here.

Slow motion controls – Video is a continuous recording of the telescopic image at a rate of 30 frames per second. Jerky controls can destroy possibly hundreds of images during the sudden interference, including the rare superb one. A good drive and electrically-operated slow motions will obviate this.

Vibration – A camcorder, with camera and tape recorder, will impart vibration to the telescope. A system with a separate tape deck overcomes this, and also allows video formats to be changed without having to change the camera head as well.

'Poor man's frame grabber' – A simple way of improving resolution and suppressing the noise in a single video frame is to take a photograph of the monitor screen with an exposure of about 1/8 second. This effectively combines about four individual frames of 1/30-second. Macfarlane uses ISO 400 film, controlling the exposure by altering the camera aperture.

Telescope type – Macfarlane's experience is that Schmidt–Cassegrains are very suitable for video work, provided that they are well-collimated. With his 280-mm telescope he has recorded magnitude 10 stars.

Focusing – An image that looks in focus on a small monitor can appear blurred when played back on a large screen. He advises to use the largest possible monitor, and to sharpen the image (using an electric focusing device) while standing about two paces from the monitor and viewing the screen with a hand-held monocular.

Readers may be unfamiliar with the 'tri-Schiefspiegler' system used in Terry Platt's 318-mm reflector. It is a type of Cassegrain telescope, with the mirrors tilted so that the secondary is clear of the principal light-path. Therefore there is no diffraction from the secondary or its supports, and in theory the Schiefspiegler, whether using two or three mirrors, should combine refractor-type image contrast with reflector-type achromatism. Strangely, very few such instruments have been constructed in the United Kingdom, though they are known in several European countries and in the United States. — Ed.

Footnote – Since writing this chapter the author has learned that the Philips 56470 Series of camera-modules has been taken out of production. Some units are still to be found on the market but this is unlikely to continue for very much longer. Several other manufacturers now produce camera modules which can be adapted for astronomy by controlling the CCD readout time and the reader is advised to contact Sony Distributors (see above) or Pulnix Ltd of Pulnix House, Avery Court, Wade Rd, Basingstoke, Hants, RG24 OPL. Tel: 0256 28954. Pulnix manufacture several cameras with time integration facilities already incorporated. Types TM6 and TM520 are typical of this type of camera.

CHAPTER EIGHTEEN · *Astronomical Communications*

GUY M. HURST
EDITOR, *THE ASTRONOMER*

THE VERY NATURE OF THE SCIENCE OF ASTRONOMY DEMANDS THAT observers in widely different parts of the world can exchange results as rapidly as possible. However, it is probably discovery of a new object such as a nova, comet or supernova which calls for the fastest communication, as it is vital that independent confirmation is obtained and that follow-up results are secured as soon as possible, preferably the same day.

DISCOVERIES

I make no excuse for plunging immediately into discoveries. Surely it is in this area that amateur astronomers are so well placed to make a vital astronomical contribution? The principal areas of potential discovery are

Novae
Supernovae
Comets
Asteroids

If I had been writing this chapter even a few months ago I would have hesitated to have included asteroids, on the basis that all the 'bright' minor planets had been discovered long ago. However, on 4 October 1989 a remarkable event occurred. Brian Manning, of Stakenbridge in Worcestershire, took a photograph of a comet with a 260-mm reflector and recorded four asteroids on the same photograph. All were very faint, near magnitude 16–17, but the combination of high-quality images and the use of specially-sensitized film enabled Brian Manning, one of the world's leading amateur astrophotographers, to show that it is still possible for amateur observers to discover asteroids [see Chapter 16 – Ed].

This photograph brought out the ultimate test in communications. There was an immediate need to establish whether any of these four minor planets were already known as numbered objects with well-defined orbits. After receiving the measured positions I used electronic mail to log into the computer service in the Central Bureau for Astronomical Telegrams in Cambridge, Massachusetts, USA, and established that all four objects were 'new' in the sense that they were not among the 4000+ numbered asteroids for which reasonably accurate orbits are available. Thus the advance of communications from (historically) paper letters to modern, sophisticated e-mail services proved vital in the rapid confirmation of these objects as new asteroids, and also added a piece of history to the textbooks. These were the first asteroids to be found from the United Kingdom, with reliable follow-up positions, for 80 years! But let us start from the beginning...

The communication of early discoveries in amateur astronomy relied on letters or telegrams. Unfortunately, neither proved as reliable as could be wished, and from time to time messages would disappear in transit. Although only comets are named after the discoverer there is clearly a need to 'record' a discovery as promptly as possible, not purely in respect of who should receive credit but because it is important that the astronomical community, both amateur and professional, should receive the news quickly and plan further observations. Let us look at the procedure that should be adopted.

Amateur astronomers tend to fall into two groups: those who read about the subject but rarely go out and observe, and those who dedicate themselves to watching the sky on as many clear nights as possible. The latter are, unfortunately, a very small minority. Of this small group, the number of observers who have made a discovery is even smaller, but all agree that there is a special kind of excitement attached to the moment when a possible new object is glimpsed for the first time. The details which should be recorded are:

1. Date and time
2. Instrument used
3. Position
4. Magnitude
5. If moving, direction and speed of motion
6. Observing conditions and location

(1) It is best to use UT. Thus in the summer months in the United Kingdom it is important to avoid using British Summer Time, which is equal to UT + 1 hour. Always use the 24-hour clock system, as this avoids confusion as to whether the observation was made in the morning or the evening of the day specified.

(2) For instrument details it is conventional to quote the aperture of the telescope

as a decimal fraction of a metre. Thus a 10-inch telescope is given as 0.26 metre, and the focal ratio and telescope type should also be added, as well as the magnification used to make the discovery.

(3) The position of the new object is, of course, absolutely vital, and reference to an atlas is necessary to quote the location in terms of right ascension and declination, together with epoch. Until recently the epoch of most atlases has been 1950, but recent publications have been based on epoch 2000, and it is very important that this is made clear.

(4) The brightness of the object needs to be derived by comparison with nearby stars of known magnitude. For example, the AAVSO Atlas quotes magnitudes for many of the plotted stars, and these will serve as a guide to reducing the object's magnitude. This value can be absolutely crucial, as in the case of a nova or supernova early magnitude estimates are vital in determining the date of maximum. If you are lucky enough to catch your star on the rise, and can alert the astronomical community quickly enough, your initial estimates coupled with the analysis of the star by professionals at other wavelengths can often play a very important part in the interpretation of the astrophysics involved.

If the star is changing rapidly, make estimates at hourly intervals. Record the comparison stars used as well as the reduced magnitude in case the observation has to be re-reduced at a later date.

(5) Moving objects, such as comets and asteroids, will also require an estimate of the direction of motion (normally quoted as a value based on position angle) and the speed (usually quoted in arc-minutes per day). This could be calculated from a series of positions and times.

(6) Observing conditions should be noted, and in particular the zenithal naked-eye limiting magnitude. The telescopic limit of the field containing the new object should also be recorded.

In addition to these basic details, it is usually in the case of extragalactic supernovae to estimate the offsets from the nucleus of the parent galaxy.

The above details could apply equally to visual or photographic discoveries, but for the latter details of film speed and type should also be logged. It is also vital that you make it clear whether the object was recorded on one exposure or on several, as the former case does not rule out the possibility of a flaw simulating a new object.

Before relaying your message, make sure if at all possible that you have eliminated such culprits as planets, asteroids, variable stars, and atlas omissions.

Now to the communications. This is a very difficult area as there has always been a conflict between the need to obtain confirmation and thus avoiding circulating false alarms, and the importance of giving the astronomical community the earliest possible change of monitoring a new object. For this reason, several experienced observers across Europe have agreed to make themselves available on an all-night basis to check out possible discoveries. Since 1964, the organization behind *The*

Astronomer magazine has coordinated this effort, and appropriate details are given at the end of this chapter. Potential new discoverers are naturally quite excited when reaching for the telephone, and it does help if you first write down all the details outlined above!

Even after you have passed on your message, it is important that you continue to monitor the new object. Obtain further photographs if suitable equipment is to hand, or make a field drawing showing the new object and the comparison stars used. Always add the field scale and the direction of north and west. Make copies and send in your drawing and notes as confirmation of the telephone call, using Express Delivery. In the case of negatives it is preferable to send one only for analysis, retaining the others just in case the letter is lost in transit. If the discovery is confirmed, *The Astronomer* group arrange for the appropriate details to be relayed to the Central Bureau for Astronomical Telegrams, which is the international clearing-house for all discoveries, whether made by professional or amateur astronomers. In practice it is usual for details to be relayed by electronic mail or telex, and currently these are sent to the Smithsonian Astrophysical Observatory in Cambridge, Massachusetts, where they are handled by the Director, Dr Brian Marsden. It is important that unconfirmed reports are not sent direct, as this would overload the work at the Central Bureau. Thus the UK group act as a valuable 'filter' organization, endeavouring to ensure that only definite discoveries are passed on.

The Central Bureau then issues postcard circulars announcing the discovery, and also telex and e-mail messages to observatories worldwide.

PUBLICATIONS

Although I have so far covered the vital area of communicating discoveries, the various magazines and journals have a role. These enable amateur astronomers to publish articles on all aspects of astronomy from historical notes to papers explaining new advances in techniques and equipment. But most of all, the regular journals provide a means by which astronomers can communicate with each other. Even in a small area like the United Kingdom, observers are so widespread that they may meet each other only rarely, except perhaps on a local basis through their own astronomical society.

Apart from articles, the journals keep amateur astronomers abreast of new developments in the science and, in particular, bring news of what professionals are currently researching. This is important as there is clearly a need for more interaction between professional and amateur astronomers (PRO-AM) on various observing projects, and this implies a need for both groups to keep up to date with each other's work.

Books are also a vital tool for the amateur astronomer. There are a vast number published each year, and the journals themselves carry reviews which guide the astronomer towards the work must suited to their specialist interests. Apart from general reading, observers quickly find a need for atlases and catalogues to which they will constantly refer as they scan the sky.

ELECTRONIC COMMUNICATION

Several times in this chapter mention has been made of media other than letters and the use of the telephone.

For the rapid transmission of charts for new objects, fax is becoming popular. Discovery messages are often relayed by telex. But a new method which is becoming increasingly popular is electronic mail, or 'e-mail' as it is more commonly known. Those using this sytem tend to look on the ordinary post as very slow, and in fact have dubbed it 'snail-mail'!

Many readers will have come across Prestel, which is just one of many viewdata services which are accessed by most users with the help of a modem linked to a telephone line. Although special home terminals can be obtained, many merely hook up a personal computer. The modem translates commands and messages into signals suitable for sending down a standard telephone line, and at the receiving end these signals are decoded. In this way, users can communicate with a mainframe computer regardless of the equipment used 'at home'. The viewdata services primarily provide screens of information which can be read on-line or downloaded to disk or tape and studied later (which saves on telephone bills!).

Prestel, Microlink, and Telecom Gold are all systems which also provide an e-mail service. Each user has a mailbox, which is a special area of the main computer memory reserved for receipt of personal messages. In effect, it is very similar to an answering machine because messages can be sent and stored there even if the recipient is unavailable. The messages are usually prepared and edited before connecting to the system (off-line), and then transmitted using special software in conjunction with the modem. The mailbox name or number is defined so that the message is stored specifically for the intended recipient, who later logs on and reads all the messages stored. This is not really a real-time system, but it has the advantage to astronomers that they can leave messages for each other even when telephone contact is not possible.

On 24 January 1987 a new service was started by *The Astronomer* group, using the Telecom Gold service. Recognizing the advantages of e-mail, the new service involved sending electronic circulars to a whole group of registered users, primarily giving news of discoveries and predictions. This has continued, and by early 1992 over 600 circulars had been issued. Each user is recommended to log on each evening just before the start of an observing session, and in this way all the up-to-date information is available to help observers concentrate on any new objects needing their attention.

The e-mail service described above is now also proving vital for investigation of possible discoveries. It is easy to route a request automatically to a large group of e-mail users requesting confirmation of a new object.

In addition to Telecom Gold, permission has recently been given for some amateur astronomers to use Janet (Joint Academic Network) and Starlink. These electronic services are provided specifically for the academic community, with Starlink being primarily for professional astronomers. These UK services also have gateways to similar systems all around the world, and it is therefore possible for messages formerly confined to Telecom Gold to be sent to Janet and thence to other astronomers overseas. We have come a long way since the days of letters!

MEETINGS

The value of meeting and exchanging notes with fellow amateur astronomers cannot be overemphasized. It is possible to learn far more from informal chats than from reading books, because at your local astronomical society gathering you have the change to ask questions. Many local societies organize observing sessions, and perhaps one day astronomers in the United Kingdom will regard the American-style 'star party' as commonplace here. [Knowing our weather, I doubt it! — Ed.]

The national organizations such as the British Astronomical Association (BAA), *The Astronomer* (TA), and the Junior Astronomical Society (JAS) all have regular gatherings. The BAA in particular have monthly meetings in London, and often invite professional astronomers to lecture on subjects of current interest. Although the Royal Astronomical Society (RAS) includes many professional astronomers, they also cater for the more serious amateur who joins their ranks. Apart from meetings, the RAS has a magnificent library in Burlington House, Piccadilly, available to assist Fellows in their research.

The Federation of Astronomical Societies (FAS) plays a valuable role in helping with the communication between local societies, and publishes an annual handbook which provides considerable information on astronomical contacts.

MEDIA

This short summary of communications in astronomy would not be complete without mention of the media. As many amateur astronomers know, there have been many cases where stories of new comets and other objects have become exaggerated in the process (Comet Kohoutek being a case in point); but the newspapers do at least keep the public informed of major events, and perhaps make astronomy more appealing to the general public than the serious amateur astronomer can. They can remember the need for plain English rather than the obscure technical jargon of which we are all, occasionally, guilty.

Television has played a major role in bringing astronomy into the living-room, and one could not fail to mention that very long-running success on UK television, 'The Sky at Night'. Many of the 'Horizon' programmes have also provided a reliable account of modern research, conveying it in terms which can be understood by the wider audience they serve. For those whose interest extends to research in space, programmes covering the launch of the Space Shuttle and manoeuvres in space to repair satellites all appeal, and one can never forget the famous walks (and rides!) on the surface of the Moon.

CONCLUSION

The current state of communications in astronomy is very healthy, and the developments in the last decade have moved us away from an insular approach to our hobby. It is very important that the spirit of international cooperation between amateur astronomers is fostered, and that the recent initiatives on liaison between professionals and amateurs are developed further. I hope that this summary gives an insight into current developments, and invite those interested to contact organizations catering for their interests. The names and addresses of major societies and information sources are given in pages 391–4, but details of *The Astronomer* services are given below.

Telephone or Fax:	(National) 0256 471074
	(International) +44 256 471074
Telex:	9312111261 (TA G)
E-mail:	Telecom Gold: 10074: MIK2885
	GMH at
	UK.AC.RUTHERFORD,STARLINK.ASTROPHYSICS
	Janet: STARLINK: RLSAC:: GMH

The Astronomer (monthly) offers the following services to amateurs:

1. Paper circulars announcing discoveries.
2. E-mail services to those on Telecom Gold, Janet, Starlink, Compuserve and overseas systems providing constant, instant news for observers.
3. Annual meetings at various venues throughout the United Kingdom.
4. A fax service is provided in conjunction with the BAA for those who do not have access to e-mail.

About the Author

Guy Hurst has been interested in astronomy all his life, but began in earnest as an active observer in 1971 with the systematic visual study of open star clusters, the results of which were published in the Webb Society Handbooks. In 1975 he took over as Editor of *The Astronomer* (TA) from James Muirden and has continued in this role to the present day.

During his time as Editor of the 'TA' he has fostered international co-operation amongst amateur astronomers worldwide, and more recently has served on the UK Professional-Amateur Liaison Committee. Since 1976 he has also coordinated the

UK Nova/Supernova Patrol whose members search for these important objects. In 1987 he launched the first astronomical electronic mail service so that a very fast system was available to alert fellow astronomers to newly-discovered objects.

Perhaps the most demanding of all is the role of investigating discoveries from around the world and verifying their authenticity prior to relaying them to the professional clearing house at the Central Bureau for Astronomical Telegrams in Cambridge, Mass., USA. In acknowledgement of his help to Dr Brian Marsden, who coordinates the Central Bureau, asteroid (3697) 'Guyhurst' was named after him.

All this is fitted into 'spare time', as by day he is an area bank manager, and it would be impossible without the support of his wife Ann and their three boys.

Contact address – 16 Westminster Close, Kempshott Rise, Basingstoke, Hants RG22 4PP, UK.

Notes and Comments

Asked about amateur communications in the United States, Stephen Edberg (La Canada, California) contributed the following notes. — Ed.

One can group communications by their media: paper, organizational, verbal, and electronic.

Paper – Magazines like *Sky & Telescope (S & T), Astronomy, The Astronomer,* and *Astronomy Now* are obvious. Less obvious are telegrams and circulars from the Central Bureau for Astronomical Telegrams, the International Comet Quarterly, the Shallow Sky Bulletin and its sister publication the Comet Rapid Announcement Service, and John Bortle's comet announcement cards.

Organizational – Most organizations communicate with their membership via paper media, usually journals or circulars. Organizations with active amateur–professional programs include the AAVSO, the ALPO, International Amateur–Professional Photoelectric Photometry (IAPP), International Occultation Timing Association (IOTA), the International Meteor Organization, and numerous others. Of course, local astronomical societies supply amateur–amateur communications as do umbrella organizations like the Astronomical League (AL) and Western Amateur Astronomers (WAA).

Verbal – Seminars and symposia are useful for communications. I do not refer so much to large amateur astronomy conventions like the Riverside Telescope Makers Conference, Stellafane, or AL and WAA conventions as to meetings such as the Electronics Oriented Astronomy, the Astrophoto seminars and the IAPP symposia, as well as the Symposium on Research Amateur Astronomy. The annual meetings of IOTA, ALPO and AAVSO also serve in this capacity to a lesser degree.

Electronic – Some small one-way communications occur via radio time-signal stations such as WWV and WWVH, and WWV's telephone transmission. IOTA updates asteroidal occultation predictions over the telephone, and many local societies and planetaria maintain telephone recordings (as does S&T).

Another electronic medium is the computer bulletin board service. CompuServe maintains an Astronomy Forum, the IAPP has a bulletin board, S&T has a bulletin board, and various materials can be accessed from a CBAT service.

CompuServe memberships can be purchased from Radio Shack stores and other computer dealers. It is a major national network with forums on space exploration, space education, and astronomy. The *Sky & Telescope* on-line service provides quick news of comet and nova discoveries as well as program listings from the Astronomical Computing department. The address is CompuServe Inc., 5000 Arlington Centre Blvd., Columbus, OH 43220, USA. — Ed.

Societies and Publications

Some of these organisations, mostly amateur observational societies, are mentioned in the text, but are grouped together here for convenience. — Ed.

American Association of Variable Star Observers
25 Birch Street
Cambridge
MA 02138
USA

American Astronomical Society
2000 Florida Avenue NW
Suite 300
Washington
DC 2009
USA

Association Française d'Observateurs d'Étoiles Variables
16 Rue de Plobsheim
67100 Strasbourg
France

Association of Lunar and Planetary Observers
8930 Raven Drive
Waco
TX 76712
USA

Astronomical League
6235 Omie Circle
Pensacola
FL 32504
USA

Astronomical Society of South Africa
South African Astronomical Observatory
PO Box 9
Observatory 7935
South Africa

Astronomical Society of the Pacific
390 Ashton Avenue
San Francisco
CA 94112
USA

Astronomishes Büro
Hasenwertgasse 32
1238 Wien
Austria

Berliner Arbeitsgemeinschaft für
Veranderliche Sterne
Munsterdamm 90
D-1000
Berlin 4
Germany

British Astronomical Association
Burlington House
Piccadilly
London W1V 0NL
UK

European Group of Star Observers
12 rue Bézant
F-75014
Paris
France

Federation of Astronomical Societies
c/o Christine Sheldon
Whitehaven
Maytree Road
Lower Moor
Pershore
Worcs. WR10 2NY
UK

IAU Central Bureau for Astronomical
Telegrams
Smithsonian Astrophysical Observatory
Cambridge
MA 02138
USA

International Amateur–Professional
Photoelectric Photometry
Organisation
Dyer Observatory
Vanderbilt University
Nashville,
TN 37255
USA

International Meteor Association
Pijnboomstraat 25
B-2800
Mechelen
Belgium

IOTA: International Occultation
Timing Association
David W. Dunham
6 N 106 White Oak Lane
St Charles
IL 60174
USA

IOTA/ES
Hans Bode
Bartold-Knaust Strasse 8
D-3000 Hanover 91
Germany

Irish Astronomical Association
The Planetarium
Armagh BT61 9DB
Northern Ireland
UK

Japan Astronomical Study Association
Natural Science Museum
Veno Park
Taito-Ku
Tokyo
Japan

Junior Astronomical Society
36 Fairway
Keyworth
Nottingham NG12 5DU
UK

Koninklijt Sterrenkundig Genootschap van Antwerpen
Boerhaavestraat 94 Bus 1
2008 Antwerp
Belgium

Minor Planet Center
Smithsonian Astrophysical Observatory,
Cambridge
MA 02138
USA

National Association of Planetary Observers
PO Box 2
Riverwood
NSW 2210
Australia

Nederlandsche Vereniging voor Weer en Sterrenkunde
Bureau 'De Koepel'
Sterrenwacht Sonnenborg
Zonnenburg 2
3512 Utrecht
Netherlands

New South Wales Branch of the British Astronomical Association
PO Box 103
Harbord
NSW 2096
Australia

Norsk Astronomisk Selskap
Postboks 677
4001 Stavanger
Norway

Oriental Astronomical Association
c/o Yamamoto Observatory
289 Kamitanakami-Kiryutyo otu
Sigaken 520-21
Japan

Royal Astronomical Society
Burlington House
Piccadilly
London W1V 0NL
UK

Royal Astronomical Society of Canada
McLaughlin Planetarium
100 Queens Park
Toronto
Ontario M5S 2C6
Canada

Royal Astronomical Society of New Zealand
PO Box 3093
Geerton
Tawanga
New Zealand

Scandinavian Variable Star Observers
SF-36280
Pikonlinna
Finland

Schweizerische Astronomische Gesellschaft
c/o Andreas Tarnutzer
Hirtenhofstrasse 9
6005 Luzern
Switzerland

Sky Watcher's Handbook

Sky Observers Association
155 Fuk Wing Street
4th Floor, Room 6
Shamshuipo
Kowloon
Hong Kong

Sky & Telescope
PO Box 9111
Belmont
MA 02178-9111
USA

Société Astronomique de France
3 rue Beethoven
75016 Paris
France

Society of Amateur Radio Astronomers
c/o John Weiss
PO Box 2632
Montgomery
AL 36105
USA

Svenska Astronomiska Sallskapet
Stockholms Observatorium
13300 Saltsjöbaden
Sweden

The Astronomer
16 Westminster Close
Kempshott Rise
Basingstoke
Hants, RG22 4PP
UK

Unione Astrofili Italiani
Universita degli Studi di Padova
Dipartmento di Astronomia
Vicolo dell'Osservatorio 5
35122 Padova
Italy

Ursa Astronomical Association
Laivanvarustajankatu 3
00140 Helsinki
Finland

Vereinigung der Sternfreunde
Volkssternwarte
Anzinger Strasse 1
8000 München 80
Germany

Werkgroep Veranderlijke Sterren
Postbus 800
9700 AV Groningen
Netherlands

Index

WHILE INDEXING THIS BOOK I HAVE DISCOVERED THAT ALTHOUGH indexing is supposed to be science, it is really an art. Like all art forms, the perfect index exists only in the mind. The indexer does not know the predilections, needs, or expectations of the reader. In addition, the best index is only an approximation to the content of the book it is supposed to unfold.

Realising this has made me feel confident enough to state my aims, thereby also revealing my shortcomings. They have been:

1. To try to short-cut the reader's journey by supplying ample cross-references.
2. To give the reader a comprehensive idea of the range of topics covered in each chapter.
3. To avoid irritating strings of bland page numbers after a single entry.
4. To give particular guidance to entries that may not be exactly where the reader expects to find them.

Expanding on (4), as an example, the separate features in a section dealing exclusively with Martian features have not been separately indexed, but features mentioned elsewhere in the chapter have (or should have been) indexed.

I have also tended to include the subjects of illustrations (not, on the whole, tables) in the index.

(Note the use of 'should', 'tended', and 'on the whole', and remember that I am talking about an art form.)

All these precepts, and others, have surely been followed before, and more consistently. But in one respect, this index breaks new ground: it contains the answer to a puzzle. Therefore, to avoid spoiling the challenge, the reader is strongly advised to read the book from cover to cover and decide what the puzzle is, before attempting to use the index.—Ed.

Index